Universal Behavior Computing for Security and Safety

Cheng Wang • Hangyu Zhu

Universal Behavior Computing for Security and Safety

Springer

Cheng Wang
Department of Computer Science
and Technology
Tongji University
Shanghai, China

Hangyu Zhu
Department of Computer Science
and Technology
Tongji University
Shanghai, China

ISBN 978-981-97-9013-5 ISBN 978-981-97-9014-2 (eBook)
https://doi.org/10.1007/978-981-97-9014-2

This Springer imprint is published by the registered company Springer Nature Singapore Pte Ltd.
The registered company address is: 152 Beach Road, #21-01/04 Gateway East, Singapore 189721, Singapore

Preface

Behavior computing plays a key role in various fields, enhancing the performance and effectiveness of computational systems through the integration of behavioral data. It involves analyzing and understanding the behaviors of agents in different application scenarios and using behavioral data to drive decision-making and problem-solving processes. By incorporating behavioral insights into computational models and algorithms, researchers can develop accurate and context-aware models to adapt to changing environments and user needs. However, with the rapid development of network technologies, a vast amount of behavioral data is generated in cyberspace, interacting with the physical world. In addition to humans, billions of smart devices also act as initiators of behaviors. Traditional behavior computing focuses on human behaviors and may not fully utilize behavioral data from a broader range of agents, such as devices, systems, and ports capable of initiating actions. Analyzing the behaviors of these agents is crucial for understanding behaviors. Therefore, researching new technologies under universal behavior computing holds practical significance. Generally, research on this topic involves three aspects: behavior analysis, behavior modeling, and behavior simulation.

The purpose of this book is to present the latest research advancements and new methods tailored for universal behavior computing for security and safety applications. The content of the book can be divided into five parts. The first part (Chaps. 2 and 3) introduces innovative solutions for behavior modeling using adversarial data augmentation. The second part (Chap. 4) discusses how to introduce information density enhancement techniques to pave the way for more accurate models. The third part (Chaps. 5, 6, and 7) explains how introducing graph-based models changes the way we analyze behavioral data, providing deeper insights into behavior correlations. The fourth part (Chap. 8) explores the significance of behavior computing models in various security authentication applications and analyzes current research and challenges. The fifth part (Chap. 9) introduces how behavior

---

The original version of this book has been revised. An erratum to this book can be found at https://doi.org/10.1007/978-981-97-9014-2_10

simulation techniques open up new avenues for simulating complex social systems and gaining a deeper understanding of behavior.

This book is a research monograph that introduces the topic of security and safety based on universal behavioral computing. It proposes a unified perspective that combines data, information, models, and utility to enhance the detection performance of security and safety models. This book provides efficient behavior modeling solutions for security and safety and indicates behavioral simulation as the potential research direction. Addressing common challenges in typical domains can support the paradigm of universal behavior computing in a wide range of practical applications. The book covers up-to-date research developments and novel methodologies, offering valuable insights for both researchers and professionals in the security and safety field. Readers are expected to have a background in behavior analysis, behavioral modeling, behavioral simulation, machine learning, neural networks, and related fundamentals.

Shanghai, China

Cheng Wang
Hangyu Zhu

Contents

Chapter 1
Overview of Universal Behavior Computing

Abstract The study of behavior traces back to the observation and analysis of human behavior in specific environments. To further understand and simulate behavior, researchers have integrated disciplines such as computer science, data science, and behavioral science, establishing complex models that represent and simulate behavior. Behavior computing encompasses three main aspects: behavior analysis, behavior modeling, and behavior simulation. This book investigates behavior analysis, modeling, and simulation, showcasing their synergistic nature in advancing our understanding of complex behavioral phenomena. With the rapid development of mobile networks, humans are increasingly immersed in a hybrid space of virtual and real, generating and collecting a vast amount of behavioral data. Meanwhile, research on behavior is no longer limited to serving humans alone; billions of smart devices can also be viewed as initiators of behavior. We conduct a comprehensive survey of recent representative research in the field of universal behavior computing. Specifically, we first review traditional methods of behavior analysis. Then, we introduce emerging key technologies and frameworks for constructing behavior models. Finally, we discuss further opportunities for utilizing behavior simulation in future research. We believe that this book will garner increasing attention, stimulate productive discussions, and inspire further research directions in behavior computing.

1.1 Introduction to Behavior Computing

Behavior, whether conscious or subconscious, voluntary or involuntary, refers to the reflection of the behavioral agent on the environment [1]. The behavioral agent can be living organisms, systems, or artificial intelligence agents, among others. Behavior is universal, ranging from human activities in society to animal movements in the biological realm, and even the flow of air in the natural world [2]. Extensive research has been conducted from various perspectives to investigate behavior, leading to a widespread consensus on the importance of modeling, analyzing, and mining behavior in various downstream applications [3–5]. For example, retailers can track the purchase history, browsing records, and click behavior of customers in

C. Wang, H. Zhu, *Universal Behavior Computing for Security and Safety*,
https://doi.org/10.1007/978-981-97-9014-2_1

both physical stores and online platforms to provide personalized recommendations and customized shopping experiences [6]. Banks and financial institutions can analyze transaction patterns, spending habits, and behavioral characteristics of customers to identify potential fraud and implement appropriate risk management measures [7]. City managers can optimize traffic flow, improve transportation planning, and provide smart city services by analyzing vehicle trajectories, public transportation passenger flows, and pedestrian behavior [8]. The utilization of behavioral information, such as representation, modeling, analysis, and data mining, has become increasingly useful, important, and challenging in various scenarios. All these research efforts have culminated in a new opportunity known as behavior computing.

Behavior computing encompasses the analysis and utilization of behavior, with a focus on unlocking the value of data through its utilization. Many researchers leverage intelligent methods within specific scenarios, leading to a range of efficient behavior computing methods that have been widely applied in business scenarios such as social media, recommendation systems, and anti-fraud systems [6, 9, 10]. In order to achieve efficient behavior models, traditional solutions have adopted customized models that are specifically tailored to the business characteristics of downstream tasks. For example, in fraud detection tasks in the financial domain, deep learning models like recurrent neural networks (RNNs) or transformers can be employed to process sequential transaction data [5, 11]. These models can capture temporal dependencies and patterns within transaction sequences to identify potential fraudulent behavior. For recommendation system tasks in the retail industry, collaborative filtering-based models such as matrix factorization and latent factor models can be used. These models utilize interaction information between users and items to provide personalized recommendations for each user.

With the rapid advancement of technologies such as mobile networks, the boundaries of behavior have expanded, and behavior in virtual worlds has gradually become a significant component of behavior research [3, 9, 12]. In this trend, behavior data has witnessed a rapid increase in volume, and the behavior of various agents has become increasingly diverse and complex [2]. Existing solutions have achieved significant success, but current behavior computing still faces several challenges: (1) Customized models for individual agents struggle to leverage universal behavioral data. (2) Massive amounts of behavioral data reveal new potential features at the information level. (3) Overreliance on known behavioral data overlooks the exploration of behavioral randomness.

To address the aforementioned challenges, we review the state-of-the-art methods in these fields and propose a new perspective, universal behavior computing, to explore potential solutions for tackling these challenges (as shown in Fig. 1.1). It overcomes the challenges faced by traditional behavior computing in terms of data, information, models, and utility. The data is represented in different forms and then applied to specific behavioral models to perform utility on various downstream tasks. The objective of universal behavior computing is to extend the behavior subject to encompass any entity, including various sensors, smart devices, and human users, in order to achieve real-time, comprehensive, and intelligent perception,

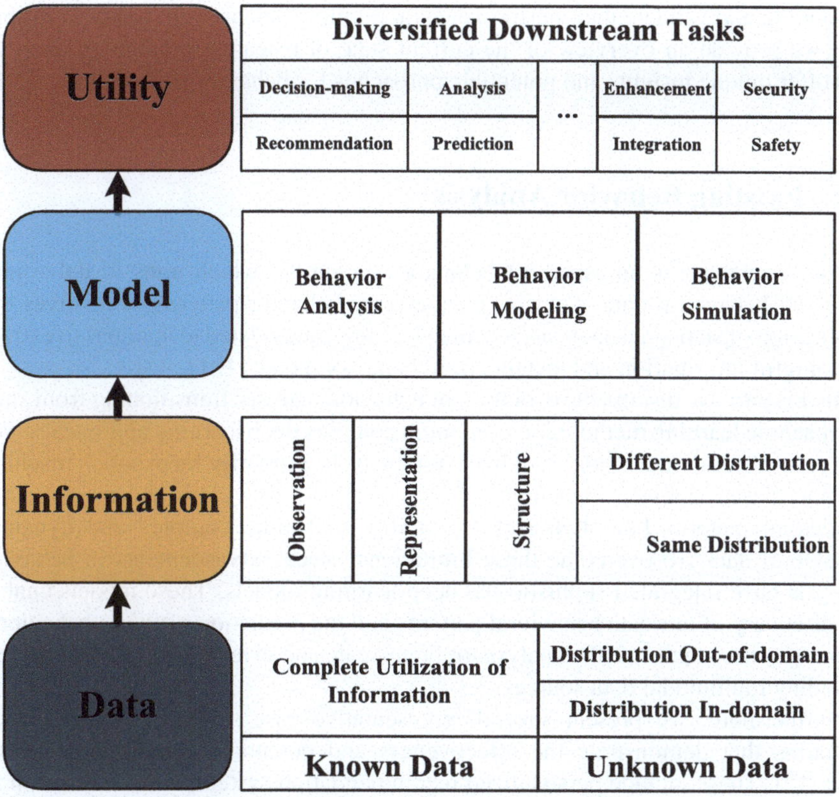

Fig. 1.1 The technology stack of universal behavior computing

analysis, and understanding of all behaviors. In particular, we gradually unravel the secrets behind behaviors by examining the data, information, models, and utility involved. At the information level, introducing unknown data, especially data that differs from the known data distribution, holds promising potential to improve the performance of behavior research on specific tasks. In the face of complex and diverse behavioral data, behavior structure serves as a novel means of behavior description that goes beyond behavioral observation and representation. It possesses stronger capabilities to express massive amounts of behavioral data and provides opportunities for the emergence of information. At the model level, a unified framework for behavior modeling and simulation is proposed to simultaneously mine behavior information from known and unknown data. This integrated behavior modeling scheme provides formatted outputs containing sufficient information for downstream tasks, making it adaptable to a wide range of highly differentiated business scenarios.

In this book, our focus lies on the research of modeling methods for behavioral data. We present a series of effective and promising solutions for universal behavior computing and validate their performance on multiple downstream tasks. Additionally, we provide an overview of the current state of research in behavior analysis and offer unique insights and potential approaches for behavior simulation.

1.2 Existing Behavior Analysis

Behavior analysis is an aspect of behavior computing, which aims to delve into observed behavioral data to uncover underlying behavior patterns. It involves the application of statistical analysis, machine learning, and related techniques to extract meaningful information and insights from behavioral data [1, 11, 13].

In this part, we discuss the evolution of behavior analysis, transitioning from simple machine learning methodologies to more complex deep learning approaches. We first review traditional methods of behavior analysis, primarily focusing on machine learning-based analytical techniques. These methods provide initial insights into behavioral patterns but often lack the ability to handle complex and dynamic behavioral data. To overcome these limitations, recent advancements in behavior analysis have integrated sophisticated deep learning models. These models enable the discovery of intricate behavioral patterns and the prediction of future behaviors. Moreover, they facilitate the analysis of large-scale and diverse behavioral datasets, including multimodal data sources.

Furthermore, we present several representative case studies and application scenarios that demonstrate the effectiveness and potential of behavior analysis [4, 6, 12]. These include personalized recommendation systems, fraud detection in financial transactions, and other relevant domains.

In conclusion, behavior analysis has evolved from simple statistical techniques to more complex methodologies capable of handling complex and dynamic behavioral data. The integration of deep learning and related techniques has enabled the discovery of intricate behavioral patterns and improved predictive capabilities [14, 15]. However, challenges remain, and further research is needed to address data heterogeneity, scalability, interpretability, and other pertinent issues in behavior analysis.

1.2.1 Behavior Analysis for Recommendations

One of the prominent application areas in behavior analysis is user behavior recommendation, particularly in the context of the Internet and mobile applications. A vast amount of user behavior data, including clicks, browsing activities, purchases, search queries, and more, is collected and recorded. Analyzing this data provides valuable insights into user preferences, interests, and behavioral patterns, supporting applications such as personalized recommendations, targeted advertising, and user profiling.

Early researchers in the field utilized collaborative filtering models to predict the behavior of target users based on the collaborative actions of all users [16]. In collaborative filtering, items were represented solely based on user behavior patterns, neglecting semantic relevance. To enhance the representation of items and users, many researchers went beyond user-item interactions and incorporated auxiliary data. Auxiliary data can be categorized into content-based information and context-aware data. Some researchers explored the possibility of using neural models to automatically discover complex high-order feature interactions for click-through rate prediction and recommendation [17]. Current relevant work on this topic can be classified into three categories: implicit MLP structures, explicit high-order modeling, and tree-based boosting models.

With the emergence of social networks, users tend to express their item preferences on online social platforms and share their interests through social relationships. Social recommendation has emerged in these platforms, aiming to model social influence and social relevance among users to improve recommendation effectiveness. The fundamental reason for social recommendation lies in the social influence between social neighbors, which leads to the correlation of user interests within the social network. Social recommendation models can be roughly categorized into two types: social relevance-enhanced models and models based on graph neural networks (GNNs) [18, 19]. The former treats users' social behaviors as the social domain and item preference behaviors as the item domain. Social relevance-enhanced and regularized models attempt to integrate the two types of behaviors from both domains into a unified representation. The latter focuses on capturing the dynamic recursive effects that reflect the social diffusion process, where each user is influenced recursively by the global social network graph structure.

1.2.2 Behavior Analysis for Cybersecurity

Another prominent application area of behavior analysis is cybersecurity. In a cyber environment, malicious activities and attack incidents are constantly increasing and evolving, posing a significant challenge for security practitioners [20, 21]. By analyzing behavior data such as network traffic, login activities, and file access, it is possible to detect and identify abnormal activities, intrusion attempts, and potential security vulnerabilities, thereby enhancing the capability of network security.

Behavior-based approaches are commonly used methods for detecting cyber anomalies on network services [22–24]. Typically, most behavior-based methods explore the associations between behavioral features and labels, extracting effective features to detect deceptive behaviors. The historical information on behavior sequences can further assist security personnel in distinguishing abnormal behaviors from benign ones. Therefore, some researchers have introduced attention mechanisms to capture the influence of historical events on current events and utilize convolutional neural networks to capture short-term contextual information. These methods partially capture the relationships between different events, but they still do not fully consider the interactions among events. Researchers have further recognized the powerful representational capacity of graphs and have used graphs to discover anomalous behaviors. Some works propose using deep network embedding methods to detect rarely occurring anomalous events in networks.

1.2.3 Potential Development Trends

The development of behavioral analysis is constantly evolving and expanding, and the following are some main trends (but not limited to):

- Multimodal Behavior Analysis: Traditional behavior analysis primarily relies on data from a single modality, such as text, images, or sensor data. However, combining data from multiple modalities for behavior analysis can provide more comprehensive and accurate insights. For example, integrating textual and image data for sentiment analysis or event recognition can better understand user behavior and intentions.
- Real-Time Behavior Analysis: Real-time behavior analysis refers to the immediate analysis and response to behavior data generated in real time. With the widespread use of the Internet and the development of Internet of Things (IoT) technologies, the volume of generated behavior data continues to increase, requiring fast analysis and decision-making. Real-time behavior analysis can help detect anomalies, predict trends, and respond accordingly when behaviors occur.
- Privacy Protection: Behavior analysis involves a significant amount of personal data, making privacy protection and ethical considerations crucial. Future behavior analysis methods need to pay more attention to technologies and mechanisms that address data privacy and protect individual rights, ensuring compliance and fairness.
- Cross-Disciplinary Collaboration: Behavior analysis spans multiple disciplines, including computer science, data science, psychology, and more. Future developments require more cross-disciplinary collaboration, integrating expertise and technologies from different fields to collectively advance the development and application of behavior analysis.

1.3 Advanced Behavior Modeling

1.3.1 Behavior Modeling on Observation

The purpose of behavior observation is to capture and analyze behavior data in order to derive behavior patterns that describe the behavior agent. There is a long history of research employing behavior observation techniques to learn behavior patterns. In *Science*, Song et al. [25] conducted observations and analysis on the 3-month travel behavior records of 50,000 anonymous mobile phone users, and exploited the discovered user behavior patterns to predict user travel behavior, which achieved a high accuracy of 93%. Kosinski et al. [26, 27] demonstrated the viability of leveraging user behavior data for the inference, prediction, and modeling of user attributes, interests, and personality traits in their studies. Besides, they pointed out that, with more appropriate training data, the revelation of other attributes becomes more probable. Based on the varying roles of behavior data during behavior observation, we categorize observation-based methods into feature engineering and shallow learning.

1.3.1.1 Feature Engineering

Feature engineering (FE) processes and transforms raw behavioral data to select, extract, or construct the most representative behavioral features. For different scenarios or tasks, the following operations are usually involved, such as feature selection, feature transformation, feature construction, feature scaling, outlier and missing value processing [28], etc. From the perspective of its role in improving data, we divide feature engineering into information optimization and information construction.

Information optimization typically encompasses feature selection, outlier processing, and feature transformation. In the raw data, not all information is useful, so information needs to be filtered out. In [29], Keserwani et al. discussed prevalent issues within intrusion detection datasets, such as duplicate records and outliers, which potentially introduce bias into intrusion detection models constructed using such data. Accordingly, they summarized some feature optimization methods, including statistical-based methods [30–32] and metaheuristic-based optimization methods [33]. Nazir et al. [34] emphasized the importance of excellent feature selection methods to improve model classification accuracy and reduce model computational complexity. Shafiq et al. [20] proposed an effective feature selection scheme that combines bijective soft sets and an accuracy measurement method based on wrapping technology, which improves the low classification accuracy of malicious traffic identification models attributed to imprecise feature selection. Outlier processing enhances the quality of the data, allowing for a better reflection of its true characteristics. Liu et al. [35] designed an identity verification system based on mandibular vibration features. To overcome the abnormal vibration signals caused by hardware defects, they used a median absolute deviation to cover the outliers with the average of normal values, thereby improving the model's ability to extract vibration features. Zhang et al. [36] proposed to exclude erroneous records or unexplained outliers during the preprocessing of fatigue driving behavior signals, thereby enhancing the effectiveness of their detection model training. In the presence of inconsistent feature dimensions in the original data, feature transformation enables a more effective standardization of feature scales, facilitating a more comprehensive utilization of the available feature data [37]. Lopes et al. [21]

employed numericalization and feature scaling techniques to enhance the quality of input data, mitigating issues posed by nonnumeric features and variations in feature scales. Malik et al. [38] normalized features with significant differences between their maximum and minimum values in the dataset to ensure stable convergence of weights and biases and to prevent issues such as vanishing and exploding gradients.

Information construction is based on combining, transforming, or domain knowledge to create new features. These strategies have been used by a number of studies [22, 23, 39, 40]. Bahnsen et al. [22] extended the strategy of aggregating transactions and created a set of new features to effectively observe customer spending behavior patterns when constructing a credit card fraud detection model. These new features for model training resulted in an average performance improvement of 252% compared to the raw model. Zhang et al. [41] proposed a Multidimensional Feature Fusion and Stacking Ensemble Mechanism (MFFSEM) that considers the correlation and relevance among fundamental feature datasets, enabling accurate exploration of the intrinsic information within traffic data. Li et al. [42] proposed a sensor-based continuous identity verification system, FusionAuth, which enhances authentication performance and system reliability by fusing the behavioral patterns captured by three sensors (accelerometer, gyroscope, and magnetometer), achieving a lower false alarm rate compared to non-fusion strategies. Li et al. [43] proposed a feature transformation method that combines linear and nonlinear spatial transformations, where generated new features improve the efficiency and performance of certain classifiers. Bahnsen et al. [44] pointed out that phishing attackers often follow certain common patterns, which experienced experts can typically discern through some features. Therefore, they introduced external expert knowledge to create 14 customized features that describe URLs based on their lexical and statistical analysis.

1.3.1.2 Shallow Learning

Shallow learning can help behavior models break through the limitations of expert knowledge and extract rich and meaningful features from raw behavioral data [24]. For instance, in the context of abnormal network traffic detection, feature engineering often involves the manual selection, extraction, and construction of a set of anomaly behavioral features. However, it is challenging to encompass all potential abnormal patterns, despite diligent efforts in feature engineering. Shallow learning methods, such as linear regression, can automatically learn normal behavioral patterns and produce outputs that deviate from normal thresholds when encountering abnormal behaviors, thereby identifying abnormal behaviors. Certainly, some complex problems have to establish relationships between multiple features and dependent variables. From the perspective of the number of features utilized, shallow learning can be divided into the following categories: single-feature strategy and multifeature strategy.

Single-feature strategy focuses on utilizing a single feature to learn the relationship between an input variable and the target outcome. They (e.g., univariate linear regression or univariate logistic regression) offer a clear understanding of how specific features affect downstream tasks. Galeano et al. [45] proposed transforming multivariate time series into univariate time series through linear combinations, which enables more effective detection of outliers in time series data. Salem et al. [46] introduced a medical insurance fraud detection framework based on anomaly detection techniques, which models the relationship between hospitalization expenses and medical reimbursements using robust regression models. Barbhuiya et al. [47] proposed a linear regression-based DDoS attack detection technique to reduce false alarm rates and maintain the accuracy of attack detection, which is based on hypothesizing a positive correlation between the mean and standard deviation of network throughput in window-based time series. The above strategies do appear concise and efficient in some scenarios; however, the simplicity brings inherent limitations. Especially in scenarios involving interactions among multiple features, this strategy may struggle to capture subtle patterns that require consideration of a wider range of input variables.

Multifeature strategy considers the complex relationship between multiple input variables and output results to make up for the shortcomings of the single-feature strategy, which can make it better understand the potential patterns that affect the results. Roy et al. [48] proposed a lightweight malware detection framework that employs a random forest classifier, achieving an ROC-AUC score of 98.87% across over 2000 applications using 50 features. Ghosh et al. [24] proposed a neural network-based module that enables end-to-end feature aggregation in fraud classification tasks. It exhibits robust performance compared to manually aggregated features and other automatic feature extraction methods in classification tasks. Lakhina et al. [49] proposed a PCA-based detector for anomaly detection on multiple time series, utilizing the top ten principal components to construct feature subspaces. In previous studies on anomaly detection in wireless sensor networks [50, 51], it has been highlighted that solely considering outliers in individual attributes of sensor data is insufficient for accurately determining anomalies. It is necessary to analyze multidimensional features to make judgments comprehensively. High-dimensional feature analysis often encounters the problem of the "curse of dimensionality," and some works have successfully alleviated this issue through dimensionality reduction methods. In the study conducted by Deng et al. [52], they proposed a method called the one-class support Tucker machine (OCSTuM) specifically designed for handling large-scale Internet of Things (IoT) sensor data. Tensor decomposition reduces the feature dimension, and the genetic algorithm simultaneously ensures the selection of the best feature subset, thereby improving the accuracy of the anomaly detection model. In the study conducted by Qiao et al. [53], an OCSVM model was proposed to address the challenges of high dimensionality and large-scale data in IoT anomaly detection.

1.3.2 Behavior Modeling on Representation

1.3.2.1 Behavior Learning with Latent Information

Deep learning methods have demonstrated their effectiveness in extracting meaningful representations from data. Unlike traditional approaches that rely on handcrafted features, deep learning models learn abstract representations of behavior directly from the available explicit behavior information. In terms of information utilization, these models can be categorized into two groups: those that utilize local information and those that utilize global information.

Utilization of Local Information Deep learning models can extract features such as local structures and correlations, resulting in excellent performance. For instance, the commonly used convolutional neural networks (CNNs) are adept at capturing local structural information in input data through convolutional operations. Wang et al. [54] introduced a CNN-based approach for classifying malicious software traffic. It eliminates the need for manually engineered features by representing traffic data in matrix form as input to the classifier. Zhang et al. [55] proposed an effective deep neural network architecture utilizing multiple gated-CNNs to learn important and relevant features extracted from each API call record. Yang et al. [56] preprocessed network traffic data into RGB images as input to a CNN model, achieving good performance on the vehicular network traffic detection dataset. Ansari et al. [57] proposed a deep learning model based on gated recurrent unit (GRU) for intrusion alert prediction, which addressed the issue of vanishing gradients in recurrent neural networks (RNNs), thereby improving long-term dependency learning. Villarreal et al. [58] proposed an anomaly detection framework based on Long Short-Term Memory (LSTM) models to address prolonged internal threat attacks. In summary, traditional CNNs, while capable of capturing local structural information in input data, may fail to fully grasp the holistic characteristics due to their limited local perception. LSTM is effective in capturing dependencies within sequential data. However, it struggles with capturing long-term dependencies and incurs significant computational overhead.

Utilization of Global Information Deep learning models that leverage global information excel at capturing the overall structure and correlations within input data from a global perspective. The attention mechanism, in particular, dynamically calculates the relevance of each position in the input sequence with respect to all other positions. This enables the model to better grasp the semantic information of the entire sequence, rather than focusing solely on local segments. Xie et al. [59] introduced the self-attention mechanism into traffic classification tasks, utilizing self-attention related to different positions of traffic packets to compute representations of sequences, thereby further improving the model's performance. Javed et al. [60] proposed a method comprising a multistage attention mechanism integrated with a Long Short-Term Memory (LSTM)-based CNN, which facilitates learning global information about the input data stream. Nedelkoski et al. [61]

proposed an anomaly detection framework utilizing a multi-head self-attention mechanism to learn vector representations of logs, enhancing the generalization capability toward novel, unseen log samples. Wang et al. [62] proposed an IoT attack behavior detection method based on the fusion of spatiotemporal features. It utilizes CNN and LSTM separately to learn spatial and temporal features, and then integrates the two using the attention mechanism. Finally, the combined attack features are fed into a deep neural network (DNN) for learning. Chai et al. [63] proposed a multimodal hierarchical attention model (MMHAM) capable of learning fraud cues from multiple-modal data to enhance the detection capability of phishing websites.

1.3.2.2 Behavior Learning with Explicit Information

In real-world scenarios, various behaviors often exhibit explicit connections, such as online payment behavior, fraudulent behavior, and system call behavior. These connections can follow specific orders or rules, forming structures that contain valuable information. The connections among behaviors can be classified into two formats: sequence-based structures and graph-based structures.

Sequential connections can be represented as a linear structure that encapsulates a specific order among behaviors and signifies temporal, spatial, or logical relations among behaviors. It is similar to natural language, in which each behavior is a word in the sentence. Given rapid progress in the natural language processing area, sequence-based methods could be applied for the embedding of a behavior sequence. For instance, Pi et al. [5] introduced an industrial solution for modeling extended-length sequential user behaviors, extending to thousands of data points. Zheng et al. [6] proposed a cold-start sequential recommendation framework based on meta-learning to address the recommendation cold-start problem by capturing dynamic user preferences based on historical behavior sequences.

The existing methods perceive sequential data as a collection of sequences, learning patterns, and associations within to enable prediction or classification. Prevalent machine learning techniques include hidden Markov models, recurrent neural networks, Long Short-Term Memory Networks, convolutional neural networks, and others. The hidden Markov model is a probability-based sequence model consisting of a hidden state sequence and a corresponding observation sequence. It infers the most probable hidden state sequence through the observation sequence. Recurrent neural networks introduce a hidden state while processing inputs at each time step, enabling continuous information flow to model the entire sequence. Long Short-Term Memory Networks addressed gradient vanishing and exploding issues in recurrent neural networks by introducing memory cells and gating mechanisms, effectively capturing and retaining dependencies within lengthy sequences. Initially prominent in image processing, convolutional neural networks demonstrated remarkable performance in sequential data analysis. They extract local features through sliding windows and reduce feature dimensionality via pooling operations. In summary, the structural relations and learning methodologies among behavior data based on sequences are relatively mature. Existing sequential data processing techniques can be directly applied to one-dimensional structures.

Graph structure can express complex relationships among behavioral data. Most popular supervised learning methods in machine learning expect their inputs to be fixed-size vectors, often leading to the flattening of behavioral data into tabular formats, disregarding potentially valuable relational information within the data [64]. Recent research has discovered significant informational gains by representing relationships among behaviors through the establishment of a latent graph structure [65]. The creation of artificial structures involves two pivotal steps: firstly, determining the type of graph to model and, secondly, establishing the rules for the graph construction.

The data can be structured into three graph forms: homogeneous graphs, heterogeneous graphs, and hypergraphs. In the context of homogeneous graphs, instance graphs [66, 67] and feature graphs [68, 69] can be constructed by using data instances or features as nodes, respectively. Forming heterogeneous graphs, You et al. [70] connect data instances with their corresponding features (alongside other metadata), thereby creating bipartite/multipartite graphs. Additionally, Guo et al. [71] consider different feature values as distinct types of edges, which delineate various relationships among data instances, thus constructing a multi-edge/multi-relational graph. Within the framework of hypergraphs, Cai et al. [72] connects table elements with similar attributes via edges, allowing any number of table elements to be linked within the hypergraph. For instance, data instances sharing common feature values can be connected by an edge within the hypergraph.

After determining the nodes, it is crucial to establish complete relationships among them to construct a comprehensive graph structure. In the field of edge construction, there are generally four approaches: rule-based, learning-based, search-based, and knowledge-based methods. Among these, the rule-based and learning-based methods are widely adopted [65]. Rule-based methods rely on data dependencies between data instances and features [64, 70, 71], or they employ common heuristic approaches such as K-nearest neighbors [67, 73] or fully connected structures [69, 74]. Learning-based methods treat edges as learnable parameters to generate edges for nodes [9]. Search-based approaches leverage neural network search techniques to discover optimal graph topologies for representation learning [75], or establish edges based on information retrieval techniques to identify similar relevant data instances [76]. Knowledge-based methods necessitate domain experts to provide relevant knowledge between data instances, enabling meticulous construction of the graph [77].

In behavioral graphs, nodes often represent an entity, and behaviors are depicted as edges signifying the connections between entities. Next, we will introduce two examples of applying behavior graphs: malware detection and fraud detection.

Behavior Graph in Malware Detection Kolbitsch et al. [78] proposed a method for detecting malicious software. It extracts system call behavior graphs for each instance, including the dependency relationships between system calls and their parameters. Given that many new malware samples are created through code or binary mutation, one approach to determine the maliciousness of a sample is by assessing its similarity to previously identified malicious software programs. Traditionally, malware detection methods treated malware programs as sequences of bytes, neglecting their higher-level internal structures, such as basic blocks and function calls. However, with the introduction of the behavior graph of malware, there has been a surge of research efforts focusing on leveraging this graph representation for malware analysis and detection. Ding et al. [3] construct a common behavior graph for each malware family and employ a graph-matching algorithm based on the maximum weight subgraph to detect malicious code. Alasmary et al. [79] build a detection mechanism for Internet of Things malware utilizing control flow graphs.

Fraud Detection in the User Behavior Graph Beutel et al. [4] integrated graph analysis tools with anomaly and fraud detection to model user behavior. Cheng et al. [12] utilized graph neural networks to learn transactional graph features based on time and location. Zhang et al. [80] developed separate models for normal behavior and fraudulent behavior by utilizing certain normal behaviors as weak supervisory information. They proposed a fraud detection system based on competitive graph neural networks. As graph neural networks continue to advance, researchers have been able to explore more valuable information related to user behavior in fraud detection graphs and model the distributions of normal behavior and fraud behavior separately.

In the aforementioned examples, behavior graphs are constructed to capture information about anomalous or fraudulent behaviors. A deep understanding and analysis of behavior graphs enable researchers to comprehensively capture the desired information. The structural aspects of behavior have gained increasing importance and have become a hot topic in this field. However, existing research on behavioral structures primarily focuses on the interactions between behaviors at a coarse level, while further exploration into finer details remains an area that warrants more in-depth investigation.

1.3.3 Behavior Modeling on Structure

From an informational perspective, as the characterization of behavior gradually elevates dimensions, the behavioral information available to us becomes more comprehensive. A natural consideration is whether we can further explore latent information within behaviors to provide greater support for behavior computation. Recent research has conducted in-depth analysis on the dimensions of behavior [2], offering insights into exploring latent information within behavior. Behavior dimensions are defined as the minimum number of past features required to make the maximum information-rich predictions about the future. Each feature of a behavior dimension can be considered as an attribute constituting the granularity of behavior atoms. Diverging from exploring complex relationships between behaviors through interactions among entities, we present two approaches that consider behavior structures from an atomic-level perspective.

1.3.3.1 Atomic-Level Interactions Within the Single Molecules

The attribute, as the smallest unit composing behavioral data, shares similarities with the fundamental atoms forming molecules in the real world. In molecules, the types of atoms and their specific structures form distinct molecules. Recent studies have revealed that attributes within behaviors possess similar underlying structural properties [7]. The associative relationships among attributes constitute valuable structural information. Employing graph neural networks for learning models allows effective exploration of the internal structural information, akin to the interactions among atomic levels within a single molecule.

Compared to the structural relationships among behaviors in behavior representation, the internal structure within a single behavior represents fine-grained information. In the context of the downstream tasks, the internal structure of behaviors accomplishes graph-level tasks, while the interbehavioral structure serves node-level tasks. So the methods outlined in the previous discussion, which manually construct interbehavioral structures, are not suitable for modeling the internal structure of behaviors. One of the pressing challenges lies in devising methodologies to construct molecular-level structures for behavioral data, enabling the representation of inherent structural information within behavioral data.

We propose a conceptual framework, molecular-level behavioral structure modeling. In the physical world, scientists often assume the existence of chemical bonds between closely spaced atoms, thus constructing molecular graphs to describe real-world molecules, which support many downstream learning and analysis tasks. When exploring the inherent structure of behavioral data, we treat the attributes of behavioral data as nodes for constructing behavior molecular graphs, and certain associations between attributes as edges of the behavior molecular graphs. As a result, each behavioral data corresponds to a specific behavior molecular graph. The obtained collection of behavioral molecular graphs can be subjected to graph-level learning tasks.

1.3.3.2 Atomic-Level Interactions Across Molecules

Notable advancements have been made in comprehending the intricate relationship between molecular structures and their functional synthesis in high-dimensional spaces. By incorporating representation learning from the field of machine learning, the fusion of data-driven molecular property and reactivity prediction allows researchers to improve the prediction of structure-property relationships through enriched local encoding of chemical environments and enhanced information exchange between reactive components. Inspired by this, we speculate that enhancing the expression of local contextual behavioral information can provide valuable support for behavior modeling, particularly in understanding the atomic-level interactions between reactive behavior molecules.

For tasks involving contextual information such as natural language processing and recommendation systems, the concept of atomic-level interactions across molecules can be analogously applied to models. This interactivity can be modeled in machine learning models through the following approaches:

- Attention mechanisms allow models to assign different weights to different elements based on contextual information. Similar to interactions between molecules, attention mechanisms calculate the relevance between different elements in the input, allocating more "attention" (weights) to atoms relevant to the downstream task.
- Long Short-Term Memory (LSTM) is a variant of recurrent neural networks (RNNs) suitable for sequential data. LSTM incorporates memory units and gate mechanisms to capture long-term dependencies in sequences. Similar to atomic interactions in molecules, LSTM can establish long-range dependencies in sequential data.
- Graph neural networks (GNNs) are neural network models specifically designed for processing graph data. In graph data, connections between nodes can represent interactions between molecules. GNNs can capture interactions between molecules by aggregating neighbor information of nodes and relationships between nodes.

These machine learning methods, by modeling the interaction between contextual information and data, can better capture complex correlations in the data, thereby improving model performance and representation.

We propose a behavior-molecule reactivity framework with dual domains (molecular domain and atomic domain). In the atomic domain, a single behavior molecule structure carried by the graph structure provides an effective way to embed self-spatial information of the behavior, enriching the expressive power of the behavior. Instead of being nodes in the graph, behaviors are represented as local subgraphs. In the molecular domain, the behavior interaction module is introduced to effectively learn the collaborative control between multiple behavior interactions, thus improving the information exchange of contextual behaviors in behavior molecule modeling. The interaction of behavior information in both the molecular and atomic domains is handled by graph neural networks, while different behavior information in the two domains is processed by attention layers, where the weight values are trained to capture the contributions of behavior information in different domains. Subsequently, a pooling layer is used to convert behavior structures with multiple atomic-level representations into one-dimensional reaction vectors. The predicted behavior interaction-related vectors are concatenated as the final one-dimensional vector, which is then passed through the last linear layer to provide the prediction value.

1.4 Sophisticated Behavior Simulation

1.4.1 Distribution In-Domain Learning: Unleashing the Power Within Known Data

Distribution in-domain learning involves leveraging the existing knowledge and patterns present in the data to gain insights and make predictions. By delving deep into the known data, we can unlock its inherent potential and extract valuable information that can drive decision-making and problem-solving processes. Furthermore, in-domain distribution analysis enables us to identify patterns, correlations, and dependencies among different variables. This knowledge can be used to build models and algorithms that accurately capture the underlying dynamics of the domain.

1.4.1.1 In-Distribution Simulation

The first stage of simulation methods primarily operates within idealized conditions, focusing on relatively simple systems. Besides, they generally utilize a simple mathematical formula to simulate the interaction of objects. We review the application of the first stage of simulation methods in various realms. McCann et al. [81] utilized a single-valued nonlinear function to simulate servomotors. Gibby [82] utilized mathematical relationships to describe amplitude-modulated data systems. Ritson et al. [83] utilized a simulator to solve the equation of the hydraulic remote position controller. The first stage of simulation was also applied to medical and health problems. Respiratory sinus arrhythmia was described by dynamic mathematical relationships which were driven by analog computer simulation [84]. These simulation methods were constrained by limited computing speeds and storage capacities. The first stage of simulation methods builds a system where everything is entirely deterministic and computable. As a result, it always converges to one or a few specific states, which can be seen as a deterministic discrete distribution.

The second stage of simulation methods primarily operates within lots of assumptions, focusing on complex systems. Compared to the first stage of the simulation methods, the second stage of the simulation methods generally utilizes a mathematical model to simulate complex systems. They have advantages in data processing tasks and multitasks. The second stage of simulation methods could be heavily applied to geographical research. Cui et al. [85] proposed a novel method to simulate magnitude-8 earthquakes and help reduce potential loss of life and property. Kolb et al. [86] utilized graphics hardware (GPU) to simulate a dynamically growing particle system. Huggins et al. [87] developed a mathematical model to simulate the surface runoff from watersheds. The second stage of simulation was also applied to the biological research. Daniel et al. [88] utilized a mathematical model to simulate the motion of oxygen in brain cells. Jones et al. [89] utilized mathematical relationships and mathematical models to simulate the changing population of the potato varieties. Kang et al. [90] proposed Biocellion which simulates biological systems through mathematical models and computer simulation. All these applications still heavily depend on prior assumptions. The second stage of simulation methods assumes that the system's behaviors strictly follow a particular distribution, which can be expressed as mathematical models.

The third stage of simulation methods primarily operates within data-driven algorithms, with a specific focus on complex systems. In contrast to the second stage of simulation methods, the third stage of simulation methods utilizes highly intricate models that are challenging or even impossible to describe using mathematical models alone. These complex models can be generated through algorithms such as clustering analysis and deep neural networks. When data is hard to acquire, the third stage of simulation methods can generate data for research. In IOT (Internet Of Things), Hui et al. [8] proposed a knowledge-enhanced GAN (generative adversarial network) model to generate large-scale IOT traffic. In disease prevention and control, Yuan et al. [91] utilized a model based on generative adversarial imitation learning to generate human activity trajectories. Moreover, in human daily activities, Yuan et al. [92] proposed a knowledge-enhanced generative adversarial imitation learning model to generate human daily activity data. The third stage of simulation methods can also be applied to simulate the interaction of the individuals and then, furthermore, infer the changes in the system. Shi et al. [93] utilized graph network-based simulation (GNS) to simulate crowd, which effectively prevent stampedes. In urban management, Zhang et al. proposed [94] PIML which utilized physical and neural network models to simulate crowds, and this model helped a lot in traffic management, urban planning, and emergency management. In social systems, Lorig et al. [95] utilized ABSS (agent-based social simulations) to simulate social systems. The third stage of simulation methods drives individual behaviors entirely by historical data. Although it is possible to simulate dynamic and complex systems, individual behaviors are still limited by a fixed distribution of historical behavioral data. In this sense, it is still an in-distribution simulation.

1.4.1.2 Bottlenecks of Current In-Distribution Simulation

The inherent bottleneck in in-distribution simulation stems from the limitations of the tools employed, which constrain the ability to generate outcomes for out-of-distribution samples. All tools utilized in in-distribution simulation learn distributions solely from the provided historical data. Consequently, when encountering out-of-distribution data, the performance of models trained using in-distribution simulation methods tends to decrease.

The emergence of out-of-distribution data is inevitable. This poses a significant challenge for in-distribution simulation in systems where dynamic data distribution exists. For instance, complex human systems, encompassing chaos and human strategies, exemplify such systems. Chaos, as a component of chaotic systems, is characterized by sensitivity to initial conditions, transient stability, long-term unpredictability, and the presence of nonperiodic state sequences [96]. Human strategies, on the other hand, are influenced by the behavior of other individuals [97], implying that humans do not consistently behave as they have in the past. This variability in data distribution is a key factor in high-order chaotic systems.

Current in-distribution simulation methods are incapable of eliminating the bottleneck. The presence of out-of-distribution data in the systemic data of dynamic distribution further emphasizes the urgent need for a new methodology, and an advanced simulation technology holds promise.

1.4.2 Out-of-Distribution Learning: Expanding Horizons Beyond Boundaries

Out-of-distribution (OOD) problem refers to the situation where the model's performance degrades when the training and the testing set do not satisfy the independent and identically distributed (i.i.d.) assumption [98]. For instance, the perception system of a self-driving car can produce erratic predictions when it encounters a new test sample with a different illumination or weather condition not seen during training [99]. In some practical production environments, the degradation of model performance is unacceptable, especially in safety-critical scenarios.

1.4.2.1 Alleviating Out-of-Distribution Problem

Given the significant importance of OOD problems, various methods have been proposed to alleviate them. Below, we introduce two related fields that alleviate the OOD problem: domain adaptation (DA) and OOD generalization (OODG).

The goal of DA is to transfer the knowledge learned in the source domain effectively to the target domain [100]. Assuming the existence of two domains, each with its own unique data distribution, DA enables a model trained in the source domain to achieve good generalization performance in the target domain. The current mainstream DA methods can be categorized into three types: discrepancy-based, adversarial-based, and reconstruction-based. Related information about all works mentioned below has been given in Table 1.1.

Discrepancy-based methods focus on the discrepancy distance between source and target distributions [111]. Early works [101–103] used Maximum Discrepancy Loss (MMD) to calculate discrepancy distance, which is viewed as a loss function, to train the model. In addition to using MMD to measure the distance between two distributions, Sun et al. [104, 105] proposed using correlational alignment (CORAL) to measure and minimize the distance between two distributions by aligning the second-order statistics. Courty et al. [106] suggested a solution to improve domain adaptation in joint distribution optimal transport with optimal transport (OT).

Table 1.1 Summary of domain adaptation methods

Type	Author	Year	Dataset	Samples	Categories	Accuracy
Discrepancy-based	Ghifary et al. [101]	2014	Office 31	4652 images	31 classes	39.3%
Discrepancy-based	Tzeng et al. [102]	2014	Office 31	4652 images	31 classes	–
Discrepancy-based	Long et al. [103]	2015	Office 31	4652 images	31 classes	–
Discrepancy-based	Sun et al. [104]	2016	Office 31	4652 images	31 classes	69.4%
Discrepancy-based	Sun et al. [105]	2016	Office 31	4652 images	31 classes	72.1%
Discrepancy-based	Courty et al. [106]	2017	Caltech-office	2533 images	10 classes	80.04%
Adversarial-based	Ganin et al. [107]	2015	MNIST MNIST-M	149,001 images	10 classes	89.01%
Adversarial-based	Zhang et al. [108]	2016	Office 31	4652 images	31 classes	82.4%
Reconstruction-based	Ghifary et al. [109]	2016	MNIST USPS	9000 images	10 classes	91.8%
Reconstruction-based	Bousmalis et al. [110]	2016	MNIST MNIST-M	149,001 images	10 classes	83.2%

Adversarial-based methods leverage adversarial learning to train an encoder that is shared between the source and target domains and is expected to extract domain-invariant representations [112]. Ganin et al. proposed a domain-adversarial neural network (DANN) [107], whose basic idea is generating features from a feature extractor and then using the domain classifier to determine whether they are features from the source domain or the target domain. A gradient reversal layer is introduced in the middle to establish an adversarial relationship, thereby extracting domain-invariant features. Based on DANN, the collaborative and adversarial network (CAN) [108] added several domain classifiers on multiple CNN feature extraction blocks, in which each domain classifier is connected to the hidden representations from one block.

Reconstruction-based methods focus on learning the transformation from the source domain to the target domain and training a classifier using the source domain data. In order to make the feature space more aligned with the samples, Ghifary et al. [109] proposed training a classifier to distinguish between samples corresponding to features from the source domain, while using a decoder to reconstruct the target domain samples as accurately as possible. Bousmalis et al. [110] explicitly learn to extract representations that are divided into two subspaces: one private to each domain and one shared across domains. By incorporating a similarity loss between the shared subspace components, the alignment between the domains is improved.

It is worth noting that DA can get a small amount of target domain data during training. The three methods mentioned above all involve incorporating various forms of prior knowledge from the target domain during the training phase to optimize the model's performance. OODG focuses on a situation similar to DA [113], but differing from DA, OODG does not have any prior information about the target domain in the model training phase. Its significance lies in ensuring the robustness of the model in practical applications, enabling it to handle data from new domains or different distributions, thereby enhancing the model's practicality and applicability. Existing OODG methods can be mainly classified into two categories: unsupervised OODG and supervised OODG [114].

Unsupervised OODG methods ensure that the representation learned in an unsupervised way possesses certain attributes that may facilitate out-of-distribution generalization [114]. Zhang et al. [115] first introduce the concept of unsupervised domain generalization (UDG) and propose DARLING, a method that learns generalizable models using less unlabeled data compared with ImageNet. Harary et al. [116] present an approach based on self-supervised learning of a Bridge Across Domains (BrAD): an auxiliary bridge domain accompanied by a set of semantics-preserving visual (image-to-image) mappings to BrAD from each of the training domains. Besides UDG, disentangled representation learning also has good performance. In the research of Dittadi et al. [117], disentanglement is found to be a good predictor for out-of-distribution (OOD) task performance. Zhang et al. [118] propose a primal-dual algorithm for joint representation disentanglement and domain generalization that can achieve competitive OOD performance.

Supervised OODG methods design and train models using data with labels, unlike unsupervised learning methods. Peters et al. [119] introduce the concept of invariant causal prediction (ICP), assuming that the underlying causal structure remains unchanged across different environments or domains. This implies that even when there are changes in data distribution, the model is expected to perform well by correctly identifying causal relationships. Subsequently, the proposal of invariant risk minimization (IRM) [120], which is the basis of invariant learning, relaxes causality to invariant features in different environments and makes ICP more practical and general. Inspired by Adversarial Training (AT), Xin et al. [121] propose Domain-wise Adversarial Training (DAT) to improve OOD generalization ability. Rahman et al. [122] proposed a correlation-aware adversarial domain adaptation and generalization method that can be applied to address the OOD generalization issue. Yuan et al. [123] suggest a Meta-learning Causal Feature Selection (MCFS) model for OOD generalization. Cha et al. [124] introduce a method named Stochastic Weight Averaging Densely (SWAD) to find flat minima, which can lead to better OOD generalization performance.

Several theories have been proposed to enhance existing methods, with a primary focus on distributionally robust optimization (DRO). DRO tackles out-of-distribution scenarios by incorporating uncertainty modeling. The main idea behind DRO is to find a model that performs robustly even in the worst-case scenario, effectively addressing the out-of-distribution generalization (OODG) problem [125]. In the context of DRO, Duchi and Namkoong [126] have made significant contributions by proving finite-sample minimax upper and lower bounds for f-divergence-based DRO. This work provides theoretical foundations and insights for the application of DRO in various domains.

1.4.2.2 Navigating Out-of-Distribution Challenges

We revisit the OOD issue from a behavioral perspective, asserting that behavior represents both the intrinsic and extrinsic factors of state changes and the patterns of state changes are embedded in behavior data. The behavior of agents, along with external feedback, drives the continuous evolution of their states, and this change is reflected in behavior data. So we believe that there are objectively existing patterns in behavior data. To obtain these patterns, it is necessary to have a sufficient amount of behavior data, and the model must have the ability to extract patterns from this data. For some complex systems, such as chaotic systems [96], the inherent uncertainty and complexity make it more challenging. From this perspective, the OOD problem essentially refers to how to effectively extract patterns from limited behavioral data.

In this book, we try to explore OOD solutions through key insights of foundation models. Agents empowered by foundation models demonstrate powerful learning and generalization abilities, giving us inspiration to solve OOD problems. The development of intelligence has gone through a process from symbolic agent [127, 128], which is based on symbolic logic, to reactive agent [129], which focuses on interacting with the environment, and RL-based agent [10], which learns from interaction to agent with transfer learning and meta-learning [130, 131], and finally to the recently emerged LLM agent [132]. The emergence of LLM signifies a revolutionary breakthrough in the field of artificial intelligence. Thanks to unprecedented training scales and model parameters, the LLM agent achieves humanlike performance in areas such as understanding, language synthesis, and common-sense reasoning.

LLM agents have powerful abilities in language and intention understanding, reasoning, memory, and even empathy, which can play an excellent role in decision-making and planning. Trained on an extensive corpus of text, the LLM agent acquires a robust foundation of common sense knowledge, showcasing remarkable generalization abilities [133]. Combining with the reflection mechanism, the LLM agent can make continuous adjustments to decisions and optimize execution sequences based on real-time feedback from the environment. Moreover, the LLM agent may exhibit intricate personalities, such as cognition, emotion, and character, which will affect the behavioral responses of the LLM agent [134]. In summary, LLM agents excel at accurately simulating individuals within realistic environments, proving crucial for effective out-of-distribution simulation.

The simulation framework built upon the LLM agent integrates personalized knowledge architecture and learning mechanisms. It has the potential to overcome OOD challenges through dynamic multilayer simulations. We believe that the knowledge architecture and the learning mechanism are the key technologies to realize sophisticated agent and behavior simulation (as shown in Fig. 1.2). Simulated

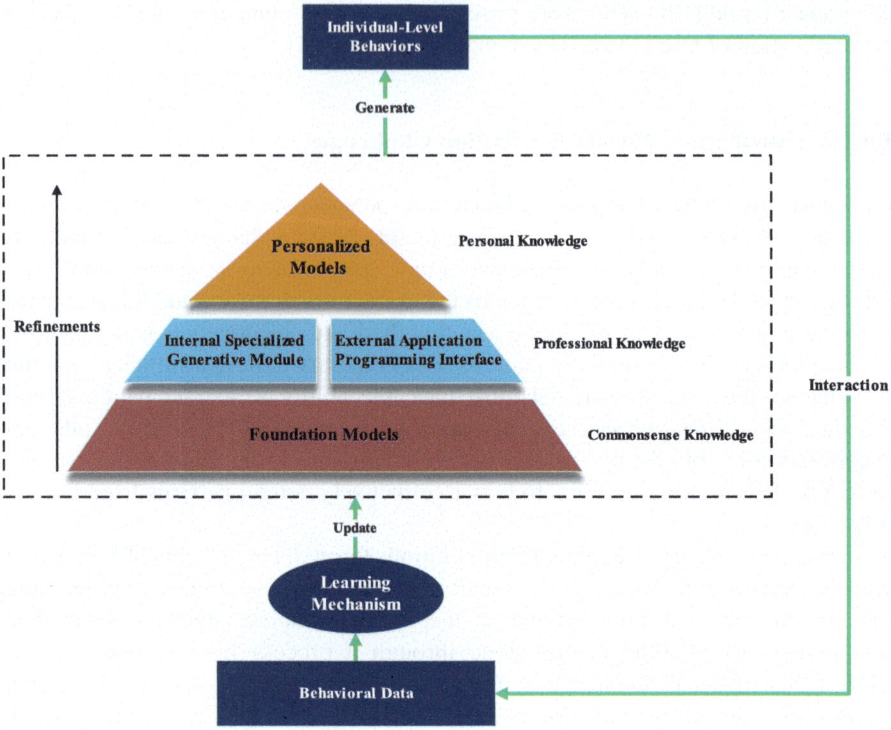

Fig. 1.2 Demonstration of agent's behaviors and learning in behavior simulation

behaviors are generated through the knowledge architecture of multi-floor models. These models will be updated dynamically based on interactions behaviors feedback according to the learning mechanism.

Hierarchical Knowledge Architecture We introduce a self-organizing multi-floor architecture (that we call the model tower) to achieve humanlike knowledge architecture. We briefly describe each layer from the bottom up. At the bottom floor, the LLM agent serves as the basis for common-sense general reasoning and sets the overall direction for the refinement of subsequent layers. To mitigate hallucinations and improve professional reasoning ability, we combine the internal specialized generative module and external APIs collectively to form the second floor. The third floor is the personality model, which establishes a comprehensive profile of the agent that contains external factors like social norms and internal factors like personality, enabling more refined behavior simulation.

Behavior Learning Mechanism While agents interact with each other and the environment, a behavioral learning mechanism enables them to adjust and refine their strategies according to their increasing observations. The behavioral learning mechanism enhances the credibility of agent behavior simulation through autonomous interactive learning methods, further improving the credibility of simulated behavior.

In conclusion, the model tower offers every individual a unique and personalized knowledge system, which is very helpful for behavior simulation.

At the system level, behavior simulation provides a unique stage for interactions among individuals. The interactions within the system form a complex and rich network, which may give rise to emergent phenomena [135], exhibiting collective characteristics beyond the simple behaviors of individuals. Similar to how the gravitational interaction in the three-body problem [136] can lead to extreme changes in celestial trajectories, interactions among individuals within the system may generate large amounts of behavioral data. The learning mechanism enables agents to use these data to adjust their knowledge systems and behaviors, enhancing the credibility of behavior simulation.

Behavior simulation demonstrates unique and efficient advantages in addressing OOD problems, especially in data utilization and pattern extraction. Firstly, in terms of data utilization, we do not need to obtain behavior data for the entire simulated system. Only collecting data that supports each individual within the system is sufficient to simulate system-level behavioral features. This data utilization strategy not only reduces the difficulty of data collection but also enhances the effectiveness and availability of the data. Secondly, in pattern extraction, behavior simulation goes beyond merely observing the overall behavioral performance from the external perspective of the system. Instead, it concerns the internal workings of the system, focusing on the individual behaviors of each intelligent agent. This in-depth exploration enables us to capture more subtle and complex patterns. In summary, behavior simulation not only simplifies the data collection process but also provides us with a more precise and comprehensive understanding of the system's behavior.

1.5 Exploring Behavior Theory on Expressive Power

The study of behavior extends beyond exclusively human actions and encompasses a broad range of phenomena observed in the natural world. Traditionally, behavior computing has relied on known data to understand and replicate existing patterns. However, recognizing that behaviors exist beyond our existing knowledge, it becomes imperative to explore the realm of the unknown. We delve into the concept of expanding from the known to the unknown in behavior theory, shedding light on the process of learning and simulating behaviors from previously unobserved data across various domains.

Behavior theory posits that all entities, from living organisms to complex systems, exhibit behaviors influenced by intrinsic characteristics, environmental factors, and interactions with their surroundings. Traditional computing techniques have predominantly focused on analyzing historical data to uncover patterns, relationships, and causal factors. While valuable insights have been gained, this approach assumes that past observations encapsulate the entirety of the behavioral landscape, neglecting the potential for novel and uncharted behaviors to emerge.

Expanding from the known to the unknown in behavior theory acknowledges the need to transcend the boundaries of existing data and explore unobserved territories. It recognizes that behaviors are not confined to static patterns but rather possess a dynamic nature capable of adapting and evolving. By venturing into uncharted realms, researchers can gain a deeper understanding of the underlying mechanisms that drive behavior and enhance their ability to model and simulate actions in unfamiliar contexts.

In this book, we discuss the preliminary analysis of known and unknown behavioral data in data and model theory, examining the expressive power of both types of data. It summarizes the characteristics and trends of existing research while providing a theoretical research idea for analyzing universal behavior computation.

1.5.1 Behavioral Data and Model Theory

Behavior is a broad concept that encompasses a wide range of actions, from human behavior to phenomena in the natural world. Behavior data refers to the recorded or represented information about behavior, providing specific values of limited dimensionality that can be used as input for machine learning algorithms. However, the limited dimensionality and partial nature of behavior data may not fully capture the complexity and diversity of the original behavior, potentially leading to information loss or misrepresentation and imposing limitations on understanding and modeling behavior. Therefore, it is essential to carefully consider the limitations of behavior data and recognize that it is an abstraction and approximation of complex behavior when utilizing it. The dimensionality estimation and sample complexity are commonly employed methods for measuring data inferences.

1.5.1.1 The Measures of Behavior Data

The process of dimensionality estimation, carried out through statistical and mathematical methods, involves inferring the quantity and complexity of features or variables encompassed within a dataset. This method is typically utilized for comprehending the structure of data. In neural image classifiers, the intrinsic dimension of the data manifold was studied by Ansuini et al. [14] using the TwoNN method. They demonstrated that the ID is much smaller than the dimension estimated via linear methods such as PCA, among other interesting results. Other authors have established a connection between ID and noisy labels [137], and demonstrated that neural models can effectively identify a low-dimensional manifold in a larger ambient space [15]. It would be interesting to understand the relationship between the data manifold and neural circuits [138], and how the manifold changes when non-robust features are eliminated [139]. Recent work

[140] relates data dimensionality and dataset size scaling exponents for kernel methods. The intrinsic dimension of the neural network parameter space has also been discussed in [141].

Sample Complexity The quantity of samples or observed data required for a learning algorithm to achieve a certain level of accuracy or performance is referred to as sample complexity. It quantifies the amount of data needed for the model to generalize effectively from the training set to unseen instances, playing a crucial role in understanding the learning behavior and limitations of the algorithm.

We study algorithms parameterized by a set $\mathscr{P} \subseteq \mathbb{R}^d$. We use $\mathscr{X}$ to denote the set of problem instances the algorithm takes as input. We measure the performance of the algorithm parameterized by $\boldsymbol{\rho} = (\rho[1], \rho[2], \cdots, \rho[d]) \in \mathbb{R}^d$ via a utility function $u_\rho : \mathscr{X} \rightarrow [0, H]$, with $\mathscr{U} = \{u_\rho : \rho \in \mathscr{P}\}$ denoting the set of all such functions. We assume there is an unknown, application-specific distribution $\mathscr{D}$ over $\mathscr{X}$. The goal is to find a parameter vector in $\mathscr{P}$ with high performance in expectation over D.

A related line of research has designed algorithms that replace some steps of classic worst-case algorithms with a machine-learned oracle that makes predictions about structural aspects of the input. Balcan et al. [142] introduce the concept of data-driven algorithm design, aiming to optimize algorithm parameters using domain-specific datasets for improved performance. It proposes a theoretical framework to establish guarantees for such algorithms, considering their performance characteristics and generalization potential. The approach is applied to various types of parameterized algorithms across domains, yielding enhanced or previously established outcomes.

Deep models provide more powerful representation capabilities to fit and approximate more complex data distributions and to deal with more complex circumstances and tasks. However, the theoretical analysis of deep models always lags behind their practical applications, among which the possible reasons are the highly nonlinear nature of the models. A large number of model parameters tend to exist in high-dimensional spaces, the complexity of the data, and the black-box nature of the models. Thus, several of the above elements lead to poor interpretability of deep learning. In order to improve the interpretability of deep learning, the theory behind deep learning needs to be further investigated, of which the most typical research is the expressive power of the model. Model expressive power refers to the complexity of the class of functions that a model can represent. In other words, how many different types of functions can be learned by a given model?

1.5.1.2 The Measures of Model Expressive Power

The model can split the input space into multiple smaller linear regions, and each linear region fits a category function. The theoretical support behind the number of linear regions is that the model splits the inputs of different classes using decision boundaries formed by hyperplanes. If we want to perfectly approximate the decision boundary between different classes, the model will use an infinite number of linear regions. Assuming we have a finite number of pieces, the number of linear regions is a good proxy for how close the model is to that decision boundary if we can learn their optimal slopes perfectly. In other words, the number of linear regions can represent the ability of the model to approximate any smooth shape. Pascanu et al. [143] first proposed to use the number of linear regions to calculate the model expressive power. Moreover, this work combined with Zaslavsky's theorem [144] for comparing the number of linear regions for deep and shallow models with a family of segmented linear functions. Next, we follow this seminal work by showing some studies on the number of linear regions of the proposed models.

Trajectory length was first proposed by Raghu et al. [145] as a novel measure of a model's expressive power. Trajectory length measures how the output of the model changes as the input follows a one-dimensional path. Specifically, for a given sample, its input is passed through each layer of the network, transformed by the activation function and weight matrix, and finally, the output is obtained. The length of the trajectory is the number of steps along this propagation path. By examining the propagation trajectories of different samples through the network, information about the structure and complexity of the network can be obtained.

VC dimension reflects the maximum complexity a classification model can handle, indicating the size of the largest sample set it can fit. It serves as a theoretical metric for analyzing the expressive and generalization performance of machine learning models, providing an upper bound estimate. A larger VC dimension implies a more expressive model but increases the risk of overfitting, commonly applied to binary classifiers or function classes. In deep learning, evaluating the VC dimension for typical neural networks becomes challenging due to their complexity. Rademacher complexity, on the other hand, assesses a function class's sensitivity to random noise by introducing random labels to the training set. A lower Rademacher complexity indicates less sensitivity to noise and better generalization performance, facilitating comparisons between function classes, model complexity assessment, and model selection. Despite its theoretical significance, Rademacher complexity is underutilized in practical deep learning applications.

1.5.2 Expressive Power in Known Behavioral Data

The concept of expressive power in known data refers to the ability of existing data to capture and represent the full range of behaviors exhibited by entities within a given context. Known data, typically obtained from historical records or observations, provides a foundation for understanding patterns, correlations, and causal relationships. The expressive power of known data lies in its capacity to reveal insights and enable the development of models that accurately describe and predict behaviors based on the available information. This analysis allows researchers to uncover the underlying mechanisms and regularities that govern behaviors within the known domain.

The expressive power of known data can be considered as the foundation of computational learning theory, which often adheres to the assumption of independent and identically distributed (I.I.D) data. This assumption states that the training and testing data belong to the same distribution. The primary focus of computational learning theory is to understand the learning process and its capabilities when the training and testing data follow the same distribution [146].

One key concept in computational learning theory is the notion of a hypothesis space [147], which represents the set of possible models that the learning algorithm can learn from the known data. The expressive power of the hypothesis space determines the complexity of the models that can be learned. A more expressive hypothesis space allows for more complex models, but it also increases the risk of overfitting.

Computational learning theory also explores the sample complexity, which refers to the number of training examples required for a learning algorithm to achieve a certain level of accuracy. By analyzing the sample complexity, researchers can gain insights into the amount of data needed to effectively learn from the known data and make accurate predictions on unseen instances.

Additionally, computational learning theory addresses the trade-off between bias and variance in learning algorithms. Bias refers to the assumptions and simplifications made by the learning algorithm, which can lead to systematic errors or underfitting. Variance, on the other hand, measures the sensitivity of the learned models to variations in the training data, which can result in overfitting. Finding the right balance between bias and variance is crucial for achieving good generalization performance.

In summary, while expressive power in known data provides a starting point for understanding patterns and behaviors, computational learning theory takes a more formal and rigorous approach. It investigates the mathematical foundations of learning algorithms, analyzing their capabilities, limitations, and generalization properties. By bridging the gap between known and unseen data, computational learning theory aims to develop robust and reliable learning algorithms that can make accurate predictions and decisions in real-world applications.

1.5.3 Expressive Power in Unknown Behavioral Data

In contrast with known data, the expressive power in unknown data pertains to the potential for discovering new and previously unobserved behaviors beyond the boundaries of existing knowledge. Unknown data represents uncharted territory, where behaviors may manifest in novel and unexpected ways. By venturing into the realm of unknown data, researchers can expand their understanding of the full spectrum of behaviors and explore the inherent complexity and diversity that exists in various contexts. The expressive power in unknown data is crucial for capturing the richness and nuances of behaviors that have yet to be observed, providing new insights and opportunities for advancing behavior theory and modeling.

The domain generalization (DG), i.e., out-of-distribution (OOD) generalization, can be seen as an exploration of the patterns and underlying information in unknown data. DG refers to the ability of our model to make predictions and generalize effectively even when there is a distribution shift between the training and testing data [148]. The challenge lies in the fact that the testing data does not adhere to the independent and identically distributed (I.I.D) assumption of the training data [149]. For example, in the fields of biological proteins [150] and drug discovery [151], the prediction of protein structures and molecular biochemical properties is often trained on a limited set of available experimental data. However, the model needs to make predictions on an unknown, diverse, and large combination of candidate molecules.

Definition 1.1 (Domain [152]) Let $\mathscr{X}$ denote a non-empty input space and $\mathscr{Y}$ an output space. A domain is composed of data that are sampled from a distribution. We denote it as $S = (x_i, y_i)_{i=1}^n \sim P_{XY}$, where $x \in \mathscr{X} \subset \mathbb{R}^d$, $y \in \mathscr{Y} \subset \mathbb{R}$ denotes the label, and P_{XY} denotes the joint distribution of the input sample and output label. $\mathscr{X}$ and $\mathscr{Y}$ denote the corresponding random variables.

Definition 1.2 (Domain Generalization [152]) In domain generalization, we are given M training (source) domains $\mathscr{S}_{\text{train}} = \{\mathscr{S}_i | i = 1, \ldots, M\}$ where $\mathscr{S}_i = (\mathbf{x}_j^i, y_j^i)_{j=1}^{n_i}$ denotes the i-th domain. The joint distributions between each pair of domains are different: $P_{XY}^i \neq P_{XY}^j$, $1 \leq i \neq j \leq M$. The goal of domain generalization is to learn a robust and generalizable predictive function $h : \mathscr{X} \to \mathscr{Y}$ from the M training domains to achieve a minimum prediction error on an unseen test domain $\mathscr{S}_{\text{test}}$ (i.e., $\mathscr{S}_{\text{test}}$ cannot be accessed in training and $P_{XY}^{\text{test}} \neq P_{XY}^i$ for $i \in \{1, \ldots, M\}$):

$$\min_h \mathbb{E}_{(\mathbf{x}, y) \in \mathscr{S}_{\text{test}}} [\ell(h(\mathbf{x}), y)] \tag{1.1}$$

where $\mathbb{E}$ is the expectation and $\ell(\cdot, \cdot)$ is the loss function.

In [153], the authors have made groundbreaking contributions to the establishment of a rigorous theoretical framework for OOD generalization. They propose a mathematical formula to describe the learnability of OOD generalization problems, as follows.

Definition 1.3 (Learnability [153]) Let Φ be the feature space and ρ be a distribution distance. We say an OOD generalization problem from E_{avail} to E_{all} is learnable if there exists an expansion function $s(\cdot)$ and $\delta \geq 0$, such that for all $\phi \in \Phi$ satisfying $I_\rho(\phi, E_{\text{avail}}) \geq \delta$, we have $s(V_\rho(\phi, E_{\text{avail}})) \geq V_\rho(\phi, E_{\text{all}})$. If such $s(\cdot)$ and δ exist, we further call this problem $(s(\cdot), \delta)$-learnable. If an OOD generalization problem is not learnable, we call it unlearnable.

Furthermore, [153] demonstrated the generalization bounds under specific conditions as follows:

$$\text{err}(f) \geq C_0 \cdot s(V_{\text{sup}}(h, E_{\text{avail}})) \quad (1.2)$$

and provided a guarantee for OOD generalization errors. Subsequently, based on this, a model selection criterion is designed to simultaneously examine model variations and validation accuracy.

In [154], the authors provided a theoretical framework of learning under subpopulation shift through label propagation. Specifically, the authors make an assumption as follows.

Assumption 1.1 (from [154]) Assume the source and target distributions have the following structure: $\text{supp}(S) = \bigcup_{i=1}^{m} S_i$, $\text{supp}(T) = \bigcup_{i=1}^{m} T_i$, where $S_i \cap S_j = T_i \cap T_j = S_i \cap T_j = \emptyset$ for all $i \neq j$. Assume the ground truth class $g^*(x)$ for $x \in S_i \cup T_i$ is consistent (constant), which is denoted as $y_i \in \{1, \ldots, K\}$. Abuse the notation to let S_i, T_i also denote the conditional distribution (probability measure) of S, T on the set S_i, T_i, respectively. In addition, we make the following canonical assumptions:

1. The teacher classifier on S_i is informative of the ground truth class y_i by a margin $\gamma > 0$, that is,
$P_{x \sim S_i}[g_{tc}(x) = y_i] \geq P_{x \sim S_i}[g_{tc}(x) = k] + \gamma, \forall k \in \{1, \ldots, K\} \setminus \{y_i\}$.
2. On each component, the ratio of the population under domain shift is upper-bounded by a constant r, i.e., $\frac{\mathbb{P}_T[T_i]}{\mathbb{P}_S[S_i]} \leq r, \forall i \in \{1, \ldots, m\}$.

Based on the fact that when the above assumptions hold, the following two theorems are proved.

Theorem 1.1 (Bound on Target Error with Multiplicative Expansion [154]) *Suppose Assumption 1 holds, and* $\frac{1}{2}(S + T)$ *satisfies* $\left(\frac{1}{2}, c\right)$*-multiplicative expansion. Then, the classifier* g *satisfies* $\varepsilon_T(g) \leq \max\left(\frac{c+1}{c-1}, 3\right)\frac{8r\mu}{\gamma}$.

Theorem 1.2 (Bound on Target Error with Constant Expansion [154]) *Suppose Assumption 1 holds, and* $\frac{1}{2}(S + T)$ *satisfies* (q, μ)*-constant expansion. Then, the classifier* g *satisfies* $\varepsilon_T(g) \leq (2\max(q, \mu) + \mu)\frac{8r\mu}{\gamma}$.

1.6 Outline of This Book

In this book, we focus on introducing some representative security and safety technologies related to behavior modeling in universal behavior computing and explore the relevant key technologies and development trends of behavior simulation. The content structure of this book is shown in Fig. 1.3.

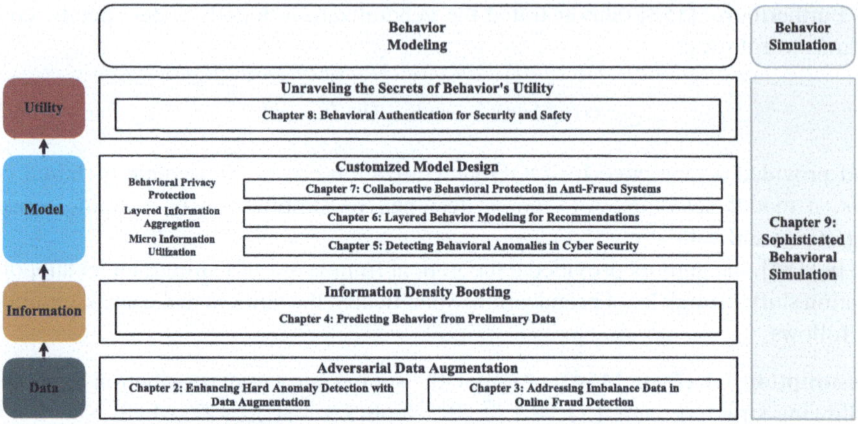

Fig. 1.3 Architecture of this book

To begin with, we mainly introduce the security and safety solution for behavior modeling by adversarial data augmentation.

- Using Known Data to Improve Boundaries Between Categories (Chap. 2).
- Exploring Unknown Data and Breaking Through Traditional Behavioral Hypotheses (Chap. 3).

Then, we introduce the information density-boosting technologies on security and safety solutions to further improve the performance of behavioral models.

- We propose a novel multistage data representation scheme, called AI2Vec (Applicant Information Vectoring), as an information density booster, to enable behavior prediction on preliminary data (Chap. 4).

Next, we improve the performance of behavior models in tasks such as security protection by designing customized strategies to address some typical challenges in security and safety scenarios.

- We propose a new graph-based behavioral modeling paradigm for the BAD problem, named *behavioral identification graph* (BIG), which has distinct advantages over existing methods by mining deeply the *property-level* (as an enhancement to the *event-level*) associations in behavioral data (Chap. 5).
- We propose an approach, named Micro Event Graph (MiEG), which seamlessly integrates side information while maintaining computational efficiency and employs layered computations to project fine-grained attribute information onto a temporal event graph (Chap. 6).
- We design a privacy-preserving distributed graph learning framework (the *association reconstruction mechanism* (ARM)) that ensures critical association repairs by merging parameter sharing and data sharing (Chap. 7).

In this way, we further provide a comprehensive examination of the background and preliminaries of behavioral authentication (Chap. 8).

Finally, we investigate the effectiveness and preference of behavior simulation technologies based on the analyses of scientific paradigms and problems (Chap. 9).

The key technologies and methods of behavior computing for security and safety include the following aspects:

- Hard anomaly detection. The hard anomaly has two characteristics in terms of the differences between abnormal and normal samples: (1) the boundary is exceedingly unobvious, and (2) the distribution is extremely imbalanced, both of which result in the difficulty of anomaly detection. It is really hard for regular classifiers to find the boundary between abnormal and normal samples here. We propose a quantitative definition of the hard anomaly according to detection difficulty and devise a GAN-based (generative adversarial network) method, called HadGAN, which produces both abnormal and normal samples with a similar distribution. We then pre-train a base anomaly detection model using the data generated by HadGAN. More details will be provided in Chap. 2.
- Enhancing imbalance data. In real business scenarios, fraudulent transactions, usually, only account for a small portion of normal transactions, resulting in extremely imbalanced data. We present the GAN-based approach for generating training data and construct a base pre-training fraud detection model to solve the imbalanced data problem. Then, we apply transfer learning to fine-tune the base fraud detection model by using the real datasets which can leverage the knowledge learned from the base fraud detection model on the synthetic datasets. More details will be provided in Chap. 3.
- Applicant information vectoring. Fraud *prediction* is an especially critical step to control loss risk and improve processing efficiency in online lending services. Unfortunately, it is definitely challenging since the ex ante prediction actually needs to be made only based on the most basic information of applicants. Accordingly, we propose a novel multi-stage data representation scheme, called AI2Vec (*Applicant Information Vectoring*), as an *information density booster*. It can gradually boost information density of associations by simultaneously decreasing the scale of information carriers and increasing the amount of useful information. More details will be provided in Chap. 4.
- Devised event-property composite model. The so-called behavioral anomaly detection (BAD) is expected to solve effectively a variety of security issues by detecting the deviances from normal behavioral patterns of protected agents. Actually, it generally has stricter requirements on behavioral data than those of other network services where behavioral data are required to meet two conditions: (1) *behavioral distinctness* for reflecting the personality of agents and (2) *behavioral stability* for detecting the abnormality of behaviors. We propose a new graph-based behavioral modeling paradigm for the BAD problem, named *behavioral identification graph* (BIG), which has distinct advantages over existing methods by mining deeply the *property-level* (as an enhancement to

the *event-level*) associations in behavioral data. More details will be provided in Chap. 5.

- Micro event graph (MiEG). Existing advanced sequential recommendation models, including transformer-based and dynamic graph-based approaches, often prioritize macroscopic event sequences or node relationships, overlooking fine-grained information within user and item attributes. This gap poses challenges in constructing stable models, particularly in sparse data scenarios. To address this, we propose an approach that seamlessly integrates side information while maintaining computational efficiency. It employs layered computations to project fine-grained attribute information onto a temporal event graph. More details will be provided in Chap. 6.
- Association reconstruction mechanism (ARM). Anti-fraud engineering for online credit loan (OCL) platforms is getting more challenging due to the developing specialization of gang fraud. The inherent difficulty that can be foreseen is the *data isolation* caused by mistrust between multiple platforms and data control legislations for privacy preservation. To maintain the advantage owned by the platforms, we design a privacy-preserving distributed graph learning framework that ensures critical association repairs by merging parameter sharing and data sharing. More details will be provided in Chap. 7.
- Behavioral appropriateness for security and safety. In a broad sense, for the security and safety issue, behavioral authentication is not only an innovative and promising method due to its inherent advantages but also a critical and fundamental problem due to the ubiquity of behavior generation and the necessity of behavior regulation in any system. By this classification, we provide a comprehensive examination of the background and preliminaries of behavioral authentication. We further summarize existing research based on their respective focus areas and characteristics. The challenges confronted by current behavioral authentication methods are analyzed, and potential research directions are discussed to promote the diversified and integrated development of behavioral authentication. More details will be provided in Chap. 8.
- Model-tower-driven sophisticated agent and simulator architecture. Simulation technologies have been widely utilized in many scientific research fields such as weather forecasting, fluid mechanics, and biological populations. They act as the best tool to handle problems in complex systems where closed-form expressions are unavailable and the target distribution in the representation space is too complex to be fully represented by data-driven learning models, such as deep learning (DL) models. We summarize existent simulation technologies aligning with the scientific paradigms and propose the concept of behavioral simulation (BS), and further sophisticated behavioral simulation (SBS). They represent a higher degree of paradigm integration based on foundation models to simulate complex social systems involving sophisticated human strategies and behaviors. Through this work, we look forward to more powerful BS and SBS applications in scientific research branches within social science. More details will be provided in Chap. 9.

References

1. C. Stachl, Q. Au, R. Schoedel, S.D. Gosling, G.M. Harari, D. Buschek, S.T. Völkel, T. Schuwerk, M. Oldemeier, T. Ullmann, et al., Proc. Natl. Acad. Sci. **117**(30), 17680 (2020)
2. W. Bialek, Proc. Natl. Acad. Sci. **119**(18), e2021860119 (2022)
3. Y. Ding, X. Xia, S. Chen, Y. Li, Comput. Secur. **73**, 73 (2018)
4. A. Beutel, L. Akoglu, C. Faloutsos, in *Proceedings of the 22nd ACM SIGSAC Conference on Computer and Communications Security* (2015), pp. 1696–1697
5. Q. Pi, W. Bian, G. Zhou, X. Zhu, K. Gai, in *Proceedings of the 25th ACM SIGKDD International Conference on Knowledge Discovery & Data Mining* (2019), pp. 2671–2679
6. Y. Zheng, S. Liu, Z. Li, S. Wu, in *Proceedings of the AAAI Conference on Artificial Intelligence*, vol. 35 (2021), pp. 4706–4713
7. C. Wang, H. Zhu, IEEE Trans. Depend. Secure Comput. **19**(1), 301 (2022)
8. S. Hui, H. Wang, Z. Wang, X. Yang, Z. Liu, D. Jin, Y. Li, in *Proceedings of the ACM Web Conference 2022* (2022), pp. 3336–3346
9. Y. Dou, Z. Liu, L. Sun, Y. Deng, H. Peng, P.S. Yu, in *Proceedings of the 29th ACM International Conference on Information & Knowledge Management* (2020), pp. 315–324
10. C. Isbell, C.R. Shelton, M. Kearns, S. Singh, P. Stone, in *Proceedings of the Fifth International Conference On Autonomous Agents* (2001), pp. 377–384
11. M. Masaeli, J.G. Dy, G.M. Fung, in *Proceedings of the 27th International Conference on Machine Learning (ICML-10)* (2010), pp. 751–758
12. D. Cheng, X. Wang, Y. Zhang, L. Zhang, IEEE Trans. Knowl. Data Eng. **34**(8), 3800 (2020)
13. L. Yu, H. Liu, J. Mach. Learn. Res. **5**, 1205 (2004)
14. A. Ansuini, A. Laio, J.H. Macke, D. Zoccolan, in *Advances in Neural Information Processing Systems*, vol. 32 (2019)
15. R. Basri, D.W. Jacobs, in *International Conference on Learning Representations* (2016)
16. J.L. Herlocker, J.A. Konstan, L.G. Terveen, J.T. Riedl, ACM Trans. Inform. Syst. **22**(1), 5 (2004)
17. S. Wang, L. Hu, Y. Wang, L. Cao, Q.Z. Sheng, M. Orgun, in *28th International Joint Conference on Artificial Intelligence, IJCAI 2019* (International Joint Conferences on Artificial Intelligence, 2019), pp. 6332–6338
18. J. Tang, X. Hu, H. Liu, Soc. Netw. Anal. Mining **3**, 1113 (2013)
19. C. Gao, Y. Zheng, N. Li, Y. Li, Y. Qin, J. Piao, Y. Quan, J. Chang, D. Jin, X. He, et al., ACM Trans. Recommender Syst. **1**(1), 1 (2023)
20. M. Shafiq, Z. Tian, A.K. Bashir, X. Du, M. Guizani, Comput. Secur. **94**, 101863 (2020)
21. I.O. Lopes, D. Zou, I.H. Abdulqadder, S. Akbar, Z. Li, F. Ruambo, W. Pereira, Comput. Secur. **135**, 103465 (2023)
22. A.C. Bahnsen, D. Aouada, A. Stojanovic, B. Ottersten, Expert Syst. Appl. **51**, 134 (2016)
23. S. Bhattacharyya, S. Jha, K. Tharakunnel, J.C. Westland, Decis. Support Syst. **50**(3), 602 (2011)
24. K. Ghosh Dastidar, J. Jurgovsky, W. Siblini, M. Granitzer, Knowl. Inform. Syst. **64**(3), 831 (2022)
25. C. Song, Z. Qu, N. Blumm, A.L. Barabási, Science **327**(5968), 1018 (2010)
26. W. Youyou, M. Kosinski, D. Stillwell, Proc. Natl. Acad. Sci. **112**(4), 1036 (2015)
27. M. Kosinski, D. Stillwell, T. Graepel, Proc. Natl. Acad. Sci. **110**(15), 5802 (2013)
28. T. Hastie, R. Tibshirani, J.H. Friedman, J.H. Friedman, *The Elements of Statistical Learning: Data Mining, Inference, and Prediction*, vol. 2 (Springer, Berlin, 2009)
29. P.K. Keserwani, M.C. Govil, E.S. Pilli, Neural Comput. Appl. **35**(7), 4993 (2023)
30. N. Hoque, D.K. Bhattacharyya, J.K. Kalita, Expert Syst. Appl. **41**(14), 6371 (2014)
31. F. Esposito, D. Malerba, G. Semeraro, J. Kay, IEEE Trans. Pattern Anal. Mach. Intell. **19**(5), 476 (1997)
32. P. Bermejo, J.A. Gámez, J.M. Puerta, Pattern Recogn. Lett. **32**(5), 701 (2011)
33. I. Boussaïd, J. Lepagnot, P. Siarry, Inform. Sci. **237**, 82 (2013)

34. A. Nazir, R.A. Khan, Comput. Secur. **102**, 102164 (2021)
35. J. Liu, W. Song, L. Shen, J. Han, K. Ren, IEEE Trans. Mobile Comput. **22**(11) (2022)
36. Z. Zhang, H. Ning, F. Zhou, IEEE Trans. Intell. Transp. Syst. **23**(11) (2022)
37. D. Singh, B. Singh, Pattern Recogn. **122**, 108307 (2022)
38. J. Malik, A. Akhunzada, I. Bibi, M. Talha, M.A. Jan, M. Usman, IEEE Sensors J. **21**(14), 15859 (2020)
39. A.C. Bahnsen, A. Stojanovic, D. Aouada, B. Ottersten, in *Proceedings of the 2014 SIAM International Conference on Data Mining* (SIAM, 2014), pp. 677–685
40. A. Dal Pozzolo, O. Caelen, Y.A. Le Borgne, S. Waterschoot, G. Bontempi, Expert Syst. Appl. **41**(10), 4915 (2014)
41. H. Zhang, J.L. Li, X.M. Liu, C. Dong, Fut. Gener. Comput. Syst. **122**, 130 (2021)
42. Y. Li, B. Zou, S. Deng, G. Zhou, IEEE Internet Comput. **24**(2), 49 (2020)
43. T. Li, G. Kou, Y. Peng, Inform. Syst. **91**, 101494 (2020)
44. A.C. Bahnsen, E.C. Bohorquez, S. Villegas, J. Vargas, F.A. González, in *2017 APWG Symposium on Electronic Crime Research (eCrime)* (IEEE, 2017), pp. 1–8
45. P. Galeano, D. Peña, R.S. Tsay, J. Am. Stat. Assoc. **101**(474), 654 (2006)
46. O. Salem, A. Guerassimov, A. Mehaoua, A. Marcus, B. Furht, in *2013 IEEE International Conference on Communications (ICC)* (IEEE, 2013), pp. 4373–4378
47. S. Barbhuiya, P. Kilpatrick, D. S. Nikolopoulos, in *Proceedings of the 2021 13th International Conference on Machine Learning and Computing* (2021), pp. 568–574
48. A. Roy, D.S. Jas, G. Jaggi, K. Sharma, Proc. Comput. Sci. **173**, 345 (2020)
49. A. Lakhina, M. Crovella, C. Diot, ACM SIGCOMM Comput. Commun. Rev. **35**(4), 217 (2005)
50. Y. Zhang, N. Meratnia, P. Havinga, IEEE Commun. Surv. Tutorials **12**(2), 159 (2010)
51. Z. Huan, C. Wei, G.H. Li, Sensors **18**(12), 4328 (2018)
52. X. Deng, P. Jiang, X. Peng, C. Mi, IEEE Trans. Ind. Electron. **66**(6), 4672 (2018)
53. Y. Qiao, K. Wu, P. Jin, IEEE Trans. Knowl. Data Eng. **35**(1), 404 (2021)
54. W. Wang, M. Zhu, X. Zeng, X. Ye, Y. Sheng, in *2017 International Conference on Information Networking (ICOIN)* (IEEE, 2017), pp. 712–717
55. Z. Zhang, P. Qi, W. Wang, in *Proceedings of the AAAI Conference on Artificial Intelligence*, vol. 34 (2020), pp. 1210–1217
56. L. Yang, A. Shami, in *ICC 2022-IEEE International Conference on Communications* (IEEE, 2022), pp. 2774–2779
57. M.S. Ansari, V. Bartoš, B. Lee, Fut. Gener. Comput. Syst. **128**, 235 (2022)
58. M. Villarreal-Vasquez, G. Modelo-Howard, S. Dube, B. Bhargava, IEEE Trans. Depend. Secure Comput. **20**(1), 451 (2021)
59. G. Xie, Q. Li, Y. Jiang, T. Dai, G. Shen, R. Li, R. Sinnott, S. Xia, in *Proceedings of the Workshop on Network Meets AI & ML* (2020), pp. 14–20
60. A.R. Javed, M. Usman, S.U. Rehman, M.U. Khan, M.S. Haghighi, IEEE Trans. Intell. Transp. Syst. **22**(7), 4291 (2020)
61. S. Nedelkoski, J. Bogatinovski, A. Acker, J. Cardoso, O. Kao, in *2020 IEEE International Conference on Data Mining (ICDM)* (IEEE, 2020), pp. 1196–1201
62. H. Wang, X. Di, Y. Wang, B. Ren, G. Gao, J. Deng, IEEE J. Sel. Areas Commun. (2023)
63. Y. Chai, Y. Zhou, W. Li, Y. Jiang, IEEE Trans. Depend. Secure Comput. **19**(2), 790 (2021)
64. M. Cvitkovic, arXiv preprint arXiv:2002.02046 (2020)
65. C.T. Li, Y.C. Tsai, J.C. Liao, in *2023 IEEE 39th International Conference on Data Engineering (ICDE)* (IEEE, 2023), pp. 3589–3592
66. Y. Liu, Y. Zheng, D. Zhang, H. Chen, H. Peng, S. Pan, in *Proceedings of the ACM Web Conference 2022* (2022), pp. 1392–1403
67. E. Rocheteau, C. Tong, P. Veličković, N. Lane, P. Liò, arXiv preprint arXiv:2101.03940 (2021)
68. M. Kim, H.S. Choi, J. Kim, arXiv preprint arXiv:2204.03225 (2022)
69. Z. Li, Z. Cui, S. Wu, X. Zhang, L. Wang, in *Proceedings of the 28th ACM International Conference on Information and Knowledge Management* (2019), pp. 539–548

70. J. You, X. Ma, Y. Ding, M.J. Kochenderfer, J. Leskovec, Adv. Neural Inform. Process. Syst. **33**, 19075 (2020)
71. X. Guo, Y. Quan, H. Zhao, Q. Yao, Y. Li, W. Tu, arXiv preprint arXiv:2108.09127 (2021)
72. D. Cai, C. Sun, M. Song, B. Zhang, S. Hong, H. Li, in *Proceedings of the 2022 SIAM International Conference on Data Mining (SDM)* (SIAM, 2022), pp. 127–135
73. A. Goodge, B. Hooi, S.K. Ng, W.S. Ng, in *Proceedings of the AAAI Conference on Artificial Intelligence*, vol. 36 (2022), pp. 6737–6745
74. I. Spinelli, S. Scardapane, A. Uncini, Neural Netw. **129**, 249 (2020)
75. Y. Xie, Z. Wang, Y. Li, B. Ding, N.M. Gürel, C. Zhang, M. Huang, W. Lin, J. Zhou, in *Proceedings of the 27th ACM SIGKDD Conference on Knowledge Discovery & Data Mining* (2021), pp. 3795–3805
76. K. Du, W. Zhang, R. Zhou, Y. Wang, X. Zhao, J. Jin, Q. Gan, Z. Zhang, D.P. Wipf, Adv. Neural Inform. Process. Syst. **35**, 16373 (2022)
77. L. Du, F. Gao, X. Chen, R. Jia, J. Wang, J. Zhang, S. Han, D. Zhang, in *Proceedings of the 27th ACM SIGKDD Conference on Knowledge Discovery & Data Mining* (2021), pp. 322–331
78. X. Hu, T.c. Chiueh, K.G. Shin, in *Proceedings of the 16th ACM Conference on Computer and Communications Security* (2009), pp. 611–620
79. H. Alasmary, A. Khormali, A. Anwar, J. Park, J. Choi, A. Abusnaina, A. Awad, D. Nyang, A. Mohaisen, IEEE Internet Things J. **6**(5), 8977 (2019)
80. G. Zhang, Z. Li, J. Huang, J. Wu, C. Zhou, J. Yang, J. Gao, ACM Trans. Inform. Syst. **40**(3), 1 (2022)
81. G. McCann, C.H. Wilts, B. Locanthi, Trans. Am. Inst. Electr. Eng. **68**(1), 652 (1949)
82. R. Gibby, Bell Syst. Tech. J. **39**(3), 675 (1960)
83. F. Ritson, P. Hammond, Proc. IEE-Part II: Power Eng. **99**(72), 533 (1952)
84. M. Clynes, J. Appl. Physiol. **15**(5), 863 (1960)
85. Y. Cui, K.B. Olsen, T.H. Jordan, K. Lee, J. Zhou, P. Small, D. Roten, G. Ely, D.K. Panda, A. Chourasia, et al., in *SC'10: Proceedings of the 2010 ACM/IEEE International Conference for High Performance Computing, Networking, Storage and Analysis* (IEEE, 2010), pp. 1–20
86. A. Kolb, L. Latta, C. Rezk-Salama, in *Proceedings of the ACM SIGGRAPH/EUROGRAPHICS conference on Graphics hardware* (2004), pp. 123–131
87. L. Huggins, E. Monke, Water Resour. Res. **4**(3), 529 (1968)
88. D.D. Reneau Jr, D.F. Bruley, M.H. Knisely, in *Chemical Engineering in Medicine and Biology: Proceedings of the Thirty-Third Annual Chemical Engineering Symposium of the Division of Industrial and Engineering Chemistry of the American Chemical Society, Held at the University of Cincinnati, on October 20–21, 1966* (Springer, 1967), pp. 135–241
89. F. Jones, D. Parrott, G. Ross, Ann. Appl. Biol. **60**(1), 151 (1967)
90. S. Kang, S. Kahan, J. McDermott, N. Flann, I. Shmulevich, Bioinformatics **30**(21), 3101 (2014)
91. Y. Yuan, J. Ding, H. Wang, D. Jin, Y. Li, in *Proceedings of the 28th ACM SIGKDD Conference on Knowledge Discovery and Data Mining* (2022), pp. 4752–4762
92. Y. Yuan, H. Wang, J. Ding, D. Jin, Y. Li, in *Proceedings of the ACM Web Conference 2023* (2023), pp. 906–916
93. H. Shi, Q. Yao, Y. Li, in *Proceedings of the ACM Web Conference 2023* (2023), pp. 4200–4209
94. G. Zhang, Z. Yu, D. Jin, Y. Li, in *Proceedings of the 28th ACM SIGKDD Conference on Knowledge Discovery and Data Mining* (2022), pp. 2439–2449
95. F. Lorig, L. Vanhée, F. Dignum, in *ECCAI Advanced Course on Artificial Intelligence* (Springer, 2021), pp. 391–414
96. E.N. Lorenz, J. Atmos. Sci. **20**(2), 130 (1963)
97. D. Richards, Behav. Sci. **35**(3), 219 (1990)
98. A. Farahani, S. Voghoei, K. Rasheed, H.R. Arabnia, in *Advances in Data Science and Information Engineering: Proceedings from ICDATA 2020 and IKE 2020* (2021), pp. 877–894

99. Y. Balaji, Robust learning under distributional shifts. Ph.D. Thesis, University of Maryland, College Park (2021)
100. W.M. Kouw, M. Loog, arXiv preprint arXiv:1812.11806 (2018)
101. M. Ghifary, W.B. Kleijn, M. Zhang, in *PRICAI 2014: Trends in Artificial Intelligence: 13th Pacific Rim International Conference on Artificial Intelligence, Gold Coast, QLD, Australia, December 1–5, 2014. Proceedings 13* (Springer, 2014), pp. 898–904
102. E. Tzeng, J. Hoffman, N. Zhang, K. Saenko, T. Darrell, arXiv preprint arXiv:1412.3474 (2014)
103. M. Long, Y. Cao, J. Wang, M. Jordan, in *International Conference on Machine Learning* (PMLR, 2015), pp. 97–105
104. B. Sun, J. Feng, K. Saenko, in *Proceedings of the AAAI Conference on Artificial Intelligence*, vol. 30 (2016)
105. B. Sun, K. Saenko, in *Computer Vision–ECCV 2016 Workshops: Amsterdam, The Netherlands, October 8–10 and 15–16, 2016, Proceedings, Part III 14* (Springer, 2016), pp. 443–450
106. N. Courty, R. Flamary, A. Habrard, A. Rakotomamonjy, in *Advances in Neural Information Processing Systems*, vol. 30 (2017)
107. Y. Ganin, V. Lempitsky, in *International Conference on Machine Learning* (PMLR, 2015), pp. 1180–1189
108. W. Zhang, W. Ouyang, W. Li, D. Xu, in *Proceedings of the IEEE Conference on Computer Vision and Pattern Recognition* (2018), pp. 3801–3809
109. M. Ghifary, W.B. Kleijn, M. Zhang, D. Balduzzi, W. Li, in *Computer Vision–ECCV 2016: 14th European Conference, Amsterdam, The Netherlands, October 11–14, 2016, Proceedings, Part IV 14* (Springer, 2016), pp. 597–613
110. K. Bousmalis, G. Trigeorgis, N. Silberman, D. Krishnan, D. Erhan, in *Advances in Neural Information Processing Systems*, vol. 29 (2016)
111. A. de Mathelin, F. Deheeger, M. Mougeot, N. Vayatis, arXiv preprint arXiv:2103.03757 (2021)
112. Y. Zhang, Y. Zhang, Y. Wang, Q. Tian, arXiv preprint arXiv:1811.12751 (2018)
113. H. Li, X. Wang, Z. Zhang, W. Zhu, arXiv preprint arXiv:2202.07987 (2022)
114. Z. Shen, J. Liu, Y. He, X. Zhang, R. Xu, H. Yu, P. Cui, arXiv preprint arXiv:2108.13624 (2021)
115. X. Zhang, L. Zhou, R. Xu, P. Cui, Z. Shen, H. Liu, in *Proceedings of the IEEE/CVF Conference on Computer Vision and Pattern Recognition* (2022), pp. 4910–4920
116. S. Harary, E. Schwartz, A. Arbelle, P. Staar, S. Abu-Hussein, E. Amrani, R. Herzig, A. Alfassy, R. Giryes, H. Kuehne, et al., in *Proceedings of the IEEE/CVF Conference on Computer Vision and Pattern Recognition* (2022), pp. 5280–5290
117. A. Dittadi, F. Träuble, F. Locatello, M. Wüthrich, V. Agrawal, O. Winther, S. Bauer, B. Schölkopf, arXiv preprint arXiv:2010.14407 (2020)
118. H. Zhang, Y.F. Zhang, W. Liu, A. Weller, B. Schölkopf, E.P. Xing, in *Proceedings of the IEEE/CVF Conference on Computer Vision and Pattern Recognition* (2022), pp. 8024–8034
119. J. Peters, P. Bühlmann, N. Meinshausen, J. Roy. Stat. Soc. Series B: Stat. Methodol. **78**(5), 947 (2016)
120. M. Arjovsky, L. Bottou, I. Gulrajani, D. Lopez-Paz, arXiv preprint arXiv:1907.02893 (2019)
121. S. Xin, Y. Wang, J. Su, Y. Wang, in *Proceedings of the AAAI Conference on Artificial Intelligence*, vol. 37 (2023), pp. 10519–10527
122. M.M. Rahman, C. Fookes, M. Baktashmotlagh, S. Sridharan, Pattern Recogn. **100**, 107124 (2020)
123. Z. Yuan, X. Peng, X. Wu, B.k. Bao, C. Xu, in *2021 IEEE International Conference on Multimedia and Expo (ICME)* (IEEE, 2021), pp. 1–6
124. J. Cha, S. Chun, K. Lee, H.C. Cho, S. Park, Y. Lee, S. Park, Adv. Neural Inform. Process. Syst. **34**, 22405 (2021)
125. H. Rahimian, S. Mehrotra, arXiv preprint arXiv:1908.05659 (2019)
126. J. Duchi, H. Namkoong, arXiv preprint arXiv:1810.08750 (2018)
127. A. Newell, H.A. Simon, in *ACM Turing Award Lectures* (2007), p. 1975

128. M. Ginsberg, *Essentials of Artificial Intelligence* (Newnes, 2012)
129. S. Nolfi, Neurocomputing **42**(1–4), 119 (2002)
130. T. Brys, A. Harutyunyan, M.E. Taylor, A. Nowé, in *AAMAS* (2015), pp. 181–188
131. K. Rakelly, A. Zhou, C. Finn, S. Levine, D. Quillen, in *International Conference on Machine Learning* (PMLR, 2019), pp. 5331–5340
132. X. Liu, H. Yu, H. Zhang, Y. Xu, X. Lei, H. Lai, Y. Gu, H. Ding, K. Men, K. Yang, et al., arXiv preprint arXiv:2308.03688 (2023)
133. P. Zhao, Z. Jin, N. Cheng, arXiv preprint arXiv:2309.14365 (2023)
134. Z. Xi, W. Chen, X. Guo, W. He, Y. Ding, B. Hong, M. Zhang, J. Wang, S. Jin, E. Zhou, et al., arXiv preprint arXiv:2309.07864 (2023)
135. T.F. Varley, Entropy **25**(1), 54 (2022)
136. C. Marchal (2012)
137. X. Ma, Y. Wang, M.E. Houle, S. Zhou, S. Erfani, S. Xia, S. Wijewickrema, J. Bailey, in *International Conference on Machine Learning* (PMLR, 2018), pp. 3355–3364
138. C. Olah, N. Cammarata, L. Schubert, G. Goh, M. Petrov, S. Carter, Distill **5**(3), e00024 (2020)
139. A. Ilyas, S. Santurkar, D. Tsipras, L. Engstrom, B. Tran, A. Madry, *Advances in Neural Information Processing Systems*, vol. 32 (2019)
140. S. Spigler, M. Geiger, M. Wyart, J. Stat. Mech. Theory Exp. **2020**(12), 124001 (2020)
141. C. Li, H. Farkhoor, R. Liu, J. Yosinski, arXiv preprint arXiv:1804.08838 (2018)
142. M.F. Balcan, D. DeBlasio, T. Dick, C. Kingsford, T. Sandholm, E. Vitercik, in *Proceedings of the 53rd Annual ACM SIGACT Symposium on Theory of Computing* (2021), pp. 919–932
143. R. Pascanu, G. Montufar, Y. Bengio, arXiv preprint arXiv:1312.6098 (2013)
144. T. Zaslavsky, Mem. Am. Math. Soc. **154**, 1 (1997)
145. M. Raghu, B. Poole, J. Kleinberg, S. Ganguli, J. Sohl-Dickstein, in *International Conference on Machine Learning* (PMLR, 2017), pp. 2847–2854
146. M.J. Kearns, U. Vazirani, *An Introduction to Computational Learning Theory* (MIT Press, 1994)
147. M.H. Anthony, N. Biggs (1997)
148. N. Ye, L. Zhu, J. Wang, Z. Zeng, J. Shao, C. Peng, B. Pan, K. Li, J. Zhu, in *Thirty-Seventh AAAI Conference on Artificial Intelligence, AAAI 2023, Washington, DC, USA, February 7–14, 2023* (2023), pp. 10927–10935
149. H. Li, X. Wang, Z. Zhang, W. Zhu, IEEE Trans. Knowl. Data Eng. **35**(7), 7328 (2023). https://doi.org/10.1109/TKDE.2022.3193725
150. J. Jumper, R. Evans, A. Pritzel, T. Green, M. Figurnov, O. Ronneberger, K. Tunyasuvunakool, R. Bates, A. Žídek, A. Potapenko, et al., Nature **596**(7873), 583 (2021)
151. Y. Li, C.Y. Hsieh, R. Lu, X. Gong, X. Wang, P. Li, S. Liu, Y. Tian, D. Jiang, J. Yan, et al., Nat. Mach. Intell. **4**(7), 645 (2022)
152. J. Wang, C. Lan, C. Liu, Y. Ouyang, T. Qin, W. Lu, Y. Chen, W. Zeng, P.S. Yu, IEEE Trans. Knowl. Data Eng. **35**(8), 8052 (2023). https://doi.org/10.1109/TKDE.2022.3178128
153. H. Ye, C. Xie, T. Cai, R. Li, Z. Li, L. Wang, in *Advances in Neural Information Processing Systems*, vol. 34 (Curran Associates, 2021), pp. 23519–23531
154. T. Cai, R. Gao, J. Lee, Q. Lei, in *International Conference on Machine Learning* (PMLR, 2021), pp. 1170–1182

Chapter 2
Enhancing Data for Hard Anomaly Detection

Abstract Anomaly detection refers to the process of identifying elements, events, or observations in a given dataset that deviate from the expected patterns or exhibit distinct characteristics compared to other elements. We focus on a special and representative anomaly detection problem, called hard anomaly detection. The so-called hard anomaly has two characteristics in terms of the differences between abnormal and normal samples: (1) the boundary is exceedingly unobvious, and (2) the distribution is extremely imbalanced, both resulting in the difficulty of anomaly detection. It is really hard for regular classifiers to find the boundary between abnormal and normal samples here. To address the issue, we first propose a quantitative definition of the hard anomaly according to detection difficulty and then design a dedicated solution framework, named hard anomaly detection (HAD). Under HAD, we devise a GAN-based (generative adversarial network) method, called HadGAN, which produces both abnormal and normal samples with a similar distribution. We then pre-train a base anomaly detection model using the data generated by HadGAN. Moreover, we apply transfer learning to fine-tune the base model on real datasets. The superiority of our solution is demonstrated both by theoretical analysis and in extensive experiments.

2.1 Introduction to Data Augmentation

In the domain of data mining, anomaly detection refers to the process of identifying elements, occurrences, or observations that deviate from the expected patterns or exhibit dissimilarities compared to other elements within a given dataset. These anomalies often manifest as instances of bank frauds [1], medical anomalies [2], textual errors [3], and so on. Anomalies are also referred to as outliers, novelties, biases, and exceptions. Under the assumption that the majority of instances within a dataset are representative of normal behavior, existing methodologies for anomaly detection are typically categorized into three main groups:

- The unsupervised anomaly detection detects anomalies in unlabeled data by finding the most mismatched data instances from the rest [4].

C. Wang, H. Zhu, *Universal Behavior Computing for Security and Safety*,
https://doi.org/10.1007/978-981-97-9014-2_2

- The supervised anomaly detection necessitates the availability of labeled data instances classified as either normal or abnormal. This approach employs classifiers to learn the discriminative boundary between normal and abnormal samples [1].
- Semi-supervised anomaly detection focuses on adding unlabeled samples to supervised classification algorithm to realize boundary learning [5].

As a matter of fact, almost all existing anomaly detection methods make efforts on finding possible boundaries that could separate the abnormal and normal patterns of data to the maximum extent. However, this attempt becomes ineffective when confronted with a representative type of anomaly detection problem. We call it *hard anomaly detection*.

We start with a set of examples that motivate the issue of hard anomaly detection, and identify the key technical challenges to be addressed. In an online trading scenario, fraudsters transfer and withdraw victims' deposits by stealing his/her e-commerce platform account and then make money laundering on the Internet [1, 6]. This kind of anomalies usually shows the following characteristics:

- To make profits from fraud as long as possible, fraudsters need tricks to disable the anti-fraud systems. Usually, they try their best to disguise themselves as legitimate traders by imitating normal trading behaviors. Consequently, the distribution of fraudulent data exhibits similarity to that of legitimate data.
- Moreover, fraudulent transactions account for only a very small part of the total transaction data, which makes the data extremely imbalanced.

Similarly, in industrial equipment anomaly detection scenario, the probability that equipment appears in an abnormal condition is so small that the data is highly imbalanced [7]. With time slices, the collected temperature, pressure, or speed from sensors does not have dramatic shifts, so the boundary between normal and anomalous data distribution is quite obscure.

In general, hard anomaly detection refers to the scenario characterized by highly imbalanced data, where the distribution of positive samples closely resembles that of negative samples. These abnormal samples are accordingly called *hard anomalies* that cannot be effectively detected by existing anomaly detection methods due to the following technical challenges:

- The abnormal samples do not exhibit the characteristics as noise or outliers but rather resemble normal ones [8]. However, current studies in the industrial field still focus on boundary learning. For example, iForest which is a tree-based method [9–12] divides the points of dataset recursively and randomly until all are isolated, and determines the abnormal samples by the points with short path. One-class SVM (support vector machine), as exemplified in studies such as [13], employs a mapping technique to project data points into a high-dimensional space. By doing so, it identifies the class boundary of normal samples and effectively separates abnormal samples. Hence, most existing anomaly detection methods by detecting outliers and noise perform not well in hard anomaly detection.

- The class distribution of positive and negative samples is extremely imbalanced. For example, in online transaction data, the majority of transactions are normal, while only a small fraction of them are abnormal. Generally speaking, when dealing with imbalanced data, researchers tend to adjust the ratio of positive and negative samples to be more balanced before building a model. For instance, both oversampling and undersampling are commonly used. Other methods are based on data generation, such as SMOTE [14] and ADASYN [15], which solves the problem of imbalanced data by adding the generated minority class samples to the training set. However, both the sampling-based and the generation-based techniques easily lead to the problem of overfitting, which eventually results in the poor performance of the model. The cost-sensitive methods such as Focal loss [16] work by weighting or redefining model loss functions to solve the data imbalanced problem, for example, the Focal loss reduces the loss weight of the majority negative samples in data, so that the attention of the resulted models could focus on those minority positive samples. Although these methods can alleviate the impact of data imbalance, it does little help to detect hard anomalies, since the class distribution boundary of positive and negative samples remains fuzzy.

In this chapter, we aim at formalizing the *hard anomaly detection* problem, and providing an effective solution. Specifically, we start from the problem itself, and the two challenges that the *hard anomalies* face are as follows: (1) extremely imbalanced data and (2) similarity of positive and negative sample distribution. So we introduce two metrics: (1) degree of data imbalance (ε) and (2) similarity of positive and negative sample distribution (δ), as important measures of our formalization to *hard anomaly*. Inspired by the generative adversarial network (GAN), we propose a detection framework of *hard anomalies* based on GAN method, called hard anomaly detection (HAD). Under the framework HAD, our proposed method, HadGAN, can generate sufficient samples for the training of the base detection model. Subsequently, in order to acquire our ultimate detection model, we employ transfer learning to fine-tune the pre-trained model using real-world data as supplementary training.

Our contributions can be summarized as follows:

- We propose a quantitative definition to measure the difficulty of anomaly detection. Based on it, we formulate the definition of *hard anomaly*.
- We propose a dedicated solution framework, named *hard anomaly detection* (HAD), to detect the hard anomalies effectively.
- We propose a method, called HadGAN, which can generate similar distribution of positive and negative samples.

It is worth noting that, unlike some other methods such as AnoGAN [17] and GANomaly [18], which directly take GAN as the anomaly detection model, our HadGAN plays an auxiliary role in the detection of *hard anomalies*.

2.2 Related Work

Based on the availability of labeled data, various methodologies for conventional anomaly detection can be categorized into the following types: supervised learning, semi-supervised learning, unsupervised learning, and methods based on generative adversarial networks (GANs).

Supervised Learning In the context of a labeled dataset, a supervised learning approach strives to train a classifier that can effectively differentiate between normal and abnormal samples within the feature space. Consequently, the performance of anomaly detection heavily relies on the learning capability of the classifier. Support vector machine (SVM) is popular in anomaly detection [19]. It uses a hinge loss as the empirical risk and optimizes it with a regularization term, which could usually guarantee good anomaly detection performance. As one of the most representative tree models for anomaly detection, XGboost [11] benefits from its powerful boosting ensemble mechanism. The model capability surpasses GBDT [20] as in each step it continuously adds trees and segmentation features to fit the residual of the previous prediction. At present, deep learning algorithms [21], equipped with strong layer-by-layer feature learning ability, have been widely studied in anomaly detection, such as LSTM and CNN.

Semi-Supervised Learning Semi-supervised learning involves two sets of samples: one set is labeled, while the other set is unlabeled. The labeled data is utilized to train a classifier, while the unlabeled data is employed to iteratively optimize the classifier. This process involves augmenting the labeled samples with the inclusion of unlabeled samples to enhance the classification performance. An example of a semi-supervised log anomaly detection method is CausalConvLSTM [22], which leverages convolutional neural networks (CNNs) for parallel extraction of spatial features and utilizes Long Short-Term Memory (LSTM) networks to capture sequential relationships. Another example is SAD [23], an end-to-end deep architecture designed for general semi-supervised anomaly detection. It operates on the principle that the entropy of the normal distribution should be lower than that of the abnormal distribution.

Unsupervised Learning With the assumption that anomalies usually turn out as outliers and noise, unsupervised anomaly detection methods work by associating each sample in a dataset with a corresponding group. Clustering is a common and typical method in unsupervised anomaly detection. According to different prior assumptions, these methodologies can be broadly categorized into the following three primary classifications: (1) Normal samples belong to a cluster while abnormal samples do not belong to any cluster. In this scenario, not all samples will be grouped into a cluster center, such as DBSCAN [24]; (2) Normal and abnormal samples belong to separate clusters, a situation that algorithms with a clustering center division could be used, e.g., the k-means; (3) Normal samples belong to dense and large cluster centers, while abnormal samples belong to sparse and small cluster centers. When this happens, the density of clusters should be considered as a distance measure of clustering methods, such as DPC [25].

Few-Shot Learning We call the anomaly detection scenario with very few labeled data as few-shot anomaly detection. REPEN [26] may be the first deep model to realize anomaly detection by using few-shot anomaly data. The fundamental concept behind REPEN is to train a feature representation that ensures the anomaly instances possess a larger nearest neighbor distance compared to the normal data samples within a randomly selected subset of data. Different from REPEN, which focuses on feature representation learning of distance-based anomaly measurement, DevNet [27] network aims to perform end-to-end anomalous scoring learning with few-shot anomaly data.

Since our hard anomaly detection is a method based on GAN, we put the GAN-based method as a category here.

GAN-Based Methods With the increasing advancement and application of GANs in the domain of image analysis, GANs have also been utilized as models for anomaly detection. AnoGAN [17] is an unsupervised anomaly detection model that builds upon the DCGAN (deep convolutional GAN) architecture [28]. AnoGAN trains the model using a substantial number of negative samples and subsequently assigns an anomalous score to the test samples for the purpose of anomaly detection. Compared with AnoGAN, EGBAD [29] adds an encoder for joint training, which saves the time cost of parameter adjustment for each new image input. GANomaly [18] further improves EGBAD, and uses the structure of autoencoder at first, then encodes the restructure image once more on the normal samples, finally detects the anomaly by the defined contextual loss function.

Although existing methods have shown their powerfulness in ordinary anomaly detection, they are not feasible enough in hard anomaly detection, where not only the data is extremely imbalanced, but also the boundary of abnormal and normal samples is quite obscure.

2.3 Overview

2.3.1 Problem Statement

Hard anomalies have the following characteristics:

- The dataset exhibits an extreme class imbalance, where the number of abnormal samples is significantly smaller compared to the number of normal samples.
- There is no obvious boundary between the distributions of abnormal and normal samples; namely, the distance between the two distributions is very close.

A comprehensive depiction of the notation is provided in Table 2.1. Based on the inherent characteristics, the following formalization is presented.

Table 2.1 The description of notations

Notation	Description
$\boldsymbol{x}_i$	The attribute set of record i
y_i	The label of record i
S_p	The set of positive samples
S_n	The set of negative samples
$\rho(\cdot)$	The underlying distribution of $\cdot$
$\mathscr{DIST}(*, *)$	The distance between two distributions
$\sigma(\cdot)$	The sigmoid function

Definition 2.1 (**ε-Skewness**) Given a dataset containing m samples $\mathscr{D} = \{(\boldsymbol{x}_i, y_i)|y_i \in \{0, 1\}\}_{i=1}^m$, the set of anomaly samples $S_p = \{(\boldsymbol{x}_i, y_i)|y_i = 1\}$ and $S_p \subset \mathscr{D}$ with $|S_p| > 0$ and the set of normal samples $S_n = \{(\boldsymbol{x}_i, y_i)|y_i = 0\}$ and $S_n \subset \mathscr{D}$ with $|S_n| > 0$. For an arbitrary constant $\varepsilon > 0$, we say the dataset is ε-Skewness, if

$$\frac{|S_p|}{|S_n|} \leq \varepsilon.$$

By referring to Definition 2.1, it can be readily deduced that ε represents the ratio between positive and negative samples within the dataset. In the context of anomaly detection, the number of positive samples is typically lower than the number of negative samples. Consequently, we constrain ε to be within the interval $(0, 1)$.

Definition 2.2 (**δ-Boundary**) Given a dataset containing m samples $\mathscr{D} = \{(\boldsymbol{x}_i, y_i)|y_i \in \{0, 1\}\}_{i=1}^m$, the set of anomaly samples $S_p = \{(\boldsymbol{x}_i, y_i)|y_i = 1\}$ and $S_p \subset \mathscr{D}$ with $|S_p| > 0$ and the set of normal samples $S_n = \{(\boldsymbol{x}_i, y_i)|y_i = 0\}$ and $S_n \subset \mathscr{D}$ with $|S_n| > 0$, for an arbitrary constant $\delta > 0$, we say there exists a δ-Boundary of the positive and negative samples in the dataset, if

$$\mathscr{DIST}(\rho(S_p), \rho(S_n)) \leq \delta,$$

where $\rho(*)$ denotes the underlying distribution of a sample set $*$ and $\mathscr{DIST}(*, *)$ is a function that measures the distance between two distributions, which can be devised based on typical distance metrics, such as the JS divergence. Without loss of generality, the range of $\mathscr{DIST}(*, *)$ needs to be mapped to $(0, 1)$ by some means, such as sigmoid function.

Definition 2.3 (**(ε, δ)-Anomaly**) Given a dataset $\mathscr{D}$ and two small arbitrary constants ε and δ where $\varepsilon, \delta \in (0, 1)$. We say that the sample in $\mathscr{D}$ is a (ε, δ)-anomaly, if $\mathscr{D}$ is ε-Skewness with δ-Boundary.

Since the values of ε and δ depend on specific application scenarios and datasets, we only list a qualitative analysis about anomaly detection in Fig. 2.1.

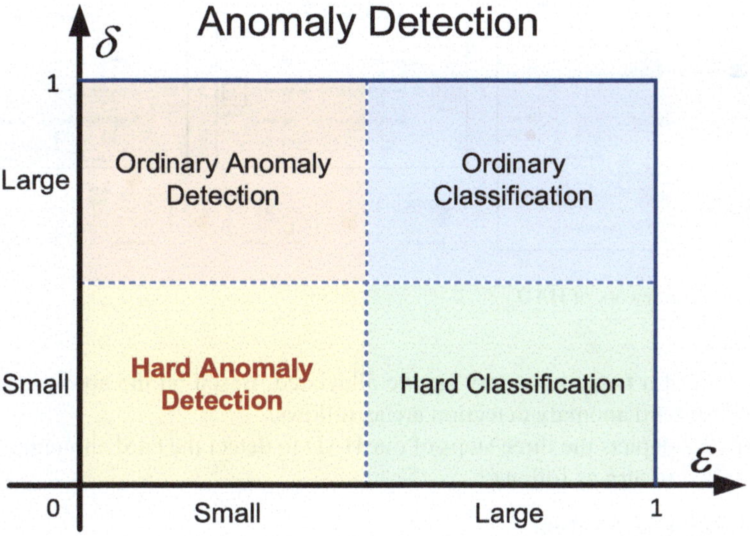

Fig. 2.1 Anomaly detection based on (ε, δ)

As we can see in Fig. 2.1, hard anomaly detection is a part of anomaly detection, and we reuse two variables ε and δ to divide the difficulty of anomaly detection. When ε is small and δ is large, (ε, δ)-Anomaly detection is transformed into an ordinary anomaly detection problem; when ε and δ are larger at the same time, it becomes an ordinary classification problem; when ε is larger and δ is smaller, it becomes a hard classification problem. This work mainly solves and discusses the anomaly detection when ε and δ are in a small state at the same time, and we call the problem hard anomaly detection.

2.3.2 Framework

Our detection framework of hard anomalies is based on transferring the problem of hard anomaly detection into an ordinary classification. According to Fig. 2.1, we consider whether the problem of hard anomaly detection can be transformed into a hard classification problem. The main difference between the two problems lies in the balance of samples. That is, as long as we solve the problem of data imbalance in hard anomaly detection, it will become a hard classification problem. One of the reasons for the fuzzy boundary of positive and negative samples in the hard classification problem is that the insufficient number of positive and negative samples leads to the inaccurate expression of their distribution, resulting in the fuzzy boundary. Therefore, if we can get a large number of positive and negative samples,

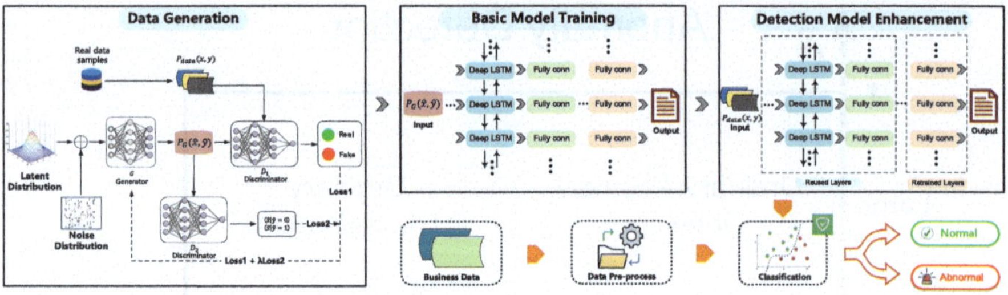

Fig. 2.2 The framework of HAD

the problem of a fuzzy boundary can be alleviated. Based on the above ideas, our solutions for hard anomaly detection are as follows.

Figure 2.2 depicts the three steps of our HAD to detect the hard anomalies. They are described in turn as follows.

Phase 1. Data Generation

The initial phase involves generating synthetic data instances that closely resemble the real data using the HadGAN framework, which encompasses three components:

- The generator, denoted as G, in the conditional generative adversarial network (CGAN) framework [30] is utilized. By incorporating the label as a conditioning factor, G is able to selectively generate synthetic samples that resemble the real data, encompassing both positive and negative instances.
- The discriminator D_1 employed in the framework is a traditional discriminator within the GAN paradigm. Its primary objective is to distinguish between synthetic samples and real samples, thereby discerning the authenticity of the generated instances.
- The discriminator D_2 serves as an additional component within the framework and is designed to differentiate between positive samples and negative samples. Its purpose is to discern the characteristics that distinguish positive instances from negative instances in the generated data.

The iteration of the generation-discrimination process concludes when the discriminator D_1 demonstrates an inability to accurately classify a substantial portion of the synthetic samples, and the discriminator D_2 fails to effectively distinguish between the generated positive and negative samples. Upon completion of the training process, the trained generator can be deployed to automatically generate a significant volume of positive and negative samples, conditioned on the input labels.

Phase 2. Basic Detection Model Training

Utilizing the converged HadGAN framework, the generator can be utilized to generate a substantial quantity of data samples comprising both positive and negative instances. These generated samples can subsequently be employed to train a foundational detection model. In the present study, our detection model incorporates a Long-Short Term Memory (LSTM) network [31].

Phase 3. Detection Model Enhancement
In this phase, transfer learning is employed to fine-tune the initial detection model using real data. This approach facilitates the utilization of knowledge acquired from synthetic data, enabling the mitigation of the highly imbalanced distribution inherent in the real data. Consequently, the performance of the base model can be significantly enhanced.

By implementing the aforementioned steps, a robust and efficient hard anomaly detection model can be obtained.

2.4 GAN-Based Framework for Hard Anomaly Detection

We first present the formal description of HadGAN, then give the training algorithm, and finally introduce the theoretic analysis.

2.4.1 *HadGAN*

In contrast to the conventional GAN model, HadGAN encompasses a generator G and two discriminators D_1 and D_2. The generator G generates a synthetic sample $\hat{\boldsymbol{x}}$ by incorporating a noise variable z conditioned on the variable y:

$$\hat{\boldsymbol{x}} = G(z|y), z \sim P_z(z); \tag{2.1}$$

here, $P_z(z)$ represents the prior distribution, and the discriminator D_1 is trained to discriminate between the real data sampled from $P_{data}(\boldsymbol{x})$ and the synthetic data generated by G from $P_G(\hat{\boldsymbol{x}})$.

It is well known that fraudsters often attempt to mimic the transaction behaviors of regular users in order to mask their fraudulent activities. In order to ensure that the generated fraudulent samples ($y = 1$) become increasingly similar to the legitimate samples ($y = 0$), an additional discriminator D_2 is integrated into the HadGAN model. This discriminator is trained to differentiate between the two categories of generated samples, namely, $P_G(\hat{\boldsymbol{x}}|y = 1)$ and $P_G(\hat{\boldsymbol{x}}|y = 0)$.

The value function of the minimax game can be represented as follows:

$$\min_{G} \max_{D_1, D_2} V(G, D_1, D_2) = V_1(G, D_1) + \lambda V_2(G, D_2), \tag{2.2}$$

where

$$V_1(G, D_1) = \mathbb{E}_{(x)\sim P_{data}(x)}\left[\log D_1(\boldsymbol{x}|y)\right] + \mathbb{E}_{(z)\sim P_z(z)}\left[\log(1 - D_1(G(z|y)))\right], \tag{2.3}$$

$$V_2(G, D_2) = \mathbb{E}_{(\hat{x})\sim P_G(\hat{x}|y=1)}\left[\log D_2(\hat{\boldsymbol{x}})\right] + \mathbb{E}_{(\hat{x})\sim P_G(\hat{x}|y=0)}\left[\log(1 - D_2(\hat{\boldsymbol{x}}))\right]; \tag{2.4}$$

here, λ is a hyperparameter responsible for balancing the two similarities: the similarity between the generated and real data and the similarity between the abnormal and normal samples. Specifically, $\lambda := f(\varepsilon, \delta)$. Next, we will thoroughly examine the relationship between λ and the parameters ε and δ.

The first value function V_1 remains consistent with that of the conventional GAN model [32], wherein the generator G endeavors to learn the distribution $P_G(\hat{\boldsymbol{x}}|y)$ over real data $P_{data}(\boldsymbol{x}|y)$. The second value function V_2 is designed to ensure the similarity between the generated positive samples and the negative samples. As the discrimination between the generated positive samples and negative samples becomes progressively challenging for the discriminator D_2 model, it indicates that the distribution of the positive samples generated by our generator approximates the distribution of the negative samples. Upon convergence of HadGAN, a sufficient number of synthetic samples can be obtained through the generator. For a proof on HadGAN's convergence, please refer to our previous work [33].

2.4.2 *Parameter Selection*

We can see that λ is a hyperparameter; here we give a strategy to select λ as follows:

$$\lambda = \frac{\sigma(\varepsilon^{-1}) \cdot e^{1-\delta}}{2}; \tag{2.5}$$

here $\sigma(\cdot)$ denotes the sigmoid function, while ε and δ correspond to the parameters utilized in (ε, δ)-Hard Anomaly, where $\varepsilon \in (0, 1)$ and $\delta \in (0, 1)$. In Eq. (2.5), a smaller value for ε and δ leads to a larger value for λ. As elucidated in Definition 2.3, a decrease in ε indicates a greater data imbalance, whereas a decrease in δ suggests a higher similarity between the positive and negative sample distributions. However, λ serves as a weighting factor for the losses of discriminators D_1 and D_2. In cases where the data quality is inferior (i.e., small values of ε and δ), a larger λ is necessary, indicating a higher weight for the discriminator D_2 loss. Nonetheless, as λ increases to a certain extent, it becomes crucial to ensure that the generated data distribution aligns with the real data distribution. Consequently, λ must approach a stable value. Equation (2.5) guarantees that λ tends toward $\frac{e}{2}$ when ε and δ are exceedingly small.

2.5 Experiments

2.5.1 Experimental Setup

2.5.1.1 Datasets

Our experiments are conducted on two private and two public datasets, which are described below.

Dataset 1 (B2C) The first dataset, denoted as B2C, is a real-world dataset pertaining to business-to-consumer transactions. It encompasses approximately three million records obtained from a commercial bank. These transaction records span a time interval ranging from April 1, 2017, to June 30, 2017, and originate from 0.1 million distinct accounts. Each record within the dataset has been manually labeled as either non-fraudulent or fraudulent. To facilitate the experimentation process, the dataset has been partitioned into two subsets. The training set consists of transaction records spanning April 1, 2017, to May 31, 2017, while the remaining records form the test set. Notably, the dataset comprises a total of 29 features. During the preprocessing stage, the categorical character features are transformed into numerical values by calculating the total count of unique categories for each character feature. Furthermore, all numerical data is normalized to serve as the input data for subsequent analysis. The specifics of the dataset are summarized in Table 2.2.

Dataset 2 (Pump) The second dataset, named Pump, comprises operational data related to the equipment of a power plant located in Thailand. This dataset encompasses the operational data of ten pumps on a per-minute basis, spanning the time period from January 1, 2009, to December 31, 2019. Each individual record within the dataset is labeled as either normal or abnormal, indicating the operational state of the pumps. During the preprocessing phase, all numerical data within the dataset is subjected to normalization, thereby transforming it into a suitable format for utilization as input data in subsequent analyses.

The public datasets are **Dataset 3 (Ionosphere**[1]) and **Dataset 4 (ECC**[2]).

Table 2.2 Description of the datasets

Dataset introduction		B2C	Pump	Ionosphere	ECC
Training set	# of positive	29,927	37,118	76	296
	# of negative	2,393,817	3,312,478	113	170,588
Test set	# of positive	11,541	18,223	50	196
	# of negative	1,003,539	1,433,425	76	113,727
Dimension		29	81	33	31

[1] https://archive.ics.uci.edu/ml/datasets/ionosphere

[2] https://www.kaggle.com/datasets/mlg-ulb/creditcardfraud

2.5.1.2 Evaluation Metrics

In practical applications, various evaluation metrics are commonly employed to assess the performance of binary classifiers. These metrics include the AUC (Area Under the ROC Curve) score, F-measure, and KS (Kolmogorov-Smirnov) score. However, these metrics do not directly capture the economic impact of the models, particularly in situations involving imbalanced data. For instance:

- In the fraud detection system, compared with fraud users, normal users occupy the vast majority of transaction records. In order to ensure the detection of fraud users, a system disturbs many normal users, which is unacceptable in reality.
- In the abnormal detection of machines, frequent false alarms will greatly waste the labor force of maintenance personnel, which is contrary to the original intention of using machine learning to bring convenience to people.

Considering the aforementioned factors, the false positive rate (FPR) serves as a fundamental criterion for evaluating a model's performance. Additionally, other metrics such as precision, recall, and F1-score are commonly reported under a fixed FPR threshold. This approach ensures that the evaluation of the model takes into account its ability to accurately identify positive instances while controlling the rate of false-positive predictions.

2.5.2 Synthetic Data Generation

Figure 2.3a and c presents the results of dimension reduction applied to a subset of samples from the B2C and Pump datasets, respectively. The dimension reduction technique employed in this study is T-SNE (t-distributed stochastic neighbor embedding) [34]. Due to the high time complexity associated with T-SNE, a random selection is made of 100 abnormal samples from the B2C dataset and 5000 normal samples from the Pump dataset. The generated samples produced by HadGAN, simulating the selected data points, are illustrated in Fig. 2.3b and d, respectively. Intuitively, according to Fig. 2.3, we have the following summaries:

- By definition, the abnormal samples in real datasets are hard anomalies.
- HadGAN can learn the distribution of real data and produce abnormal samples that are similar to normal samples.

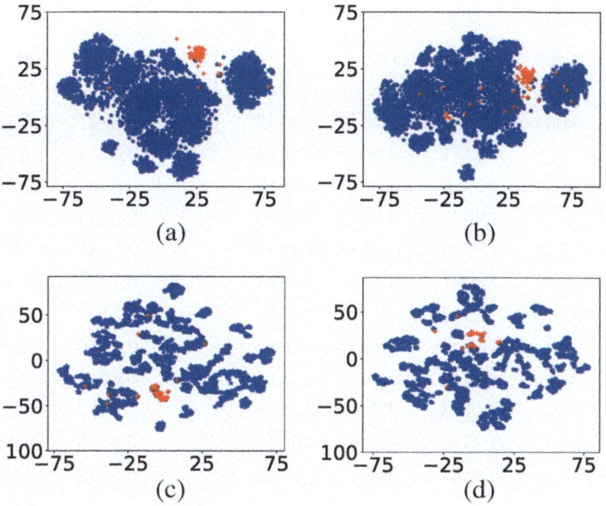

Fig. 2.3 The results of T-SNE dimensionality reduction for samples

2.5.3 Contrast Experiments

We compare HadGAN[3] with some representative and state-of-the-art models for hard anomaly detection. The comparison models we choose are as follows: Isolation Forest [10], SMOTE [14], Focal loss [16], Robust Deep Autoencoder (RDA) [35], and GANomaly [18].

Table 2.3 illustrates the performance of HadGAN and other models over datasets B2C and Pump. Notably, we present the precision, recall, and F1-score by assigning the FPR with different values. We run each model ten times, and then take the average of metrics to mitigate the impact of randomness. From Table 2.3 we can make out that HadGAN + LSTM is the best in detecting hard anomalies. GANomaly and Focal loss + LSTM are close in performance, next is RDA, then is SMOTE + LSTM, while Isolation Forest is the weakest. It seems that the performance superiority of HadGAN + LSTM is not obvious. However, when we set FPR = 0.0001, other models cannot detect any anomalies on both datasets, which demonstrates the advantages of our HadGAN. This also means that LSTM is not the key to detecting hard anomalies.

The results in Fig. 2.4a and b further verify this conclusion. The figures depict the changes of Recall and F1-score along with FPR on B2C and Pump datasets. Specifically, we have used broken lines of different colors to represent different methods.

[3] https://github.com/AnnHt/HadGAN

Table 2.3 Results of comparative experiments

	B2C ($\varepsilon = 0.0125$ and $\delta = 0.1245$)				Pump ($\varepsilon = 0.0112$ and $\delta = 0.3069$)			
Methods	FPR	Precision	Recall	F1-score	FPR	Precision	Recall	F1-score
Isolation Forest	0.001	0.8242	0.4517	0.5835	0.005	0.6841	0.6231	0.5914
SMOTE		0.8531	0.5050	0.6344		0.6311	0.7198	0.6725
RDA		0.7919	0.7326	0.7610		0.6338	0.7158	0.6723
Focal loss		0.8977	0.7630	0.8249		0.6543	0.7963	0.7184
GANomaly		0.8963	0.7516	0.8176		0.6645	0.8333	0.7394
HadGAN		0.8991	0.7748	0.8324		0.7033	0.9973	0.8249
Isolation Forest	0.0005	0.8347	0.3115	0.4537	0.001	0.8509	0.4532	0.5914
SMOTE		0.8911	0.3558	0.5085		0.8516	0.4829	0.6163
RDA		0.9246	0.6442	0.7814		0.8912	0.8335	0.8614
Focal loss		0.9563	0.9514	0.9539		0.9112	0.8635	0.8867
GANomaly		0.9546	0.9142	0.9336		0.8875	0.6638	0.7595
HadGAN		0.9581	0.9942	0.9758		0.9213	0.9851	0.9521
Isolation Forest	0.0001	–	–	–	0.0001	–	–	–
SMOTE		–	–	–		–	–	–
RDA		–	-	-		–	–	–
Focal loss		–	–	–		–	-	–
GANomaly		–	–	–		–	–	–
HadGAN		0.9812	0.4538	0.6206		0.9514	0.1647	0.2808

However, an unanswered question still remains. Given the highly similar distribution of positive and negative samples in the data, one might consider designing a conditional generative adversarial network (CGAN) to directly learn the distribution of positive and negative samples. It could be argued that if the CGAN converges, the generated positive and negative samples would also exhibit a similar distribution. This raises the question of whether D_2 is necessary in this context. The answer is that D_2 is indeed indispensable. Due to the extreme data imbalance, the limited number of positive samples is insufficient for the generator G to accurately learn the real distribution of anomalies. The additional discriminator D_2 serves a crucial role by not only leveraging the similarity information between the distributions of positive and negative samples but also utilizing the knowledge of negative samples to train HadGAN. Consequently, the generator G is able to improve its learning of the distribution of positive samples.

2.5.4 *Stability Study*

As we mentioned earlier, in the HAD framework, the anomaly detection model can be any neural network classification model; meanwhile, to further validate the role of HadGAN in hard anomaly detection, we have done the following ablation

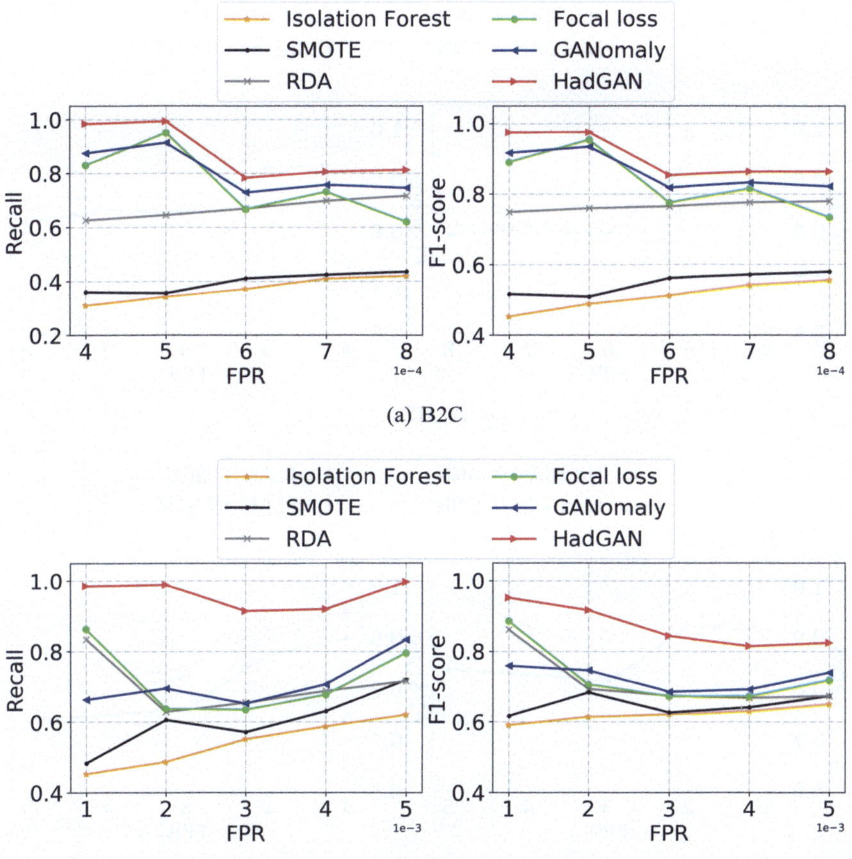

(a) B2C

(b) Pump

Fig. 2.4 Recall and F1-score change with FPR on B2C and Pump datasets

experiments. Therefore, we verify the performance of different neural network models in the HAD framework which includes artificial neural network (ANN), convolutional neural network (CNN), gate recurrent unit (GRU), and Long-Short Term Memory network (LSTM). If these models are not pre-trained with the data generated by HadGAN, since the data is extremely unbalanced, the trained models are very unstable and cannot detect abnormalities under the given FPR. We can see from Fig. 2.5 and Table 2.4 that these models have a good performance under the HAD framework. At the same time, we also verified the stability effect of our HAD framework on public datasets Ionosphere and ECC, as shown in Fig. 2.6. It can be seen that for the current mainstream neural network model, the detection ability of hard anomalies can be strengthened under the HAD framework.

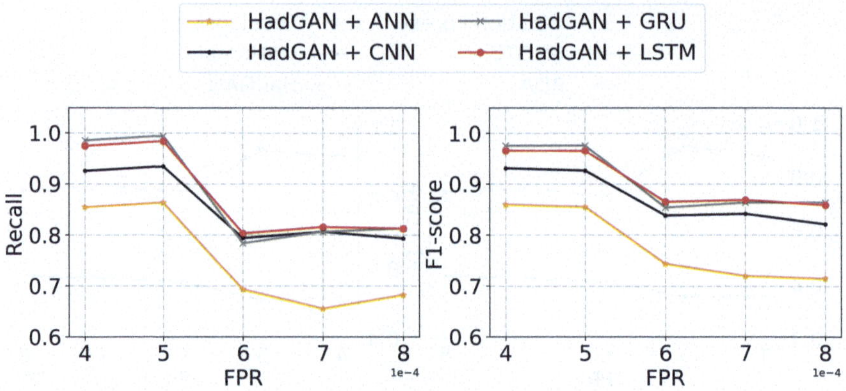

(a) B2C

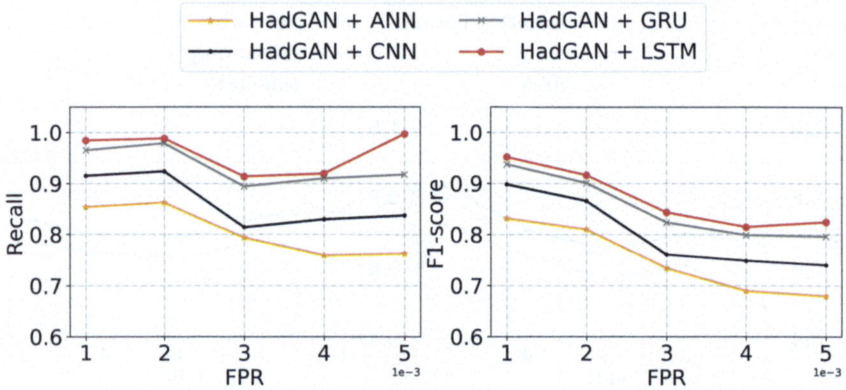

(b) Pump

Fig. 2.5 Performance of different neural networks in HAD

Table 2.4 Ablation experiments in HAD

	B2C ($\varepsilon = 0.0125$ and $\delta = 0.1245$)				Pump ($\varepsilon = 0.0112$ and $\delta = 0.3069$)			
Methods	FPR	Precision	Recall	F1-score	FPR	Precision	Recall	F1-score
HadGAN + ANN	0.004	0.8659	0.8552	0.8605	0.001	0.8113	0.8551	0.8326
HadGAN + CNN		0.9361	0.9248	0.9305		0.8807	0.9150	0.8779
HadGAN + GRU		0.9561	0.9732	0.9655		0.9109	0.9648	0.9374
HadGAN + LSTM		0.9673	0.9860	0.9755		0.9213	0.9851	0.9521
HadGAN + ANN	0.008	0.7512	0.6832	0.7156	0.005	0.6133	0.7632	0.8326
HadGAN + CNN		0.8531	0.5050	0.6344		0.6633	0.8373	0.7402
HadGAN + GRU		0.9108	0.8128	0.8594		0.7029	0.9167	0.7962
HadGAN + LSTM		0.9218	0.8135	0.8638		0.7033	0.9973	0.8249

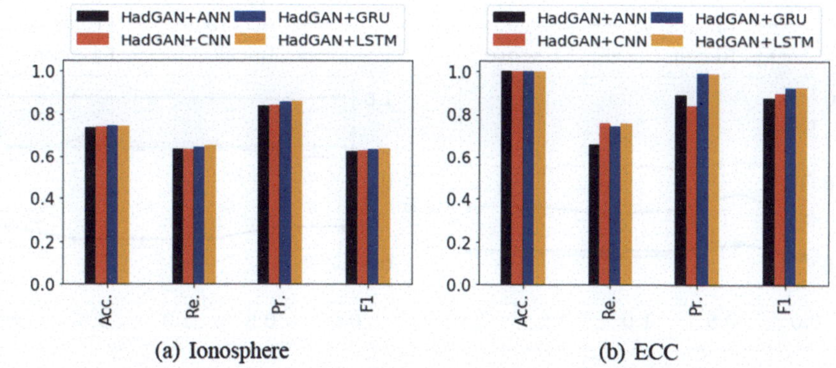

Fig. 2.6 Performance of different neural networks in HAD on Ionosphere and ECC datasets

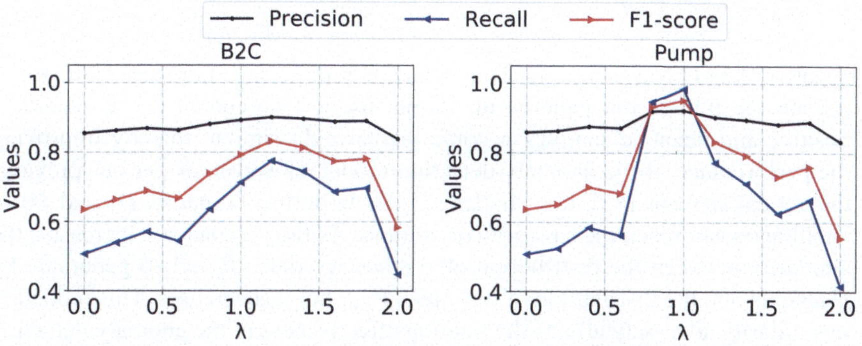

Fig. 2.7 HadGAN's performance changes with λ on B2C and Pump datasets

2.5.5 Robustness Study

To further assess the performance of HadGAN and examine the sensitivity of its performance to the parameter λ on two real-world datasets, we conducted experiments where the false positive rate (FPR) was fixed at 0.001. Subsequently, we observed the variations of different evaluation metrics with respect to λ. The corresponding results are presented in Figs. 2.7 and 2.8.

From the obtained results, we observed that the precision, recall, and F1-score initially increase and then decrease as λ increases. Notably, the maximum values of these metrics are achieved at different values of λ. It is worth noting that when $\lambda = 0$, HadGAN degenerates into a conditional generative adversarial network (CGAN). The underlying reason for these observations is as follows:

- First of all, in different datasets, the similarity of positive and negative samples is different, so their performance will achieve the optimal at different λ.
- During the training process of HadGAN, the distribution of synthesized positive data ($P_G(\hat{\boldsymbol{x}}|y = 1)$) gradually converges to resemble the distribution of

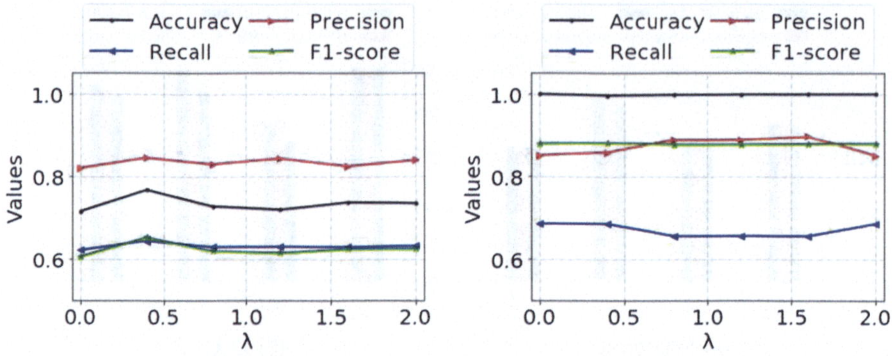

(a) Ionosphere (b) ECC

Fig. 2.8 HadGAN's performance changes with λ on Ionosphere and ECC datasets

negative data ($P_G(\hat{\boldsymbol{x}}|y = 0)$). This behavior mimics the fraudsters' attempt to imitate the transaction patterns of normal users. Consequently, the generated positive and negative samples become increasingly similar, thereby improving the performance of the anomaly detection model. However, as per our previous theoretical analysis [33], it is challenging for both discriminators D_1 and D_2 to simultaneously reach their respective minima. As the parameter λ increases, the similarity between the distribution of synthesized data ($P_G(\hat{\boldsymbol{x}}|y)$) generated by HadGAN and the distribution of real data ($P_{data}(\boldsymbol{x}|y)$) decreases. This reduction in similarity adversely affects the training effectiveness of the anomaly detection model. Consequently, it results in a decrease in precision, recall, and F1-score.

2.6 Discussion

2.6.1 Experimental Results

In the evaluation experiment, we choose some classical and representative methods as the contrast methods of our HAD framework. As a representative method of anomaly detection, Isolation Forest performs poorly in the detection of hard anomalies. Because it is an unsupervised method, it cannot use even a small amount of positive sample information. Because our data is extremely imbalanced, SMOTE is difficult to generate fake positive samples that is close to real ones based on a small amount of real positive samples, so SMOTE + LSTM has a poor performance in detecting hard anomalies. The main function of RDA is to improve the anti-noise ability of the autoencoder, so it's less effective in helping to detect hard anomalies. Although Focal loss can weaken the loss weight of a large number of negative samples and alleviate the problem of data imbalance, it cannot solve the problem that the distribution of positive and negative samples is very similar, so Focal loss

+ LSTM is weaker than HAD framework in performance. Because GANomaly is a semi-supervised anomaly detection method, it only uses negative samples in training process, thus losing a small amount of positive sample information, so the performance of this method is weaker than HAD in hard anomaly detection. In HAD framework, we first design the HadGAN method to obtain the generated data close to the real ones, and add the discriminator D_2 to make the distribution of generated positive samples similar to that of the negative ones. Then, we pre-train an anomaly detection model through the generated data, and finally fine-tune the model based on the real data. By doing this, HAD can be effective to solve the detection problem of hard anomaly.

2.6.2 *Experimental Boundary*

As a supervised anomaly detection, in recent years, there have been a surge of studies on using some limited labeled anomaly data in anomaly detection, such as methods on few-shot anomaly detection and weakly supervised anomaly detection as discussed in [36]. Here we need to explain that, first of all, hard anomaly detection is also a kind of limited anomaly detection for labeled data. In few-shot learning, the limited label reflected in both labeled positive and negative samples is few. While in weakly supervised, the limitation of label is reflected in the following: (1) incomplete supervision [37], where only a subset of training data is given with labels; (2) inexact supervision, where the training data is given with only coarse-gained labels; and (3) inaccurate supervision, where the given labels are not always ground-truth. The difference among few-shot anomaly detection, weakly supervised anomaly detection, and hard anomaly detection is as follows: (1) In hard anomaly, the labeled positive samples are also few, but the difference is a large amount of labeled negative samples exist; (2) The labeled data is complete, exact, and accurate. Based on the above considerations, we do not compare our HAD with few-shot and weakly supervised anomaly detection in experiments.

2.7 Conclusions

We formulate the definition of hard anomaly and then propose a novel framework hard anomaly detection (HAD), with a novel model named HadGAN, to detect the so-called hard anomalies. In particular, the generator G produces fake samples. Discriminator D_1 is trained to identify whether samples are real or fake, while another discriminator D_2 is trained to distinguish whether the generated samples are from positive or negative distribution. This allows the generator G to produce synthetic samples similar to the real ones. We pre-train a base anomaly detection model with enough synthetic samples generated by generator G. Finally, transfer learning is applied to fine-tune the final model with real data. Experiments show that, compared with the state-of-the-art methods to detect hard anomalies, our method based on HadGAN achieves better performance. In future work, we will further investigate the quantitative division of the anomaly detection domain from the perspective of data distribution, starting from the introduced variables ε and δ.

References

1. C. Wang, in *IJCAI 2020*, ed. by C. Bessiere (ijcai.org, 2020), pp. 4611–4618
2. T. Fernando, S. Denman, D. Ahmedt-Aristizabal, S. Sridharan, K.R. Laurens, P.J. Johnston, C. Fookes, Neural Netw. **127**, 67 (2020)
3. X. Yan, H. Zhang, X. Xu, X. Hu, P. Heng, in *Thirty-Fifth AAAI Conference on Artificial Intelligence, AAAI 2021, Thirty-Third Conference on Innovative Applications of Artificial Intelligence* (AAAI Press, 2021), pp. 3110–3118
4. H. Liang, L. Song, J. Wang, L. Guo, X. Li, J. Liang, Neurocomputing **423**, 444 (2021)
5. K. Demertzis, L.S. Iliadis, I. Bougoudis, Neural Comput. Appl. **32**(9), 4303 (2020)
6. D. Cheng, X. Wang, Y. Zhang, L. Zhang, IEEE Trans. Knowl. Data Eng. Published Online (2020)
7. X. Yi, X. Yang, Y. Huang, S. Ke, J. Zhang, T. Li, Y. Zheng, IEEE Trans. Knowl. Data Eng. Published Online (2021)
8. X. Han, X. Chen, L. Liu, in *Thirty-Fifth AAAI Conference on Artificial Intelligence, AAAI 2021, Thirty-Third Conference on Innovative Applications of Artificial Intelligence, IAAI 2021, The Eleventh Symposium on Educational Advances in Artificial Intelligence, EAAI 2021, Virtual Event, February 2–9, 2021* (AAAI Press, 2021), pp. 4090–4097
9. G. Ke, Q. Meng, T. Finley, T. Wang, W. Chen, W. Ma, Q. Ye, T. Liu, in *Advances in Neural Information Processing Systems 30: Annual Conference on Neural Information Processing Systems 2017, December 4–9, 2017, Long Beach, CA, USA* (2017), pp. 3146–3154
10. F.T. Liu, K.M. Ting, Z. Zhou, in *Proceedings of the 8th IEEE International Conference on Data Mining (ICDM 2008), December 15–19, 2008, Pisa, Italy* (IEEE Computer Society, 2008), pp. 413–422
11. T. Chen, C. Guestrin, in *SIGKDD 2016* (ACM, 2016), pp. 785–794
12. L.O. Prokhorenkova, G. Gusev, A. Vorobev, A.V. Dorogush, A. Gulin, in *Advances in Neural Information Processing Systems 31: Annual Conference on Neural Information Processing Systems 2018, NeurIPS 2018, December 3–8, 2018, Montréal, Canada* (2018), pp. 6639–6649
13. O. Kawi, K. Clawson, P. Dunn, D. Knight, J. Hodgson, Y. Peng, in *2020 International Joint Conference on Neural Networks, IJCNN 2020, Glasgow, United Kingdom, July 19–24, 2020* (IEEE, 2020), pp. 1–7
14. N.V. Chawla, K.W. Bowyer, L.O. Hall, W.P. Kegelmeyer, J. Artif. Intell. Res. **16**, 321 (2002)
15. H. He, Y. Bai, E.A. Garcia, S. Li, in *Proceedings of the International Joint Conference on Neural Networks, IJCNN 2008, part of the IEEE World Congress on Computational Intelligence, WCCI 2008, Hong Kong, China, June 1–6, 2008* (IEEE, 2008), pp. 1322–1328
16. T. Lin, P. Goyal, R.B. Girshick, K. He, P. Dollár, in *IEEE International Conference on Computer Vision, ICCV 2017, Venice, Italy, October 22–29, 2017* (IEEE Computer Society, 2017), pp. 2999–3007
17. T. Schlegl, P. Seeböck, S.M. Waldstein, U. Schmidt-Erfurth, G. Langs, in *Information Processing in Medical Imaging - 25th International Conference, IPMI 2017, Boone, NC, USA, June 25–30, 2017, Proceedings* (2017), pp. 146–157
18. S. Akcay, A.A. Abarghouei, T.P. Breckon, in *Computer Vision - ACCV 2018–14th Asian Conference on Computer Vision, Perth, Australia, December 2–6, 2018, Revised Selected Papers, Part III* (2018), pp. 622–637

19. P. Montague, J. Kim, in *2017 International Joint Conference on Neural Networks, IJCNN 2017, Anchorage, AK, USA, May 14–19, 2017* (IEEE, 2017), pp. 2843–2850
20. J.H. Friedman, Ann. Stat. **29**(5), 1189 (2001)
21. K. Lee, K. Lee, H. Lee, J. Shin, in *Advances in Neural Information Processing Systems 31: Annual Conference on Neural Information Processing Systems 2018, NeurIPS 2018, December 3–8, 2018, Montréal, Canada* (2018), pp. 7167–7177
22. S. Yen, M. Moh, T. Moh, in *18th IEEE International Conference On Machine Learning And Applications, ICMLA 2019, Boca Raton, FL, USA, December 16–19, 2019*, ed. by M.A. Wani, T.M. Khoshgoftaar, D. Wang, H. Wang, N. Seliya (IEEE, 2019), pp. 1334–1341
23. L. Ruff, R.A. Vandermeulen, N. Görnitz, A. Binder, E. Müller, K. Müller, M. Kloft, in *8th International Conference on Learning Representations, ICLR 2020, Addis Ababa, Ethiopia, April 26–30, 2020* (OpenReview.net, 2020)
24. M. Ester, H. Kriegel, J. Sander, X. Xu, in *Proceedings of the Second International Conference on Knowledge Discovery and Data Mining (KDD-96), Portland, Oregon, USA* (AAAI Press, 1996), pp. 226–231
25. A. Rodriguez, A. Laio, Science **344**(6191), 1492 (2014)
26. G. Pang, L. Cao, L. Chen, H. Liu, in *Proceedings of the 24th ACM SIGKDD International Conference on Knowledge Discovery & Data Mining, KDD 2018, London, UK, August 19–23, 2018* (2018), pp. 2041–2050
27. G. Pang, C. Shen, A. van den Hengel, in *Proceedings of the 25th ACM SIGKDD International Conference on Knowledge Discovery & Data Mining, KDD 2019, Anchorage, AK, USA, August 4–8, 2019* (2019), pp. 353–362
28. A. Radford, L. Metz, S. Chintala, in *4th International Conference on Learning Representations, ICLR 2016, San Juan, Puerto Rico, May 2–4, 2016, Conference Track Proceedings* (2016)
29. H. Zenati, C.S. Foo, B. Lecouat, G. Manek, V.R. Chandrasekhar, CoRR abs/1802.06222 (2018). http://arxiv.org/abs/1802.06222
30. M. Mirza, S. Osindero, CoRR abs/1411.1784 (2014). http://arxiv.org/abs/1411.1784
31. V. Frinken, F. Zamora-Martínez, S.E. Boquera, M.J.C. Bleda, A. Fischer, H. Bunke, in *Proceedings of the 21st International Conference on Pattern Recognition, ICPR 2012, Tsukuba, Japan, November 11–15, 2012* (IEEE Computer Society, 2012), pp. 701–704
32. I.J. Goodfellow, J. Pouget-Abadie, M. Mirza, B. Xu, D. Warde-Farley, S. Ozair, A. Courville, Y. Bengio, Adv. Neural Inform. Process. Syst. **3**, 2672 (2014)
33. H. Teng, C. Wang, Q. Yang, X. Chen, R. Li, IEEE Trans. Comput. Soc. Syst. (2023)
34. L. Van, der Maaten, G. Hinton, J. Mach. Learn. Res. **9**(2), 2579 (2008)
35. C. Zhou, R.C. Paffenroth, in *Proceedings of the 23rd ACM SIGKDD International Conference on Knowledge Discovery and Data Mining* (ACM, 2017), pp. 665–674
36. G. Pang, C. Shen, L. Cao, A. van den Hengel, ACM Comput. Surv. **54**(2), 38:1 (2021)
37. Z.H. Zhou, Natl. Sci. Rev. **5**(1), 44 (2018)

Chapter 3
Addressing Imbalance Data for Online Fraud Detection

Abstract Nowadays, the emergence of online trading greatly facilitates people's life. Meanwhile, online trading also brings hidden dangers, such as online fraudulent trading. To solve the issue, researchers have proposed many different detection models. However, in actual business scenarios, fraudulent transactions usually only account for a small portion of normal transactions, resulting in extremely imbalanced data; besides, the concealment of fraud is reflected in that the fraudsters are imitating the normal transactions of users, posing a huge challenge for fraudulent transaction detection modeling. Inspired by generative adversarial networks (GANs), we propose a GAN-based framework to detect online banking fraud on extremely imbalanced data, called BalanceGAN. A fraud detection model is first pre-trained using the data generated by the generator, and then the model is fine-tuned using transfer learning on real-world datasets, by using this approach to address data imbalances. Compared with the conventional methods for solving imbalanced data, our BalanceGAN can avoid overfitting of the model relatively, experiments on two real datasets show our BalanceGAN has more than 10% performance improvement in Precision and Recall.

3.1 Introduction to Imbalance Data

With the pervasive use of the Internet, the extension of eCommerce has accelerated the evolution of bank business, encouraging people to be more willing to carry out online trading affairs. The "Data Report 2021" published by Taobao[1] released that on average, more than 800 million users are active on Taobao every month. Among them, there are more than 300 million active users every day, and more than 60 million female users spend 3–4 hours shopping on Taobao at night. By mid-2019, EBay[2] had 182 million users worldwide. This indicates that the number of users increased by 1.11% over the previous quarter and 4% over the previous year. In

[1] https://www.taobao.com/

[2] https://www.ebay.com/

C. Wang, H. Zhu, *Universal Behavior Computing for Security and Safety*,
https://doi.org/10.1007/978-981-97-9014-2_3

2019, 172 million people purchased goods from Amazon[3] America. According to the analysis, there are more than 300 million active users on Amazon, 65% of the American population visits the Amazon website at least once a month, and most of them have become real buyers. It is the existence of these eCommerce platforms that greatly facilitate people's lives [1–4]. Online trading brings convenience to people's lives but also brings huge hidden dangers, for instance, fraud. Fraudsters transfer and withdraw users' deposits by stealing users' eCommerce platform accounts and making money laundering on the Internet. When the user perceives it, the loss has already been caused. Therefore, the identification of online anti-fraud has become a more and more widely concerned issue [5–10].

In the practical application scenario, the fraudulent transaction samples are extremely few compared with the normal transaction samples, which leads to data imbalance; besides, the fraudsters try to imitate the user's normal trading behavior as much as possible, so that their fraud means to become more and more hidden, these bring great difficulties to the identification and modeling of anti-fraud. Although there are many existing technical solutions for imbalanced learning, such as sampling methods [11–14] and data generation methods [15–18], these methods mainly aim at transforming training sets from imbalanced datasets to balanced datasets, which improve the final results in most cases. Sampling is divided into oversampling and undersampling. Oversampling duplicates multiple copies of a few categories to solve the problem of data imbalance, and undersampling excludes some sampling from the general class or selects some samples from the general class. The method based on data generation solves the data imbalance by adding the generated few samples to the few samples. The advantage of sampling is that the operation is simple, but the disadvantages are also obvious. Some samples will appear repeatedly in the dataset after the oversampling, and the trained model will be overfitting to some extent for undersampling, the final training set loses data, and the model only learns a part of the overall pattern, which may lead to the model underfitting. Even with the method based on data generation, it is difficult to avoid the risk of overfitting. We call the methods of copying data or adding the generation data to the dataset as oversampling. Different from the above sampling and data generation methods, in this chapter, we propose a new model called BalanceGAN, which can generate a large number of synthetic samples as training data for the pre-training model. Then, we use transfer learning to put the real training data into the pre-training model for further training to obtain our final detection model. The advantage of our BalanceGAN is to avoid directly adding or deleting data to real data, so as to reduce the risk of model overfitting and underfitting.

We make the following main contributions:

- We first present the GAN-based approach for generating training data and construct a base pre-training fraud detection model to solve the imbalanced data problem.

[3] https://www.amazon.com/

- For the first time, we apply transfer learning to fine-tune the base fraud detection model by using the real datasets which can leverage the knowledge learned from the base fraud detection model on the synthetic datasets.
- Compared with the conventional methods to solve data imbalance and fraud detection, the results of our BalanceGAN in real datasets are effective.

3.2 Related Work

3.2.1 Anti-Fraud Identification Model

Anti-fraud projects in many cases are that customers simply do not know what is and what is not. In other words, the definition of fraud is vague. From a small point of view, anti-fraud seems to be a two-class problem, but on the other hand, it can also be considered as a multi-class problem because you can treat each type of fraud as a separate type. Even in industries like banks and insurance companies that deal with fraud all year round, it is necessary to constantly update their detection methods rather than consider using a single model to solve the problem of anti-fraud. In addition, fraud detection generally faces the following problems: (1) Most of the data are not labeled, and various mature supervisory learning is useless [19]; (2) Even with labels, when various types of frauds are mixed, it will bring difficulties to our fraud detection [20].

Therefore, there are usually two ways to detect anti-fraud problems. One is to use unsupervised methods or anomaly detection methods to identify fraud without labeled data. The method in [21] uses unsupervised user behavior representation to detect frauds with a cold-start problem; P. Ravisankar [22] presented the feature selection and data mining techniques to detect financial statement fraud. For labeled data, the probabilistic model [23] is used to detect the fraud identification; E. L. Paula [24] adopted deep learning anomaly detection to support the investigation in Brazilian exports and anti-money laundering; sequence classification algorithm [25] is used to solve the credit card fraud; Saia et al. [26] evaluated two proactive strategies, which are based on the Fourier Transform and Wavelet Transform respectively, these strategies overcome the problem that reduces the effectiveness of the canonical retroactive state-of-the-art solutions. Zhou et al. [27] proposed a method based on PCA and improved Adaboosting to identify fraud. Unlike the above machine learning approach, there are still several data mining-based techniques to identify financial fraud [28].

In the absence of labels in existing anti-fraud methods, unsupervised learning is usually used as a model of fraud detection, but even if we have labels, there will be a huge risk, because the model learned from such historical data can detect frauds similar to historical frauds, and for variants of frauds and never seen. Our model will be quite weak for fraud, so in practice, we cannot rely on a single monitoring model to extravagantly detect all frauds.

3.2.2 Learning from Imbalanced Data

In methods on anti-fraud, there are usually few labels on data, especially for fraudulent data, this will bring up the situation of data imbalance, and solving the problem of data imbalance in machine learning is also a current difficulty. At present, we mainly divide the methods to solve the data imbalance into three categories: (1) oversampling, which includes the traditional oversampling techniques an methods involving the introduction of model-generated data [29]; (2) undersampling; (3) weighting, which includes techniques such as adjusting parameter weights and redefining loss functions [30]; and (4) GAN-based methods, utilizing generators in GANs to tackle data imbalance.

3.2.2.1 Oversampling

SMOTE [31] is a quite typical oversampling-based method; the basic idea of the algorithm is to analyze and simulate a few categories of samples, and add new samples to the dataset so that the categories in the original data are no longer seriously imbalanced. SMOTE-Borderline [32] is improved on the basic algorithm, which makes the selected representative samples more representative and the generation process more stable. ADASYN [16] uses a specific mechanism to automatically determine how many composite samples need to be generated for each minority sample, instead of synthesizing the same number of samples for each minority sample as SMOTE does. The method [33] is an improvement of the oversampling method to solve the problem of data imbalance by introducing clustering as well as kernel functions.

3.2.2.2 Undersampling

Undersampling methods are generally not recommended; thus, the amount of information will be lost, which limits the upper line of a model [34]. EasyEnsemble [35] and BalanceCascade [36] by using the ensemble learning mechanism, the negative samples are divided into several sets for different learners to use. In this way, each learner is undersampling, though important information is not lost in a global view. Unlike EasyEnsemble, which uses unsupervised methods to divide subsets, BalanceCascade uses supervised and Boosting methods.

3.2.2.3 Weighting

The methods in [37] and [38] solve the problem of data imbalance by weighting and redefining loss function. It should be noted that such methods are usually quite dependent on parameters; thus, how to determine parameters will be a crucial factor affecting the effectiveness of the model. Susto et al. [39] proposed a hierarchical structure for fraud detection that incorporates unsupervised and supervised learning, and the method extracts new features from historical data of users' online transactions to address data imbalances. R. Saia [40] addressed the data imbalance by introducing a high-dimensional space and then employing machine learning in the high-dimensional space.

3.2.2.4 GAN-Based

There are also many GAN-based methods to solve the machine learning problems with imbalanced data, such as designing different GAN frameworks [41] or constructing different loss functions [42] to make the data generation have better results, thus facilitating the training of models to better solve the machine learning problems with imbalanced data. JDGAN [43] enhances the generator via joint distribution on extremely limited data to solve the data imbalance. Hu et al. [44] solved the data imbalance by using GAN to generate minority samples and then adding them to the real dataset.

3.3 GAN-Based Framework on Extremely Imbalanced Data

3.3.1 Framework

Figures 3.1, and 3.2a, b depict the three steps of our framework to detect frauds on imbalanced data. They are described in turn as follows. We provide detailed descriptions of some important notations in Table 3.1.

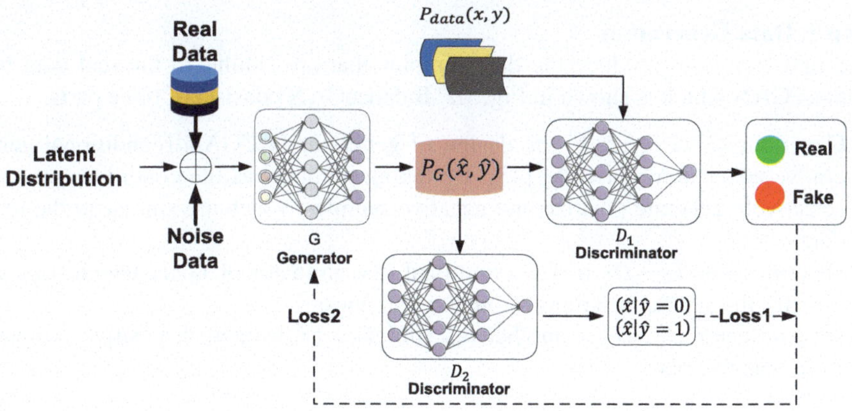

Fig. 3.1 Framework of BalanceGAN

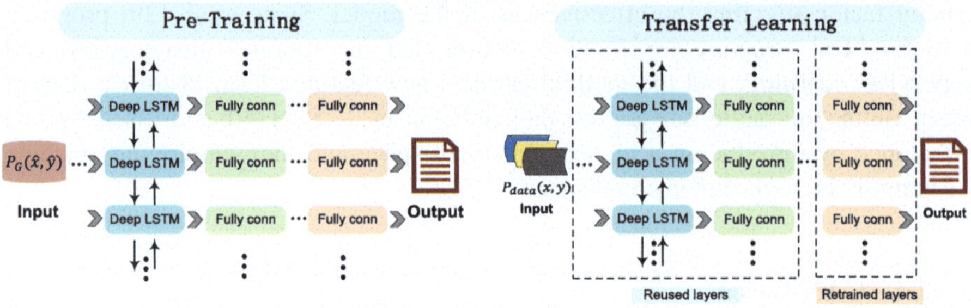

(a) (b)

Fig. 3.2 Process of pre-training and transfer learning

Table 3.1 The description of notations

Notation	Description
x_i	Attribute set of record i
y_i	Label of record i
z	Noise variable z
$G(\cdot)$	Generator in BalanceGAN
$D_1(\cdot)$	First discriminator in BalanceGAN
$D_2(\cdot)$	Second discriminator in BalanceGAN
$P_z(z)$	A prior noise distribution
$P_{data}(x, y)$	Distribution of real data
$P_G(x, y)$	Distribution of generated data
λ	Parameter to balance $D_1(\cdot)$ and $D_2(\cdot)$
Φ	Basic fraud detection model
m	Sample size contained in each batch
n	Number of reused layers in transfer learning
$n\_epoches$	Total number of times the dataset is trained
$n\_batches$	Total number of blocks to divide the dataset

Step 1. Data Generation

The first step is to produce the data samples that are similar to the real data by BalanceGAN which is shown in Fig. 3.2. BalanceGAN consists of three parts:

- Generator G. G refers to the design of generator in CGAN (conditional generative adversarial network) [45]. By taking the label as the condition, we can selectively generate positive and negative samples, which are similar to the real data.
- Discriminator D_1. D_1 is a conventional discriminator of GAN which tries to identify the synthetic samples from the real ones.
- Discriminator D_2. D_2 is another discriminator to distinguish positive samples from negative ones.

This generation-discrimination process terminates when the discriminator D_1 fails to identify a large portion of the synthetic samples and the discriminator D_2 fails to distinguish the generated positive and negative samples. Once the training is terminated, we can then apply the trained generator to automatically generate a large number of positive and negative samples conditioned on input labels.

Step 2. Base Detection Model Training
With the convergent BalanceGAN, we can employ the generator to produce a large number of data samples (both positive and negative) and use the data to learn a base detection model. In this work, our detection model is shown in Fig. 3.2a.

Step 3. Detection Model Enhancement
As shown in Fig. 3.2b, in this step, we apply the transfer learning to fine-tune the base detection model on the real data. Transfer learning allows us to leverage knowledge learned from synthetic data, so as to settle the extremely imbalanced distribution of the real data, and further improve the performance of base model.

With the above steps, an effective fraud detection model can be obtained.

3.3.2 *BalanceGAN*

By comparing Figs. 3.1 and 3.3, the specific differences are reflected in the generation model, discrimination model, and training process. BalanceGAN consists of one generator G and two discriminators D_1 and D_2. We apply the generator to generate both discrete and continuous data. The generator G generates a fake pair $(\hat{x}, \hat{y})$ from the noise variable z:

$$(\hat{x}, \hat{y}) = G(z), z \sim P_z(z), \tag{3.1}$$

where $P_z(z)$ is a prior distribution. The discriminator D_1 is trained to distinguish between the real data from $P_{data}(x, y)$ and the generated fake data from $P_G(x, y)$.

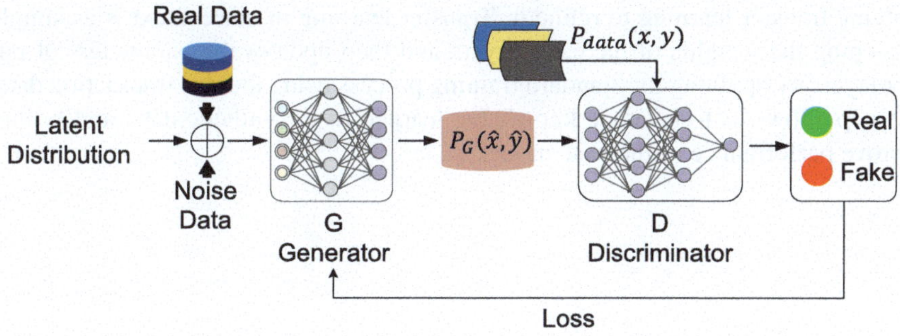

Fig. 3.3 Illustration of generative adversarial networks

As we mentioned earlier, fraudsters try to imitate the normal transaction users' behavior to conceal their fraud. In order to make the fraud samples ($\hat{y} = 1$), we generate more and more close to the non-fraud samples ($\hat{y} = 0$). Therefore, another discriminator D_2 is introduced into the BalanceGAN model and trained to distinguish the two categories of generated samples, $P_G(x|y = 1)$ and $P_G(x|y = 0)$.

The value function of the minimax game is described as

$$\min_{G} \max_{D_1, D_2} V(G, D_1, D_2) = V_1(G, D_1) + \lambda V_2(G, D_2), \tag{3.2}$$

with

$$\begin{aligned} V_1(G, D_1) = & \mathbb{E}_{(x|y)\sim P_{data}(x|y)}[\log D_1(x|y)] \\ & + \mathbb{E}_{(\hat{x}|\hat{y})\sim P_G(x|y)}[\log(1 - D_1(\hat{x}|\hat{y}))], \end{aligned} \tag{3.3}$$

$$\begin{aligned} V_2(G, D_2) = & \mathbb{E}_{(\hat{x})\sim P_G(\hat{x}|\hat{y}=1)}[\log D_2(\hat{x})] \\ & + \mathbb{E}_{(\hat{x})\sim P_G(\hat{x}|\hat{y}=0)}[\log(1 - D_2(\hat{x}))], \end{aligned} \tag{3.4}$$

where λ is a hyperparameter to balance the similarity between the generated data and the real data, as well as the similarity between the fraud samples and the non-fraud samples.

The first value function V_1 is the traditional GAN model [46], where the generator G seeks to learn the joint distribution $P_G(x, y)$ over real data $P_{data}(x, y)$. The second value function V_2 aims to make the generated fraud samples similar to the non-fraud samples. Once our D_2 model is more and more difficult to distinguish the generated fraud samples and non-fraud samples, it shows that the distribution of fraud samples generated by our generator is more and more close to the distribution of non-fraud samples. When the BalanceGAN converges, it means that we can obtain a large number of synthetic samples through the generator in BalanceGAN, and the data distribution of fraud samples is close to that of normal samples.

In this work, our detection model is a recursive neural network (RNN) [47, 48]. Figure 3.2a illustrates the process of training a base fraud detection model and applying transfer learning to refine it. Transfer learning in our context is as simple as keeping the weights of the early layers and then updates the parameters of our later layers by applying the standard training process using the real transaction data. The step allows us to leverage knowledge learned from synthetic data, and further improve performance of the base model.

Algorithm 1: Minibatch stochastic gradient descent training of Balanced-GAN

Input : Number of training iterations.
Output: Trained models D_1, D_2, and G.
for *a number of training iterations* **do**
  Sample a batch of m examples from the real data;
  $(\boldsymbol{x}|y) \sim P_{data}(\boldsymbol{x}|y)$;
  Sample a batch of m examples from the noise data;
  $(\hat{\boldsymbol{x}}|y) \sim P_G(\boldsymbol{x}|y)$;
  Update D_1 by stochastically ascending along its gradient:;
  $\nabla_{\theta_{d_1}} \frac{1}{m} \sum_{i=1}^{m} [\log D_1(\boldsymbol{x}_i|y_i) + \log(1 - D_1(\hat{\boldsymbol{x}}_i|y_i))]$;
  Update G by stochastically descending along its gradient:;
  $\nabla_{\theta_g} \frac{1}{m} \sum_{i=1}^{m} \log(1 - D_1(\hat{\boldsymbol{x}}_i|y_i))$;
  Sample a batch of m examples;
  $(\hat{\boldsymbol{x}}|y=1) \sim P_G(\hat{\boldsymbol{x}}|y=1)$;
  Sample another batch of m examples;
  $(\hat{\boldsymbol{x}}|y=0) \sim P_G(\hat{\boldsymbol{x}}|y=0)$;
  Update D_2 by stochastically ascending along its gradient:;
  $\nabla_{\theta_{d_2}} \frac{1}{2m} \sum_{i=1}^{2m} [\log D_2(\hat{\boldsymbol{x}}_i) + \log(1 - D_2(\hat{\boldsymbol{x}}_i))]$;
  Update G by stochastically descending along its gradient:;
  $\nabla_{\theta_g} \frac{1}{2m} \sum_{i=1}^{2m} [\log D_2(\hat{\boldsymbol{x}}_i) + \log(1 - D_2(\hat{\boldsymbol{x}}_i))]$;
end

3.3.3 Algorithm

The detailed training steps of BalanceGAN are described in Algorithm 1, where θ_g, θ_{d_1} and θ_{d_2} are the parameters of G, D_1, and D_2. At first, we sample a batch of m examples from the real data $P_{data}(\boldsymbol{x}|y)$ and noise data $P_G(\boldsymbol{x}|y)$, respectively (Lines 2–3). According to Eq. (3.2), we need to maximize the value of Eq. (3.3), so we update D_1 by stochastic gradient ascent (Line 4). As shown in Eq. (3.2), we need to maximize the loss of D_1, that is, minimize the objective function of Eq. (3.3), so we update G by stochastic gradient descent (Line 5). After updating G, we use it to sample a batch of m positive samples and m negative ones (Line 6). Similarly, with Eqs. (3.2) and (3.4), we update D_2 with stochastic gradient ascent (Line 7) and update G with stochastic gradient descent (Line 8). The iteration ends when BalanceGAN converges. The convergence of BalanceGAN will be proven in the following part. When we get a convergent BalanceGAN, we can apply it to generate a large number of positive and negative samples, and then use them to pre-train a base model ϕ, and then use the transfer learning to fine-tune it. For the detailed steps, refer to Algorithm 2.

For the detailed process of transfer learning, we can combine Algorithm 2 with Fig. 3.3. First, we use the BalanceGAN trained in the previous step to generate a large number of positive and negative samples, and then use the generated samples to pre-train a fraud detection model. In Algorithm 2, we first load the model (Line 1), then freeze parameters of the feature extraction layers (Lines 2–4) in the model,

Algorithm 2: Detection model enhancement

Input : $P_{data}(\boldsymbol{x}, y)$: The distribution of real data;
$\phi: \boldsymbol{x} \mapsto y$ Base model;
n: The number of reused layers.
Output: $\phi^{'}: \boldsymbol{x} \mapsto y$ Fine-tuning model.
Load model ϕ;
for *each layer of* ϕ *from 1 to n* **do**
  Set layer.trainable = False;
end
for $j = 1$ *to* $n\_epoches$ **do**
  for $k = 1$ *to* $n\_batches$ **do**
    $T \longleftarrow$ Sample t data from $P_{data}(\boldsymbol{x}, y)$;
    $\phi^{'} \longleftarrow$ Train ϕ with data T;
  end
end
return $\phi^{'}$

and next fine-tune the parameters of other layers with real data. Through constant iteration (Lines 5–9), we finally get the fraud detection model after fine-tuning.

3.3.4 *Theoretical Analysis*

We also make the similar theoretical analysis as the traditional GAN model [46]. We first consider the optimal discriminators D_1 and D_2 for any given generator G.

3.3.4.1 Optimization Proposition

Proposition 3.1 *With G fixed, the optimal discriminators D_1 and D_2 for the value function $V(G, D_1, D_2)$ are respectively*

$$D_1(x, y) = \frac{P_{data}(x|y)}{P_{data}(x|y) + P_G(x|y)},$$

$$D_2(x) = \frac{P_G(x|y=1)}{P_G(x|y=1) + P_G(x|y=0)}.$$

Proof The training criteria for the discriminators D_1 and D_2, given any generator G, are to maxmize $V(G, D_1, D_2)$:

$$
\begin{aligned}
V(G, D_1, D_2) = & \int_x \int_y P_{data}(x, y) \log D_1(x, y) \\
& + \int_x \int_y \log(1 - D_1(x, y)) \\
& + \lambda \int_x P_G(x|y = 1) \log D_2(x) \\
& + \lambda \int_x P_G(x|y = 0) \log(1 - D_2(x)).
\end{aligned}
$$

Following [46], for any $(m, n) \in \mathbb{R}^2 \backslash \{0, 0\}$, the function $y \rightarrow m \log y + n \log(1 - y)$ achieves its maximum in [0,1] at $\frac{m}{m+n}$. This concludes the proof.

3.3.4.2 Convergence Theorem

Theorem 3.1 *Given the optimal discriminators D_1 and D_2, there must be a global minimum of the virtual training criterion $C(G)$.*

Given D_1 and D_2, we reformulate Eq. (3.2) as

$$
\begin{aligned}
C(G) = & \max_{D_1, D_2} V(G, D_1, D_2) \\
= & \mathbb{E}_{(\boldsymbol{x}) \sim P_{data}(\boldsymbol{x})} \left[\log \frac{P_{data}(\boldsymbol{x}|y)}{P_{data}(\boldsymbol{x}|y) + P_G(\hat{\boldsymbol{x}}|y)} \right] \\
& + \mathbb{E}_{(\boldsymbol{x}) \sim P_G(\hat{\boldsymbol{x}})} \left[\log \frac{P_G(\hat{\boldsymbol{x}}|y)}{P_{data}(\boldsymbol{x}|y) + P_G(\hat{\boldsymbol{x}}|y)} \right] \\
& + \lambda \mathbb{E}_{(\boldsymbol{x}) \sim P_G(\hat{\boldsymbol{x}}|y=1)} \left[\log \frac{P_G(\hat{\boldsymbol{x}}|y = 1)}{P_G(\hat{\boldsymbol{x}}|y = 1) + P_G(\hat{\boldsymbol{x}}|y = 0)} \right] \\
& + \lambda \mathbb{E}_{(\boldsymbol{x}) \sim P_G(\hat{\boldsymbol{x}}|y=0)} \left[\log \frac{P_G(\hat{\boldsymbol{x}}|y = 0)}{P_G(\hat{\boldsymbol{x}}|y = 1) + P_G(\hat{\boldsymbol{x}}|y = 0)} \right] \\
= & -(2 + \lambda) \log 4 + 2JS(P_{data}(\boldsymbol{x}|y) || P_G(\hat{\boldsymbol{x}}|y)) \\
& + 2\lambda JS(P_G(\hat{\boldsymbol{x}}|y = 1) || P_G(\hat{\boldsymbol{x}}|y = 0)).
\end{aligned}
\tag{3.5}
$$

The minimax game of BalanceGAN can be reformulated as from Eq. (3.2):

$$
\begin{aligned}
C(G) = & -(2 + \lambda) \log 4 + 2JS(P_{\text{data}}(\boldsymbol{x}|y) || P_G(\hat{\boldsymbol{x}}|\hat{y})) \\
& + 2\lambda JS(P_G(\hat{\boldsymbol{x}}|y = 1) || P_G(\hat{\boldsymbol{x}}|y = 0)),
\end{aligned}
$$

where $C(G)$ represents the minimum value of the objective function of the converged generator in our BalanceGAN and $JS(\cdot)$ indicates the Jensen-Shannon divergence between two distributions. The generator reaches its minimum value when the two $JS(\cdot)$ terms converge to a global optimal point, a condition that can be guaranteed due to the convexity of the JS function.

Since the two equations of distribution $P_{data}(\boldsymbol{x}|y) = P_G(\hat{\boldsymbol{x}}|y)$ and $P_G(\hat{\boldsymbol{x}}|y = 1) = P_G(\hat{\boldsymbol{x}}|y = 0)$ cannot be achieved simultaneously in Eq. (3.5), the two JS terms can still be converged to a global optimal point due to convexity of JS. Hence, the optimum value of $V(G, D_1, D_2) = -(2+\lambda)\log 4 + \Delta$, where Δ is the minimum value when the two JS terms are converged to a point.

3.4 Experiments

In this part, we first describe our experimental platform and then show the experimental results and detailed comparative experiments.

3.4.1 Experimental Setup

Our models are implemented using Python. Specifically, the GAN and detection model are coded using Keras v.2.1.2. Our hardware platform is a server with four 2.4 Ghz Intel Xeon CPUs, four NVIDIA TITAN Xp GPUs, and 256 GB of RAM, running Centos 7 operating system with Linux kernel 3.10.

Our experimental datasets are respectively from online transaction data of two different banks. Table 3.2 shows the statistical information of two datasets. From the table, we can see that the proportion of the data of fraud and non-fraud samples is extremely imbalanced. To prevent the occurrence of time leaps (i.e., using the future data to train the model to predict the past frauds), we use the time sequence to cut the dataset to form the final training and test sets. Two different original datasets contain more than 30 features. In order to facilitate our modeling, we extract some of the same and highly correlated features for our prediction of fraud. Finally, the

Table 3.2 Statistics of datasets

	B2C1		B2C2		ECC	
Datasets	Training set	Test set	Training set	Test set	Training set	Test set
# of Fraudulent transactions	2992	1154	3711	1822	296	196
# of Non-fraudulent transactions	2,393,817	1,003,539	3,312,478	1,533,425	170,588	113727
Dimension	29		29		31	

Table 3.3 User's payment records in bank transaction dataset

Feature	Type	Explanation	Example
Bank ID	Char	The number of the main card to identify different account	1
Transaction Type	Numeric	The ways of payment for transactions	0
Transaction Time	Char	It is the time when the transaction takes place	20170401080080
Merchant Number	Char	Used to identify different payees	0200EC26243821
Check Mode	Numeric	The checking ID of the user when the card is handled by the bank	9
IP	Numeric	The value is 0 (common IP) or 1 (non-common IP)	1
Label	Numeric	The value is 0 (non-fraudulent) or 1 (fraudulent)	0

features we retain are shown in Table 3.3. In the preprocessing, for those character features, we count the total categories of characters in each character feature and digitize them, so that all our data preprocessing has been completed.

In the early-warning system of banks, if the bank finds a suspicious transaction, it immediately contacts the holder of the bank card to determine whether the transaction is self-operated or not. Therefore, an important metric in evaluating the validity of a model is the false positive rate (FPR); thus, our other metrics such as Precision (Pr.), Recall (Re.), F1-score, and AUC are calculated under given FPR.

The detailed description of our datasets is as follows.

B2C The dataset B2C is a real-world B2C transaction dataset including about three million records from a commercial bank. The transaction records have a time interval from April 1, 2017, to September 30, 2017 generated by 0.1 million accounts. All these records have been labeled as non-fraud or fraud manually. The features listed in Table 3.3 (except Bank ID and Label) are most helpful to detect the anomalies, i.e., Transaction Type (Numeric), Transaction Time (Char), Merchant Number (Char), Check Mode (Numeric), and IP (Numeric). In the preprocessing, we count the total category of characters for each character feature, digitize them to numerical values, and then normalize all the numerical data as the input data.

Dataset 1 (B2C1) B2C1 is a part of dataset B2C. We select the data from April to June as dataset B2C1, in which the transaction data in April and May are used as training data, and the data in June are used as test data.

Dataset 2 (B2C2) Similarly, B2C2 is the transaction data from July to September, in which the data in July and August are used as the training set, and the data in September are used as the test set.

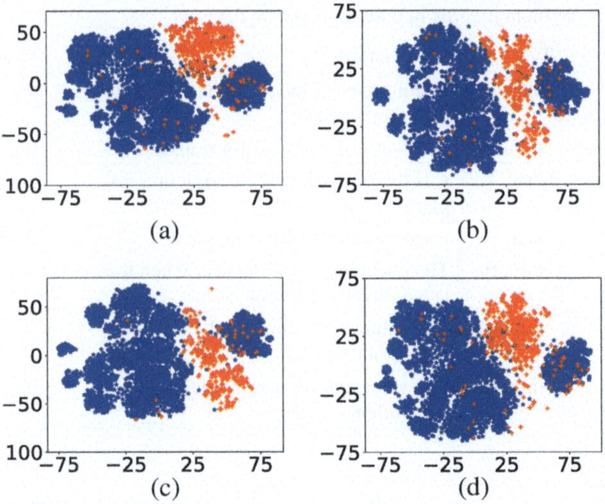

Fig. 3.4 Dimension reduction results of T-SNE

Dataset 3 (ECC) We sourced the third dataset from the European Credit Card (ECC)[4] transactions provided by the ULB ML Group. This dataset presents transactions that occurred in 2 days, where we have 492 frauds out of 284,807 transactions. The dataset is highly unbalanced, the positive class (frauds) account for 0.172% of all transactions.

For details of all training and test sets, please refer to Table 3.2.

3.4.2 Synthetic Data Generation

To visually verify the similarity between the data generated by BalanceGAN and the real data, we use the dimensionality reduction visualization method of T-SNE. We reduce the multidimensional data to a two-dimensional space, so x and y are the two new dimensional spaces after dimensionality reduction.

Figure 3.4a–d shows the dimension reduction of two real datasets and generated data, in which the dimension reduction method we adopt for the data is T-SNE [49]. Since the time complexity of T-SNE is too high, the data shown in Fig. 3.4a, c are 2% fraud and non-fraud samples randomly selected from the real datasets, and Fig. 3.4b, d are synthetic samples generated by BalanceGAN referring to the samples, respectively.

[4] https://www.kaggle.com/datasets/mlg-ulb/creditcardfraud

By comparing the fraud and non-fraud samples in Fig. 3.4a, c, we confirm that the positive sample distribution is similar to the negative sample distribution. By comparing Fig. 3.4a, c or 3.4b, d, we can conclude that our BalanceGAN can learn the data distribution of fraud and non-fraud samples in the real dataset.

3.4.3 Contrast Experiments

3.4.3.1 Implementation

We compare BalanceGAN with other learning methods to solve data imbalance. The detailed experimental process and results are as follows. Since Isolation Forest [50] is a classical unsupervised anomaly method, we use the default parameter ($\phi = 256, t = 100$) to select it as our baseline. In the method of NaiveGAN, we retain all the non-fraudulent samples; then, we use the traditional GAN methods shown in Fig. 3.3 to generate the fraud samples and add them to the real datasets as the training set, using Fig. 3.2a RNN as the detection model. In order to prevent the influence of randomness by sampling, all results are an average of ten experiments. In the method of SMOTE-Borderline [51, 52], we retain all the non-fraudulent samples; then, we use the method to generate representative fraud samples and add them to the real dataset as training set, using RNN in Fig. 3.2a as the detection model. Meanwhile, to prevent the randomness brought by the SMOTE-Borderline, all results are an average of ten experiments. In the setting of focal loss, we use an RNN neural network as our anomaly detection model and Focal loss [53] as our loss function ($\gamma = 2, \alpha = 0.25$). For more information about Focal loss, please refer to the link.[5] The GANomaly [54] is composed of an encoder, a decoder, and a discriminator. In the training phase, we retain all negative samples as the training set, and the anomalous threshold ϕ is set to the default value 0.2, that is, when the anomalous score of a sample is greater than 0.2, it is an anomaly. For detailed experimental code, please refer to the link.[6]

We train three convergent BalanceGANs for the three datasets, with $\lambda = 1.2$ for B2C1, $\lambda = 1.0$ for B2C2, and $\lambda = 1.5$ for ECC, respectively. Then, the generator G in BalanceGAN is used to generate equal number of positive and negative samples, which are used as the training set to train a base detection model. Next, through the transfer learning method, the original feature layer is retained, and the real data is used to fine-tune output layer, that is, the fine-tuned model is the final anomaly detection model.

[5] https://github.com/clcarwin/focal_loss_pytorch

[6] https://github.com/samet-akcay/ganomaly

Table 3.4 Results of comparative experiments

		B2C1		B2C2		ECC	
Methods	FPR	Pr.	Re.	Pr.	Re.	Pr.	Re.
Isolation Forset	1.00%	48.13%	62.31%	41.71%	36.98%	85.09%	45.32%
GANomaly		61.81%	87.39%	56.13%	83.33%	91.12%	86.35%
Focal loss		63.98%	93.17%	65.43%	89.35%	88.75%	66.38%
BalanceGAN + RNN		66.56%	95.15%	68.13%	92.36%	92.13%	98.51%
Isolation Forset	0.10%	82.42%	45.17%	68.41%	62.31%	89.11%	35.58%
GANomaly		89.63%	75.16%	63.31%	71.98%	92.46%	64.42%
Focal loss		89.77%	76.30%	70.33%	71.58%	95.63%	91.42%
BalanceGAN + RNN		89.91%	77.48%	83.41%	73.26%	94.46%	95.14%

3.4.3.2 Findings

Table 3.4 shows the performance of our method compared to Isolation Forest, Focal loss, and GANomaly on the three datasets, and we can see that our method has a maximum improvement of 26.42% on Pr. compared to Isolation Forest (on B2C2, FPR = 1.00%). Even with the existing state-of-the-art GAN-based method GANomaly, our performance still has a slight edge on the three datasets. Compared with the existing method Focal loss, which focuses on solving data imbalance, our performance is still superior to it.

Also, we conduct several different sets of experiments on datasets B2C1 and B2C2, respectively, to verify the performance (F1-score) of different comparison models under extreme conditions by continuously changing the value of FPR. The results of the experiments further confirm our conclusion that BalanceGAN is still ahead of other comparative methods in solving anomaly detection of imbalanced data, and the detailed results are represented in Fig. 3.5a, b.

Since both Isolation Forest and GANomaly are unsupervised anomaly detection methods, the label information of the few class samples is hardly used in the training process, which results in information loss. Therefore, the performance of these methods is worse than a majority of supervised learning anomaly detection methods; from the perspective of the loss function, Focal loss solves the problem of data imbalance by weakening the effect of the loss of a large number of negative samples on the gradient. However, a large number of negative samples play an important role in the construction of normal patterns, i.e., it is helpful for the model to identify a few class samples of anomalies. For this reason, Focal loss weakens the construction of patterns with negative samples, which causes the poor performance of the model; different from the traditional data generation-based approach, BalanceGAN only applies the generated data to model pre-training, avoiding adding the generated data directly to the real dataset, which can alleviate the overfitting of the model, and finally fine-tune the model with the real data to achieve better results.

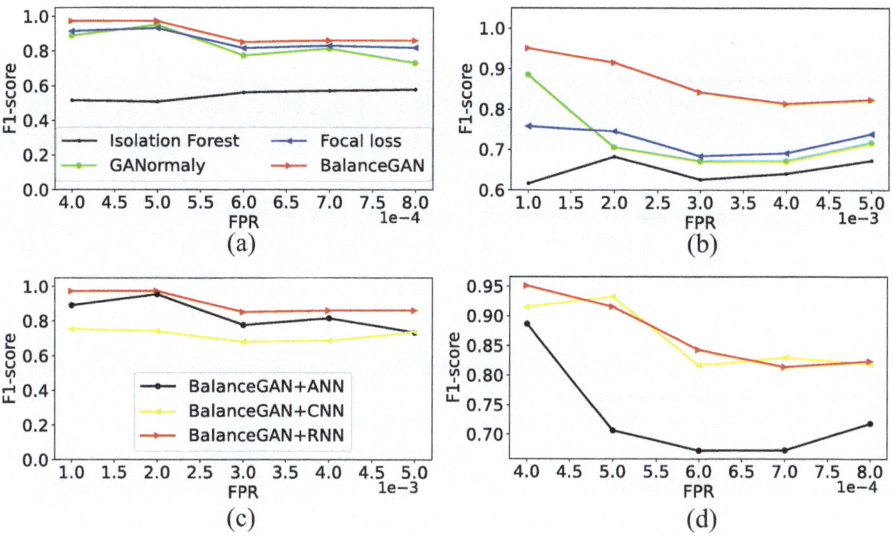

Fig. 3.5 Comparison and ablation experiments

3.4.4 Runtime Comparison

In bank transaction fraud detection, the algorithms' running time is also critical for fraud detection; thus, we also do the corresponding comparison experiments. Considering that real bank transactions are a pipeline of sequential data, and the proposed framework can be used as an offline learning mechanism, that is, in this scenario, the training of BalanceGAN, pre-training, and fine-tuning of the transfer learning model can be performed offline, so here we compare the running time (prediction) of different fraud detection models as shown in Fig. 3.6a, b. The x-axis in these figures represents the amount of data, and the y-axis represents the running time of the algorithms. By comparing the two figures, we can see there is a linear relationship between the running time and the amount of data for the different algorithms. Specifically, the Isolation Forest algorithm has the longest running time, and GANomaly has the shortest running time for the same amount of data. This is because Isolated Forest sacrifices computational time for lower memory. Since the GANomaly algorithm uses CNN as the encoder for fraud detection, the running time is shorter than our BalanceGAN + RNN combination. The Focal loss algorithm redefines the loss function and introduces more computations in the redefined loss function, which results in a longer running time than the BalanceGAN + RNN. The above is also consistent with our experimental results. Comparing Fig. 3.6a, b, the running time is longer on the ECC dataset than on the B2C1 dataset with the same algorithm and data volume, due to the larger data dimension of the ECC dataset than the B2C1 dataset.

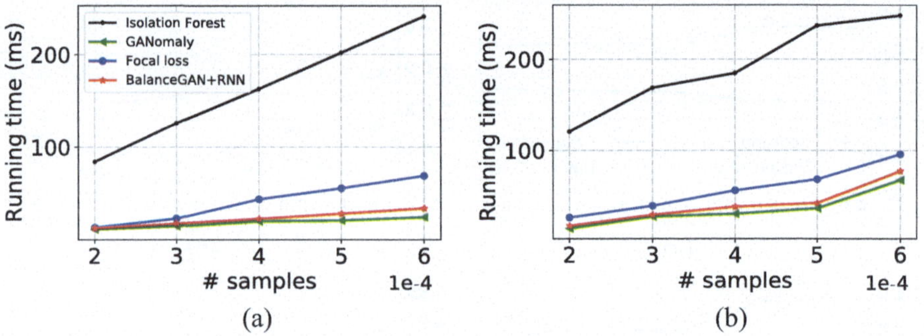

Fig. 3.6 Running time comparison on different datasets. (**a**) Running time comparison on B2C1. (**b**) Running time comparison on ECC

3.4.5 Ablation Study

Since our anomaly detection method on imbalanced data is a combination of BalanceGAN + RNN, we need to further verify that our method works better in anomaly detection because BalanceGAN plays an important role, and we have done experimental comparisons in the following two scenarios.

3.4.5.1 Scenarios 1

We replace BalanceGAN with two other data generation-based methods, NaiveGAN and SMOTE-Borderline, and use the same anomaly detection model, RNN, to do performance comparison on B2C1 and B2C2, respectively.

We compare the Precision and Recall by controlling the FPR equal to 1.00%. From the results in Table 3.5, we can make out that BalanceGAN is the best in solving the problem of imbalanced data, the performance of NaiveGAN is second, and the last is SMOTE-Borderline. At the same time, Fig. 3.7a–d further confirm our point view. We draw PR and ROC curves of three methods applied to B2C1 and B2C2.

From the observation of Table 3.5 and Fig. 3.7a–d, we can see that NaiveGAN and SMOTE-Borderline based on the oversampling method have different perfor-

Table 3.5 Ablation experiments for BalanceGAN

		B2C1			B2C2		
Methods	FPR	Pr.	Re.	AUC	Pr.	Re.	AUC
SMOTE-Borderline + RNN	1.00%	55.35%	64.80%	60.38%	44.19%	38.04%	59.27%
NaiveGAN + RNN		55.96%	82.10%	61.25%	52.95%	46.24%	60.96%
BalanceGAN + RNN		66.56%	95.15%	94.29%	68.13%	92.36%	96.78%

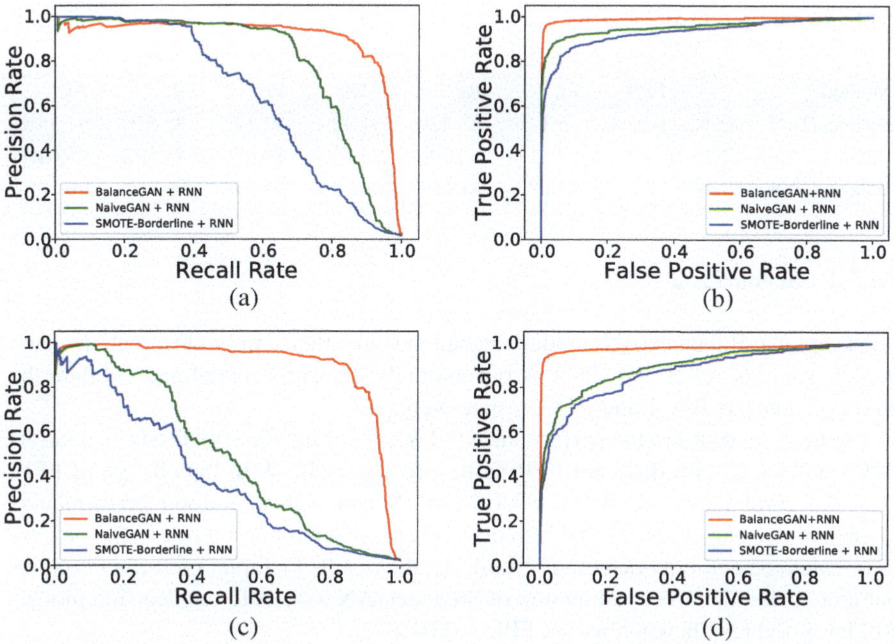

Fig. 3.7 PR and ROC curves of experiments on different datasets. (**a**) PR curves of experiments on B2C1. (**b**) ROC curves of experiments on B2C1. (**c**) PR curves of experiments on B2C2. (**d**) ROC curves of experiments on B2C2

mance in the final fraud detection model. The main reason is that the quality of data generated by NaiveGAN is better than that of SMOTE-Borderline. By comparing the methods based on BalanceGAN and NaiveGAN, there are two main reasons why the method based on BalanceGAN wins in performance: (1) The NaiveGAN method adds the generated fraud samples directly to the real data as the training data for fraud detection, which has a great probability to cause the model overfitting; on the contrary, based on the BalanceGAN method in this work, we pre-train a base fraud detection model by generating synthetic data and then use the transfer learning method to fine-tune our final model with the real data, which greatly reduces the risk of overfitting. (2) Different from NaiveGAN method, BalanceGAN adds a new discriminate model (D_2), which can make the generated fraud data distribution far away from user's transaction data distribution. By generating a large number of these synthetic data to pre-train the base fraud detection model, the model can better learn the difference between fraud and non-fraud data distribution; then, it can produce better discrimination performance.

Table 3.6 Ablation experiments for fraud detection model

Methods	FPR	B2C1			ECC		
		Pr.	Re.	AUC	Pr.	Re.	AUC
BalanceGAN + ANN	0.04%	75.12%	68.32%	91.42%	81.13%	85.51%	91.35%
BalanceGAN + CNN		85.31%	50.50%	93.75%	88.07%	91.50%	93.64%
BalanceGAN + RNN		91.08%	81.28%	94.29%	91.09%	96.48%	96.78%

3.4.5.2 Scenarios 2

We keep the BalanceGAN model unchanged and then apply different decision models such as ANN and CNN as our anomaly detection model and validate the effect on datasets B2C1 and ECC, respectively.

Figure 3.5c, d shows the performance (F1-score) of the three methods on datasets B2C1 and ECC with different FPR values, respectively. With the change of FPR, even the combination of BalanceGAN + ANN can still have good performance. Therefore, we can conclude that BalanceGAN plays an important role in the process of imbalanced anomaly detection. Similarly, the results in Table 3.6 further confirm our conclusion that the combination of BalanceGAN with different decision models still has good results when we set FPR = 0.04%.

3.4.6 *Sensitivity Study*

In order to further test the ultimate performance of BalanceGAN and the sensitivity of performance to parameter λ, we control the value of FPR no more than 0.1%, and then observe the change of Precision and Recall with the value of λ, and the curves are shown in the Fig. 3.8a, b.

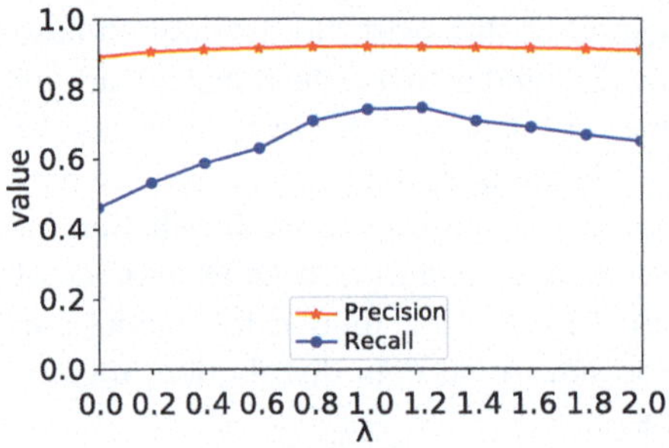

(a)

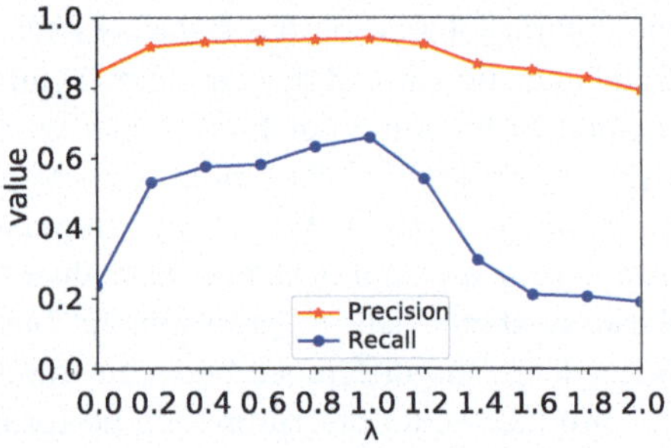

(b)

Fig. 3.8 Parameter experiments on different datasets. (**a**) BalanceGAN performance changes with λ on B2C1. (**b**) BalanceGAN performance changes with λ on B2C2

The red lines indicate that the Precision changes with λ, while the blue lines indicate Recall in Fig. 3.8a, b. Compared with Fig. 3.8a, b, we find that the values of Precision and Recall increase first and then decrease with the increase of λ, and reach the maximum values at different λ. The reason is that when we train the BalanceGAN, and with the increase of λ, the synthesized data generated fraud data distribution ($P_G(x|y=1)$) is gradually far away from the normal user's transaction data distribution ($P_G(x|y=0)$), but at the same time, it is necessary to ensure that the generated data distribution ($P_G(x,y)$) is similar to the real data distribution ($P_{data}(x,y)$). Thus, the huge amounts of training data generated by BalanceGAN can ensure that the fraud detection model can learn the difference between the distribution of fraud data and that of non-fraud data. As a result, Precision and Recall increased. According to our previous theoretical analysis (Sect. 3.3.4), it is difficult for the discriminators D_1 and D_2 to reach the minimum at the same time. When the value of λ increases to a certain extent, the similarity between the distribution of synthesized data $P_G(x,y)$ generated by BalanceGAN and the distribution of real data $P_{data}(x,y)$ decreases, which leads to the poor training effect of fraud detection model. Thus, it causes the decrease of Precision and Recall.

3.5 Conclusions

In this chapter, we propose a new fraud detection method for online banking based on extremely imbalanced data, BalanceGAN, which consists of a generator G and two discriminators D_1, D_2. Specifically, the game of G and D_1 allows G to generate data that is similar to the real data distribution, while the game of G and D_2 is used to distinguish whether the generated samples are from the fraud group or a non-fraud group. These allow G to generate fraud data distribution that far away from the normal user's transaction data distribution. In the traditional data generation-based approach, directly adding the generated data to the real data tends to cause overfitting of the model. To alleviate the above problem, we adopt the transfer learning approach to pre-train the model using the data generated by G, and then fine-tune the model with real data. Experiments show that, compared with the traditional method to solve the data imbalance, our method based on BalanceGAN has better fraud detection performance.

Despite the above advantages of our BalanceGAN, it also has some limitations. Since we introduce the parameter λ, the choice of λ will have an impact on the fraud detection effect of the model to a certain extent. Therefore, how to effectively determine the appropriate λ value will be an important issue for our future research. In future work, we will try to improve our model by studying the data utility to adjust the parameter λ in BalanceGAN adaptively.

References

1. L. Liu, N. Antonopoulos, M. Zheng, Y. Zhan, Z. Ding, ACM Trans. Embed. Comput. Syst. **15**(2), 38:1 (2016). https://doi.org/10.1145/2811264
2. J. Xu, D. Zhang, L. Liu, X. Li, IEEE Trans. Serv. Comput. **5**(1), 20 (2012). https://doi.org/10.1109/TSC.2010.33
3. Y. Xie, G. Liu, C. Yan, C. Jiang, M. Zhou, IEEE Trans. Comput. Soc. Syst. **10**(3), 1004 (2022)
4. Y. Xie, G. Liu, C. Yan, C. Jiang, M. Zhou, M. Li, IEEE Trans. Neural Netw. Learn. Syst. **35**(4), 5735 (2022)
5. F. Zhao, D. Xiang, G. Liu, C. Jiang, IEEE Trans. Comput. Soc. Syst. **9**(2), 480 (2022). https://doi.org/10.1109/TCSS.2021.3099475
6. R. Cao, G. Liu, Y. Xie, C. Jiang, IEEE Trans. Comput. Soc. Syst. **8**(6), 1291 (2021). https://doi.org/10.1109/TCSS.2021.3074175
7. Z. Li, G. Liu, C. Jiang, IEEE Trans. Comput. Soc. Syst. **7**(2), 569 (2020). https://doi.org/10.1109/TCSS.2020.2970805
8. L. Zheng, G. Liu, C. Yan, C. Jiang, M. Zhou, M. Li, IEEE Trans. Comput. Soc. Syst. **7**(5), 1304 (2020). https://doi.org/10.1109/TCSS.2020.3017013
9. S. Yin, G. Liu, Z. Li, C. Yan, C. Jiang, in *20th International Conference on Data Mining Workshops, ICDM Workshops 2020, Sorrento, November 17–20, 2020*, ed. by G.D. Fatta, V.S. Sheng, A. Cuzzocrea, C. Zaniolo, X. Wu (IEEE, Piscataway, 2020), pp. 875–882. https://doi.org/10.1109/ICDMW51313.2020.00125
10. J. Cui, C. Yan, C. Wang, IEEE Trans. Comput. Soc. Syst. **8**(3), 643 (2021). https://doi.org/10.1109/TCSS.2021.3052950
11. P. Vuttipittayamongkol, E. Elyan, Inf. Sci. **509**, 47 (2020). https://doi.org/10.1016/j.ins.2019.08.062
12. J. Oh, J.Y. Hong, J. Baek, Expert Syst. Appl. **133**, 1 (2019). https://doi.org/10.1016/j.eswa.2019.05.006
13. Z. Li, M. Huang, G. Liu, C. Jiang, Expert Syst. Appl. **175**, 114750 (2021)
14. L. Zheng, G. Liu, C. Yan, C. Jiang, IEEE Trans. Comput. Social Syst. **5**(3), 796 (2018)
15. B.S. Raghuwanshi, S. Shukla, Knowl.-Based Syst. **187** (2020). https://doi.org/10.1016/j.knosys.2019.06.022
16. H. He, Y. Bai, E.A. Garcia, S. Li, in *Proceedings of the International Joint Conference on Neural Networks, IJCNN 2008, Part of the IEEE World Congress on Computational Intelligence, WCCI 2008, Hong Kong, June 1–6, 2008* (2008), pp. 1322–1328. https://doi.org/10.1109/IJCNN.2008.4633969
17. C. Jiang, J. Song, G. Liu, L. Zheng, W. Luan, IEEE Internet Things J. **5**(5), 3637 (2018)
18. C. Yang, G. Liu, C. Yan, C. Jiang, Sci. China Inf. Sci. **64**(12), 1 (2021)
19. J. Cui, C. Yan, C. Wang, in *DSDE '21: 2021 4th International Conference on Data Storage and Data Engineering, Barcelona, February 18–20, 2021* (ACM, New York, 2021), pp. 22–30. https://doi.org/10.1145/3456146.3456151
20. Q. Yang, C. Wang, C. Wang, H. Teng, C. Jiang, IEEE Trans. Comput. Soc. Syst. **8**(2), 398 (2021). https://doi.org/10.1109/TCSS.2020.3035462
21. Q. Li, Q. Wu, C. Zhu, J. Zhang, W. Zhao, in *Advances in Knowledge Discovery and Data Mining - 23rd Pacific-Asia Conference, PAKDD 2019, Macau, April 14–17, 2019, Proceedings, Part I* (2019), pp. 222–236. https://doi.org/10.1007/978-3-030-16148-4_18
22. P. Ravisankar, V. Ravi, G.R. Rao, I. Bose, Decis. Support Syst. **50**(2), 491 (2011). https://doi.org/10.1016/j.dss.2010.11.006
23. R. Domingues, CoRR abs/1903.01730 (2019). http://arxiv.org/abs/1903.01730
24. E.L. Paula, M. Ladeira, R.N. Carvalho, T. Marzagão, in *15th IEEE International Conference on Machine Learning and Applications, ICMLA 2016, Anaheim, CA, December 18–20, 2016* (2016), pp. 954–960. https://doi.org/10.1109/ICMLA.2016.0172
25. J. Jurgovsky, M. Granitzer, K. Ziegler, S. Calabretto, P. Portier, L. He-Guelton, O. Caelen, Expert Syst. Appl. **100**, 234 (2018). https://doi.org/10.1016/j.eswa.2018.01.037

26. R. Saia, S. Carta, Future Gener. Comput. Syst. **93**, 18 (2019). https://doi.org/10.1016/j.future.2018.10.016
27. H. Zhou, L. Wei, G. Chen, P. Lin, Y. Lin, in *2019 International Conference on Intelligent Computing, Automation and Systems (ICICAS)* (IEEE, Piscataway, 2019), pp. 507–510
28. K.G. Al-Hashedi, P. Magalingam, Comput. Sci. Rev. **40**, 100402 (2021). https://doi.org/10.1016/j.cosrev.2021.100402
29. Y. Liu, Y. Liu, B.X.B. Yu, S. Zhong, Z. Hu, Pattern Recognit. **133**, 109008 (2023). https://doi.org/10.1016/j.patcog.2022.109008
30. L. Ren, A.S. Seklouli, H. Zhang, T. Wang, A. Bouras, Inf. Syst. **111**, 102122 (2023). https://doi.org/10.1016/j.is.2022.102122
31. K.W. Bowyer, N.V. Chawla, L.O. Hall, W.P. Kegelmeyer, CoRR abs/1106.1813 (2011). http://arxiv.org/abs/1106.1813
32. D. Huang, X.S. Zhang, G. Huang (eds.), *Advances in Intelligent Computing, International Conference on Intelligent Computing, ICIC 2005, Hefei, August 23–26, 2005, Proceedings, Part I. Lecture Notes in Computer Science*, vol. 3644 (Springer, Berlin, 2005). https://doi.org/10.1007/11538059
33. J. Ren, Y. Wang, Y. Cheung, X.Z. Gao, X. Guo, Pattern Recognit. **133**, 108992 (2023). https://doi.org/10.1016/j.patcog.2022.108992
34. M. Prince, P.M.J. Prathap, Comput. Syst. Sci. Eng. **44**(2), 1009 (2023). https://doi.org/10.32604/csse.2023.024277
35. C. Sun, H. Cui, W. Zhou, W. Nie, X. Wang, Q. Yuan, Int. J. Neural Syst. **29**(10), 1950021:1 (2019). https://doi.org/10.1142/S0129065719500217
36. X. Liu, J. Wu, Z. Zhou, IEEE Trans. Syst. Man Cybern. Part B **39**(2), 539 (2009). https://doi.org/10.1109/TSMCB.2008.2007853
37. H. Yu, X. Yang, S. Zheng, C. Sun, IEEE Trans. Neural Netw. Learn. Syst. **30**(4), 1088 (2019). https://doi.org/10.1109/TNNLS.2018.2855446
38. Y. Wang, L. Yang, Neurocomputing **331**, 40 (2019). https://doi.org/10.1016/j.neucom.2018.11.024
39. G.A. Susto, M. Terzi, C. Masiero, S. Pampuri, A. Schirru, in *First International Conference on Artificial Intelligence for Industries, AI4I 2018, Laguna Hills, CA, September 26–28, 2018* (IEEE, Piscataway, 2018), pp. 9–14. https://doi.org/10.1109/AI4I.2018.8665694
40. R. Saia, in *Proceedings of the 3rd International Conference on Internet of Things, Big Data and Security, IoTBDS 2018, Funchal, Madeira, March 19–21, 2018*, ed. by V.M. Muñoz, G.B. Wills, R.J. Walters, F. Firouzi, V. Chang (SciTePress, Setúbal, 2018), pp. 29–40. https://doi.org/10.5220/0006663000290040
41. U. Fiore, A.D. Santis, F. Perla, P. Zanetti, F. Palmieri, Inf. Sci. **479**, 448 (2019). https://doi.org/10.1016/j.ins.2017.12.030
42. D. Cabrera, F. Sancho, J. Long, R. Sánchez, S. Zhang, M. Cerrada, C. Li, IEEE Access **7**, 70643 (2019). https://doi.org/10.1109/ACCESS.2019.2917604
43. W. Li, L. Xu, Z. Liang, S. Wang, J. Cao, T.C. Lam, X. Cui, Neurocomputing **431**, 148 (2021). https://doi.org/10.1016/j.neucom.2020.12.001
44. J. Hu, H. Zhang, Y. Liu, R.F.E. Sutcliffe, J. Feng, Appl. Intell. **52**(6), 6723 (2022). https://doi.org/10.1007/s10489-021-02623-9
45. M. Mirza, S. Osindero, CoRR abs/1411.1784 (2014). http://arxiv.org/abs/1411.1784
46. I.J. Goodfellow, J. Pouget-Abadie, M. Mirza, B. Xu, D. Warde-Farley, S. Ozair, A.C. Courville, Y. Bengio, CoRR abs/1406.2661 (2014). http://arxiv.org/abs/1406.2661
47. N. Jin, F. Yang, Y. Mo, Y. Zeng, X. Zhou, K. Yan, X. Ma, Adv. Eng. Inf. **51**, 101442 (2022). https://doi.org/10.1016/j.aei.2021.101442
48. A.E. Coca, D.C. Corrêa, L. Zhao, in *The 2013 International Joint Conference on Neural Networks, IJCNN 2013, Dallas, TX, August 4–9, 2013* (2013), pp. 1–7. https://doi.org/10.1109/IJCNN.2013.6706747
49. J. B. Tenenbaum, V. D. Silva, J. C. Langford, A global geometric framework for nonlinear dimensionality reduction. Science **290**(5500), 2319–2323 (2000)

50. F.T. Liu, K.M. Ting, Z. Zhou, in *Proceedings of the 8th IEEE International Conference on Data Mining (ICDM 2008), December 15–19, 2008, Pisa* (IEEE Computer Society, Washington, 2008), pp. 413–422
51. L. Camacho, G. Douzas, F. Bação, Expert Syst. Appl. **193**, 116387 (2022). https://doi.org/10.1016/j.eswa.2021.116387
52. D. Liang, B. Yi, W. Cao, Q. Zheng, Expert Syst. Appl. **188**, 116051 (2022). https://doi.org/10.1016/j.eswa.2021.116051
53. T. Lin, P. Goyal, R.B. Girshick, K. He, P. Dollár, in *IEEE International Conference on Computer Vision, ICCV 2017, Venice, October 22–29, 2017* (IEEE Computer Society, Washington, 2017), pp. 2999–3007
54. S. Akcay, A.A. Abarghouei, T.P. Breckon, in *Computer Vision - ACCV 2018 - 14th Asian Conference on Computer Vision, Perth, December 2–6, 2018, Revised Selected Papers, Part III* (2018), pp. 622–637

Chapter 4
Vectoring Applicant Information for Ex Ante Fraud Prediction

Abstract In online lending services, fraud *prediction* is an especially critical step to control loss risk and improve processing efficiency. Unfortunately, it is definitely challenging since the ex ante prediction actually needs to be made only based on the most basic information of applicants. This work figures out that the essential difficulty here is the *low information density* of data associations which contain the useful information for fraud prediction. Accordingly, we propose a novel multistage data representation scheme, called AI2Vec (*Applicant Information Vectoring*), as an *information density booster*. It can gradually boost information density of associations by simultaneously decreasing the scale of information carriers and increasing the amount of useful information. The qualified performance of our AI2Vec is validated by the experiments over real-life data from a prestigious online lending platform. It can help commonly used machine learning classifiers outperform the state-of-the-art methods, including the method of pilot platform with manual feature engineering by the subject matter experts.

4.1 Introduction to Information Density Boosting

Online finance is in the ascendant, but so are new forms of accompanying frauds [1, 2]. As one of most representative functional patterns of online finance, online lending services have been developing personal loan markets, coming with the inherent risk of credit loan fraud [3]. Beyond doubt, the loss from *bad debts* is a substantial cost of online lending platforms (OLPs). The bad debts, also called *bad loans*, refer to the loans that online lending platforms cannot recover from borrowers. They are usually of two types: (1) the loan whose borrower is unable to repay the due loan for certain reasons and (2) the loan whose beneficiary deliberately denies it. The latter is the fraudulent loan, i.e., the so-called fraud, where its beneficiary, i.e., the fraudster, usually uses the forged or misused information to defraud the loan. To avoid loss, OLPs adopt a sequence of steps for a whole online loan application, as depicted in Fig. 4.1. It is noteworthy that OLPs conduct the operation of *credit limit evaluation* to predict the non-fraud bad loans under the confirmation that the loans are not frauds [4–6]. Furthermore, it takes a long delay

C. Wang, H. Zhu, *Universal Behavior Computing for Security and Safety*,
https://doi.org/10.1007/978-981-97-9014-2_4

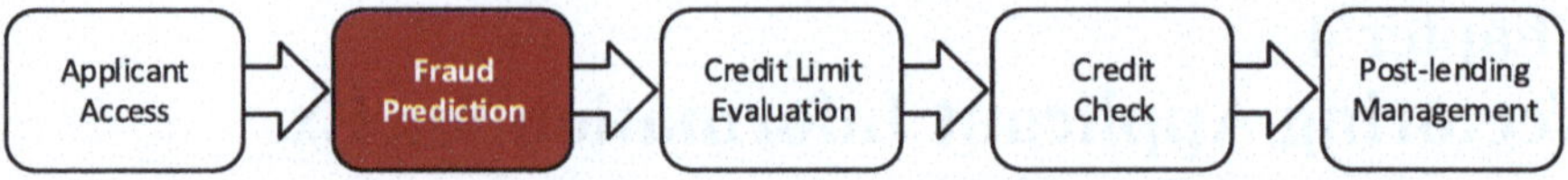

Fig. 4.1 The online lending operation flow chart

for the true feedback, i.e., a loan application is bad debt or legitimate, to be judged based on whether applicants always repay on time in *post-lending management*. As a matter of fact, such a delay is unacceptable in business, since OLPs had already suffered losses when they realized the frauds were happening [7, 8].

Therefore, *fraud prediction*, following the *applicant access* and preceding the *credit limit evaluation*, is really necessary for predicting all kinds of bad loans at the pre-loan stage, which is indeed the lifeblood of any legal OLPs. It is exactly the focus of this work.

It is profitable for OLPs to make efforts to predict dishonest applicants before these fraudsters enter *credit limit evaluation* step [9, 10]. A straightforward approach is to establish the blacklist for *applicant access* as the first line of defense to screen applicants who are at great risk of being overdue. Despite the blacklist's simplicity and effectiveness, it suffers from the limitations. It can only deal with applicants who have loan histories or risky credit histories and are incapable of predicting new applicants [11]. Actually, a *qualified*[1] online credit loan fraud prediction (OCLFP) is the most important anti-fraud barrier for OLPs [13–15].

The data-driven paradigm is an effective solution to anti-fraud engineering for different patterns of online finance at present [15–19]. Unfortunately, when data-driven methods are applied to the OCLFP issue, there is a big gap from the preliminary applicant information to qualified fraud prediction due to two special difficulties brought by data: On the one side, it is really hard to make such a qualified ex ante prediction only based on the available information in the step of fraud prediction, i.e., the preliminary information of applicants. The useful information for fraud prediction is usually insufficient due to the data limitation. On the other side, most of the features for OCLFP are hidden in the associations of applicant data. The explicit associations that are often rather sparse [20], it is naturally necessary to mine deep-hidden associations by an efficient method, e.g., network representation learning [21]. In practice, the types of implicit associations are too many to distinguish the useful ones that carry useful information for fraud prediction. The efficiency of methods keeping too many associations cannot meet the special requirements of OCLFP, e.g., the scalability and low latency. Combining these two difficulties, the problem can be reduced to a fact that the *information*

[1] Generally, for real-world OLPs, according to the de facto standards, the ex ante fraud prediction can be adopted as a *qualified* one if it has a *KS (Kolmogorov-Smirnov) value* that is not much less than 0.5 and an *average processing latency* within 10 seconds. This partly reflects the difficulty of fraud prediction. Usually, it can tolerate relatively high false positive rates [12].

density[2] of applicant data is too low for OCLFP. Therefore, the central task of this work is how to improve the information density.

Some existing methods directly conduct on the complete original information carrier to adequately preserve the useful information [19, 23, 24], which generally perform poorly since useful information is buried in a mass of useless information. A feasible way is to reduce the scale of information carrier by discarding part of information to overcome the influence of useless information [18, 25]. Obviously, this solution may lead to information incompleteness, which usually significantly decreases the performance of fraud prediction. We aim to boost the information density of associations by a two-pronged strategy under which the useful information is increased and the associations acting as information carriers are reduced. To the best of our knowledge, there are no existing methods that can solve the two aspects simultaneously [21].

In this chapter, we propose a multistage data representation method based on network representation learning, named *Applicant Information Vectoring* (AI2Vec), serving as an *information density booster*. Under AI2Vec, we first lay the groundwork for boosting the information density by restructuring the original information network into a layered network consisting of two interrelated layers: the entity layer and attribute layer. We define multiple associations to carry useful information in Stage 1. Then, in Stage 2, we refine the information carrier network (ICN), which is the *entity layer*, by incorporating exterior semantic knowledge, such as attribute similarity, into the attribute layer. In Stage 3, we transfer the information contained in all other associations into ICN. Finally, we make the representation of ICN to realize information density boosting.

Besides the design of AI2Vec, the following work is also done for or based on AI2Vec:

- We develop a Multi Stages framework (OCLFP-MS) for OCLFP, which takes AI2Vec as an efficient *data booster* to enhance the information density of data associations. After completing each stage, it will allow the remaining stages to easily achieve the desired performance of the model with a higher information density.
- An effective network representation learning algorithm, named layered mapping network embedding (LMNE), is designed to deal with the special requirements of OCLFP.
- A real-world online lending scenario is used to verify the performance of OCLFP-MS using AI2Vec with LMNE. It shows our method with the KS value of 0.5067 outperforms the state-of-the-art methods, including the method currently adopted by the platform with the KS value of 0.4426 via *manually* feature

[2] The term "information density" refers to the amount of information valid for a specific utility, e.g., fraud detection, per unit comprising the information carriers, e.g., the associations, in the most basic information-theoretic sense. Here, we borrow its basic meaning from [22], i.e., the amount of information per unit comprising the utterance.

engineering. Meanwhile, the average processing latency of our method is only 0.95 second per transaction.

4.2 Related Work

We first clarify the special challenges of the online lending anti-fraud issue through a brief description of representative anti-fraud techniques for different online services. Furthermore, we introduce the development bottlenecks of online lending anti-fraud technologies, particularly associated with low-quality limited data. Finally, we analyze typical network representation learning (NRL) methods, as an important technology to alleviate the above bottlenecks, as well as the technical weaknesses for the online credit loan fraud prediction, which supports the research motivation of our work.

4.2.1 Anti-Fraud for Online Services

Cyber fraud has become a serious social and economic problem [26]. Rule-based anti-fraud methods are incapable of handling the changeable and diversified fraud patterns. Then, researchers have turned their attention to developing data-driven anti-fraud solutions.

Machine learning, as a promising data-driven paradigm, has been widely used in anti-fraud fields [2, 14, 17] for various network services. In *online romance scams*, fraudsters craft fake profiles and manually interact with their victims, which is difficult to detect by traditional methods. Guillermo et al. [27] combined a range of structured, unstructured, and deep-learned features to build a detection system which stops scammers as they create fraudulent profiles or before they engage with potential victims. In *online payment fraud*, the typical feature is the identity theft which leads to user anomalous behavior and makes behavioral models effective for fraud detection. Unfortunately, user behavioral data are usually low-quality for behavior modeling due to the preservation of user privacy or the difficulty of data collection. Wang et al. [28] extracted fine-grained co-occurrence relationships by using a knowledge graph and realized data enhancement for diversified behavior models. *In money laundering*, criminals used the bank's services to move massive amounts of illegal money to untraceable destination accounts, and then inject their illegal money into the legitimate financial system. Li et al. [29] modeled the transactions using a multipartite graph, and detect the complete flow of money from source to destination. In *insurance fraudulent claim*, the deliberate abuse of the insurance policy could lead to heavy financial losses. It is one of the greatest challenges the insurance industry confronts. Liang et al. [30] developed an automated solution for fraud detection based on graph learning algorithms to separate fraudsters from regular customers and uncover groups of organized fraudsters. Most existing solutions were closely related to specific scenarios due to scenario-based fraud features. Different from online services mentioned above, online lending services present new anti-fraud challenges.

When lending services are carried out over the Internet, the convenience brings also huge fraud risk [18]. In general online services, the platforms often concentrate on individual-level behavioral models to detect abnormal behavior of his/her account [31]. However, the main feature of credit loan frauds in online lending services is of *abnormal associations* among different loan application transactions and users. Some studies gave the applicant credit ratings by analyzing their social connections. Babaev et al. [18] adopted a deep learning method, RNNs, on fine-grained transnational data to compute credit scores for the loan applicants, and produced significant financial gains for the banks. Boz et al. [32] utilized a set of machine learning methods to point the institution toward re-evaluating a set of pre-screened customers and proposed modeling approaches to predict the customer behavior using previous customer status records. Liang et al. [5] adopted multiple views of heterogeneous information that is generally available in online E-Commerce platforms to build a credit risk rating model. Besides, the lending risk can be obtained by checking whether the relevant information of the applicant is reliable. Thus, some studies paid attention to how to effectively verify the applicant's information. Cheng et al. [15] investigated the problem of predicting repayment delinquency in the networked-guarantee loans, and attempted on predicting bank guarantee loan defaults by adding objectively temporal network structure using an end-to-end learning framework. Tan et al. [19] introduced an end-to-end deep learning method to model potential risks simultaneously by multiple binary classification subtasks. Hu et al. [6] focused on the communicability of loan default detection and developed a novel attributed multiplex graph for integrating multiplex relations in financial scenarios.

Existing methods relatively paid more attention to model optimization than data enhancement. However, it is difficult to achieve good performance with low information density data in existing methods.

4.2.2 Network Representation Learning

NRL algorithms play a key role in our method. Here, we summarize the related literatures on previous researches. The existing NRL algorithms usually have a main objective, i.e., *network reconstruction*. They can be divided into two categories: NRL of structure maintenance [33, 34] and NRL of merging accompanying information [35, 36]. The commonly used models for learning the representation space include matrix factorization [24], random walk [25], graph neural networks [37], and their variations.

Many researchers have concentrated on structure maintenance in NRL algorithms, which is the most fundamental requirement for the NRL algorithm. Some researchers focused on preserving proximity in a homogeneous network, which only has one type of node and relationship. In this direction, many effective algorithms have been proposed, such as DeepWalk [38] and node2vec [25]. These algorithms neglect the rich relationship types, so some studies tried to obtain the embeddings of relationships, such as metapath2vec [34] and HIN2Vec [23]. Besides, the accompanying information can provide additional depiction about nodes; they shall be considered together in the representation of nodes. Some researchers have provided many algorithms to integrate and balance the link structure and accompanying information effectively, for instance MARINE [24], ASNE [37], and ANRL [36].

To transform networks from the original network space to the representation space, researchers have proposed various models to achieve this goal. Feng et al. [24] expanded two-way matrix factorization into the three-way tensor factorization to preserve the proximity information for each type of relation. Perozzi et al. [38] exploited local information obtained from random walks to learn latent representations by treating walks as the equivalent of sentences. Liao et al. [37] adopted a multilayer neural network to take advantage of strong representation and generalization ability of deep learning for capturing the complex interactions between features. However, existing NRL algorithms do not consider updating the representation dynamically as new edges keep feeding, and thus may not be perfectly suitable for fraud prediction. In the case of a rapidly evolving lending network, this problem can be even more challenging. For instance, Wang et al. proposed a framework that transformed the data into a general graph and initialized node and edge features for fraud detection and aimed to optimize the graph pre-training efficiency in the future [39]. Zhu et al. proposed a novel graph neural network that can handle the homogeneity and heterogeneity of networks simultaneously by introducing edges between homogenous mini-graphs constructively. They also were interested in extending the method to solve dynamic graphs [40]. Therefore, an NRL algorithm, which simultaneously preserves complete information for high prediction performance and reduces computational complexity for high processing efficiency, is highly desirable to design in online credit loan fraud prediction (OCLFP).

4.3 Overview

4.3.1 Scenario

The usage of online lending services has proliferated, especially in developing countries, and they have served more than 500 million people (estimated by the World Bank) [4, 41]. A stark contrast with traditional lending services is that online lending services provide users with small loans in an eased and fast way (such as the minimal paperwork and without any sort of collateral). However, banks usually offer loans based on reliable collaterals or existing customers with good reputations. If a customer wants to apply for an online loan, he/she needs first to download the respective APP from Play Stores and can then borrow money up to the limit set by the application after registration. The borrowing process is fast, with customers requesting and receiving money within minutes [42]. They improve financial inclusion, e.g., allowing users (many without formal credit history) to easily access loans online (on smart devices), but the flexibility and efficiency also bring new threats [43, 44]. In addition to financially strained customers prone to delayed repayment, there are also some experienced fraud agents who can incite high-risk borrowers to cheat and help them provide false information to pass the material evaluation process. These fraudulent agencies demand a large portion of the loan as a benefit fee, resulting in customers' inability to repay the loan.

A serious challenge faced by service providers is how to quickly assess risks for each loan application based on this low-quality application information (e.g., untrue paperwork, unreliable collaterals, and informal credit history) to minimize potential financial losses.

4.3.2 Motivation

During online loan processes, both fraudulent and normal loans may occur, where the former may result in lost principal and interest payments. The key difference between them is whether the applicant has a genuine intention to use the money. This can be predicted in advance from the applicant's ability to repay the loan.

In fraud cases, the applicants, knowing they lack ability to repay, still borrow large sums. They may use the loan for personal gain, criminal activities, or even escape. Moreover, they often fail to repay the loan after maturity. What's worse, the fraudster initially intends to apply for the maximum amount. To gain large profits, they usually forge relevant paperwork to deceive online lending platforms. Fraudsters with rich loan experiences may even incite high-risk borrowers or buy real citizen information to carry out mass fraud. They often disguise themselves by providing false information to make them "look normal" to pass the platform's review process.

Consistency exists between true/false classification tasks and fraud/normal prediction tasks. The latter aims to predict if a loan application will involve fraud in the future. Thus, true/false classification can apply to fraud prediction [6, 43]. In practice, many classification models (e.g., logistic regression, support vector machine, XGBoost) are widely used to model the credit scores of loan applicants. They detect fraud applicants by examining surface and obvious signals (like false address information and unrealistic income). However, these methods are rudimentary and not always effective.

Fraudsters use sophisticated tactics to create the seemingly normal surface of an application. But in essence, the opposite is true; their tricks to conceal fraudulent behaviors lead to inconsistencies that can be detected through data associations. In financial lending scenarios, the application information provided by the customers is rich in associations [5, 40]. For example, customers have social relationships with friends, classmates, and relatives, and may work at the same workplace as other customers. All of these associations can be beneficial for fraud prediction issues.

Unlike normal customers with various associations, there are few direct associations between fraudsters and their accomplices but abundant indirect associations. To evade regulation, fraudsters have little connection with each other. However, to maximize profits, fraudsters typically reuse a customized set of materials across multiple fraudulent applications, which will establish many implicit associations [39]. If multiple customers with no apparent associations provide identical information, this could suggest a gang-level effort to commit fraud.

4.3.3 Approach Overview

The workflow of our framework consists of four main parts: three representation stages, cooperatively acting as the information density booster, and one predictor, as shown in Fig. 4.2. In this part, we give each part a brief introduction.

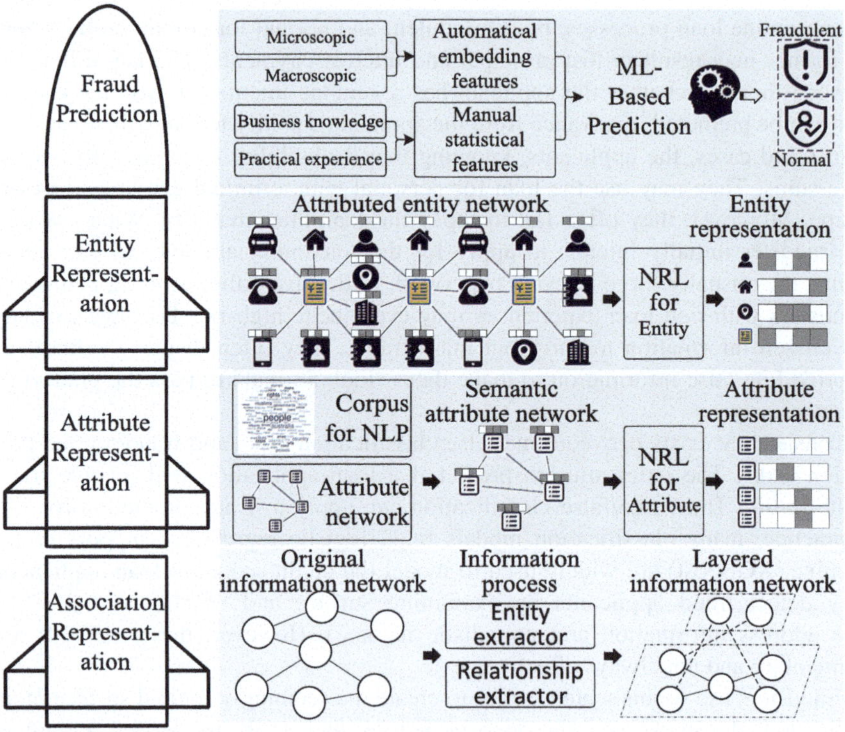

Fig. 4.2 Workflow of the framework OCLFP-MS

Stage 1 Association Representation: Given online loan application data, we can naturally generate an *original information network* by transforming the relational transactions. In the original information network, all data is considered as nodes in the network, and if different data appear in the same loan application, the edge can be established between the nodes. Normal loan applications rarely involve a large amount of material reuse, so they not only do not exhibit highly clustered connection patterns in the information network but are also typically independent of each other. This sparse connectivity (i.e., low information density of associations) results in limited information transmission between different loan applications, making it difficult to learn efficient representations based on the original information network. To alleviate the problem, we design the *information process* to divide all nodes into entities and attributes according to the principle of *unicity* and *indivisibility*,[3] and extract the relationships between nodes. Based on this classification, we build the basic architecture of *layered information network* consisting of the entity layer and attribute layer. The edges in the original information network transform into the co-occurrence associations in the layered information network. Especially, the attribute layer reserves an interface with exterior knowledge for generating semantic similarity associations [45–47]. Thus, all involved associations can be classified into five categories: *entity co-occurrence*, *attribute co-occurrence*, *attribute similarity*, *layer correspondence*, and *relationship correspondence* associations which are described in Fig. 4.3. Note that the entity associations are purely co-occurrence associations due to the above two principles.

In Stage 1, association representation lays the groundwork for boosting the information density by a two-pronged strategy, i.e., reducing association information carriers and enhancing useful association information. For the former, all nodes of the original information network are divided into entities and attributes according to the principle of unicity and indivisibility, and the entity layer with entity co-occurrence associations is selected as the final information carrier network. For the latter, the various involved associations can be classified into *co-occurrence*, *similarity*, and *correspondence* associations[4] which carry all useful association information. In particular, attribute similarity associations are responsible for absorbing exterior semantic knowledge (e.g., pre-trained word embeddings learned from open-source corpora), which will be integrated into the attribute layer through synchronously learning the co-occurrence and similarity associations between attributes. It is worth noting that the correspondence associations ensure that different forms

[3] For an object, its properties with *unicity* refer to that can uniquely identify it, and its properties with *indivisibility* refer to the properties whose components cannot provide any information about this object. For example, the ID card number of an applicant (i.e., the object) is a property with unicity and indivisibility, since it uniquely corresponds to this applicant and its components usually have no information about him/her, especially, when it is desensitized.

[4] The *co-occurrence* reflects the consistency of different objects in the data, the *similarity* reflects the semantic partial overlap of different objects, and the *correspondence* ensures the affinity of objects is positively correlated in that of different spaces.

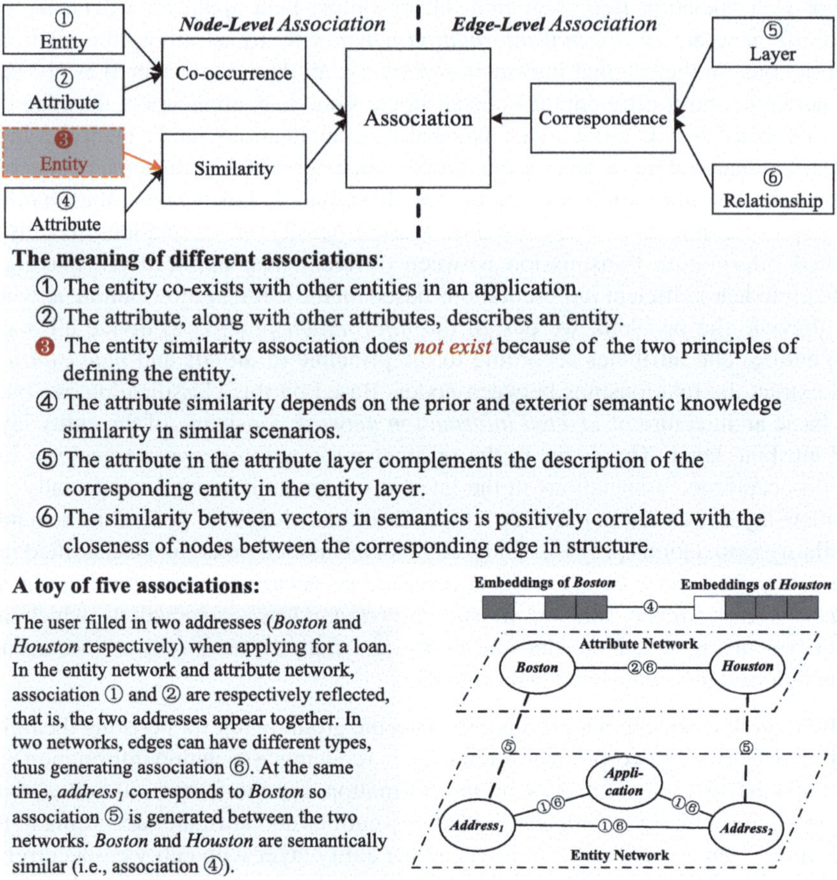

Fig. 4.3 The classification of associations

of information can be fused together (as illustrated in the left part of Fig. 4.4), i.e., layer correspondence associations act as the bridges to generate attribute co-occurrence associations in Stage 1; relationship correspondence associations act as the channels of mapping the generated attribute co-occurrence information into attribute similarity associations of the attribute layer in Stage 2; finally, both layer and relationship correspondence associations cooperatively act as the channels of mapping the enhanced attribute information back into the entity layer in Stage 3.

Stage 2 Attribute Representation: Utilizing the relationship correspondence associations, we integrate the attribute co-occurrence information into attribute similarity information obtained from the priori and exterior semantic knowledge. To this end, we generate a *semantic attribute network* based on the *attribute network* and pre-trained word vectors learned from *natural language processing (NLP)*. The attribute network is extracted from the layered information network to carry

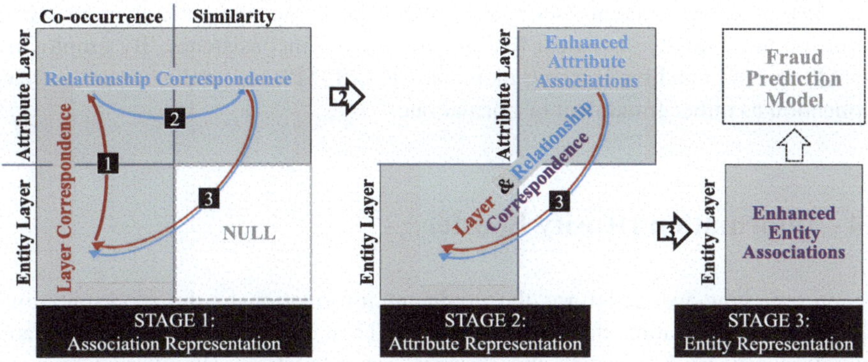

Fig. 4.4 The illustration of multistage data representation

attribute co-occurrence information, and pre-trained word vectors reflect attribute similarity information in semantics. Finally, the network representation learning (NRL) algorithm is adopted to enhance the integrated attribute information.

In Stage 2, attribute representation enhances attribute association information by fusing the information carried by data-driven co-occurrence associations and knowledge-driven similarity associations. The enhanced information is stored in the attribute layer and ready to be mapped back into the entity layer by Stage 3 (as illustrated in the middle part of Fig. 4.4).

Stage 3 Entity Representation: Utilizing both the layer and relationship correspondence associations, the *attributed entity network* is built to carry the information of entity co-occurrence associations and enhanced attribute information from Stage 2. Then, we adopt the NRL algorithm to enhance the final entity association in the attributed entity network. The enhanced entity association information maintains all involved associations described in Stage 1.

In Stage 3, entity representation loads losslessly all useful and enhanced information of co-occurrence and similarity associations into the entity layer, i.e., the final information carrier network, and outputs the multistage enhanced entity association information with sufficiently high information density. It is ready for building fraud prediction models, e.g., machine learning models for classification or clustering (as illustrated in the right part of Fig. 4.4). Note that the semantic attribute network and the attributed entity network have similar forms, and both of them utilize NRL algorithm to integrate different types of information into the corresponding representations, so the same NRL algorithm can be adopted in two stages.

In OCLFP-MS, as the core module, AI2Vec gradually boosts the information density of associations in a multistage mode, which is illustrated in Fig. 4.4.

Predictor Fraud Prediction: The inconsistency problem is crucial in online credit loan fraud prediction (OCLFP), e.g., the neighbor inconsistency, feature inconsistency, and association inconsistency [48]. We focus on the automatically embedding features and manually statistical features to find the inconsistency. By employing our AI2Vec and machine learning methods, OCLFP-MS can predict an online loan application as either fraudulent or normal one.

4.4 Information Density Booster

We present how we transform the applicant information to the representations with higher information density by AI2Vec. Then, we provide a feasible fraud prediction scheme by feeding the features generated from our method into the commonly used supervised learning methods. Finally, we give the detailed design of our customized network representation learning (NRL) algorithm, layered mapping network embedding (LMNE).

4.4.1 The Multistage Data Representation

4.4.1.1 Association Representation

To present the associations among application transactions better, the graph structure is a common way to depict the applicant information network [46]. In online lending services, a loan application transaction in the step of fraud prediction is generally characterized by a set of basic information fields that include the applicant's identity, affiliation, relation information, etc. Accordingly, we can build an *original information network* by using the common *star topology* structure for each transaction, where we denote its identifier and all field values as nodes, and the pairs of the identifier and corresponding field values as edges. Different transaction identifiers are connected by the shared field values.

In the original information network, the redundancy among some fields may lead to excessive network size and bury useful information. Some original fields are supplementary to other ones. They should not act as the independent entities alone, e.g., the address has overlapping meanings with a special city. Therefore, we select some fields with high information as the entities to represent the original transactions, like relative, company, phone, mail, address, etc. Ideally, we hope that loan application transactions with a limited number of fields can capture predictive information about future loan application transactions. That is, assuming a sequence of transaction occurrences, the ratio of the mutual information between two cases is a constant. Specifically, when the ratio is 1, the above fields can completely predict future application transactions. However, in reality, calculating mutual information rigorously may be difficult. In this work, we adopt a feasible method (i.e., information gain based on a tree model) to preliminarily screen fields that are significantly helpful for fraud prediction.

The determination of an entity depends on two principles: the unicity and indivisibility. If a unique value can be used as the indicator for a subject and has no overlapping meaning with other subjects, then it is an entity determined by the above two principles. For instance, a detailed address uniquely represents a location and does not overlap with other detailed addresses, while the provincial information is only a coarse-grained description and corresponds to multiple detailed addresses. The above two principles of entities ensure that the association between entities is purely co-occurrence association without similarity associations, which is helpful for network analysis. We put all the entities into the same layer and call it the entity layer.

In the entity layer, we divide the entity set $\mathscr{V}$ into two parts, denoted by $\mathscr{V}_t$ and $\mathscr{V}_d$, to explain in more detail how to extract the associations among entities. For each loan application transaction, we denote its identifier as $v_t \in \mathscr{V}_t$ and other entities as $v_d \in \mathscr{V}_d$. Similar to the original information network, the co-occurrence of an entity v_d and the identifier v_t is represented as the relationship $e = (v_t, v_d, r) \in \mathscr{E}$, where the type $r \in \mathscr{R}$ is the description of entity v_d in the application data, and $\mathscr{E} \subset \mathscr{V}_t \times \mathscr{V}_d \times \mathscr{R}$. For example, the relationship r indicates that, when the entity v_d is a phone, the applicant's phone is v_d in the transaction with v_t.

After the entities are chosen, the remaining information still contains rich semantic information. The semantic information can serve as the complement to ensure loan application transactions to be represented precisely. To make full use of information, we further consider the remaining fields as the attributes to provide supplementary descriptions of entities. We also create a layer for all attributes to store them, and call it the attribute layer where relationship correspondence associations link the co-occurrence and similarity of attributes.

There are associations between the entity layer and attribute layer to describe the attribute of the entity. In loan application transactions, there are multiple fields describing the same object. Thus, we can extract the fields describing the same object entity from the loan application transactions as corresponding attributes of the entity. In the original information network, the node corresponding to attribute θ_v and the node corresponding to entity v_d are both connected to the entity v_t. We formulate the common connection between attribute θ_v and entity v_d as the layer correspondence association (v_d, θ_v) that means that the attribute complements the description of the corresponding entity. Besides, other fields are directly related to entity v_t, we transform the pair v_t and corresponding field θ_v into the layer correspondence association (v_t, θ_v).

In the attribute layer, we define the attribute as θ_v^i for each entity $v \in \mathscr{V}$, where i denotes the type of the attribute. We use Θ to denote the set of attributes, i.e., $\theta_v^i \in \Theta$.

By processing the original information network, we build a *layered information network* to serve as the basic information carrier network, which is composed of two layers, the entity layer and attribute layer. The correspondence association (v, θ_v^i) between attributes and entities is important to link the entity layer and the attribute layer. In the layered information network, we consider five different associations in transactions, i.e., the *entity co-occurrence*, *attribute co-occurrence*, *attribute similarity*, *layer correspondence*, and *relationship correspondence* associations, as described in Fig. 4.3.

4.4.1.2 Attribute Representation

As a kind of important side information, attributes provide important assistance for the representation of entities. Next, we introduce how to represent the attribute information.

Different from the entity that is unique and indivisible, the attribute exists as the coarse-grained description complementing the entity. The attribute brings a more abundant description for entities, where the partial similarity of two entities can be reflected in having many same attributes. Therefore, it is important to represent the attributes with rich semantic information so as to enhance the attribute representation. In real life, we can learn the prior and exterior semantic knowledge through the natural language processing algorithms in the data of similar loan scenarios. Sometimes constrained by data availability, a simple alternative approach is to use the currently popular pre-trained word vectors as the vectorization of attributes. But this method only evaluates the similarity between attributes and ignores the fact that the semantic information is not the same in different scenarios.

To learn the semantics of attributes in a specific scenario, we first extract an attribute network $G'_{\Theta} = (\Theta, \mathscr{E}_{\Theta}, \mathscr{R}_{\Theta})$ from the attribute layer in the layered information network, to reflect the co-occurrence between attributes. Specially, we denote the co-occurrence between the attribute θ_{v_1} and θ_{v_2} as the relationship when the two attributes correspond to the same entity in the entity layer. Different from the entity layer, the co-occurrence between attributes may appear multiple times, i.e., the two attributes appear in multiple transactions. We count the number of co-occurrences as weight w of the relationship between attributes. We formally denote the relationship between attributes as e', $(\theta_{v_1}, \theta_{v_2}, r', w) \in \mathscr{E}_{\Theta}$, where $r' \in \mathscr{R}_{\Theta}$ depends on the type of the two attributes θ_{v_1} and θ_{v_2}, and $\mathscr{E}_{\Theta} \subset \Theta \times \Theta \times \mathscr{R}_{\Theta}$. For example, the relationship between the city and district attributes should be treated differently from the relationship between the city attribute and province attribute.

In addition to the relationships between attributes, there are also similarities between the rich semantic information contained in attributes. Thus, we attach a pre-trained word vector to each attribute to reflect the semantic similarity of the attribute. The generation of the semantic network is illustrated in Fig. 4.5.

In this work, we use the widely used pre-trained vectors as the vectorization of attributes, and define **a** as the pre-trained vector. The pre-trained word vectors[5] are obtained in a large number of other common corpus data with strong generality, and can be applied by fine-tuning or directly [49]. Then, the semantic attribute network can be formulated as $G_{\Theta} = (\Theta, \mathscr{E}_{\Theta}, \mathscr{R}_{\Theta}, \mathscr{A})$, where $\mathscr{A}$ is the set of the corresponding pre-trained vectors in word embedding.

---

[5] The pre-trained word vector is available via https://fasttext.cc/docs/en/english-vectors.

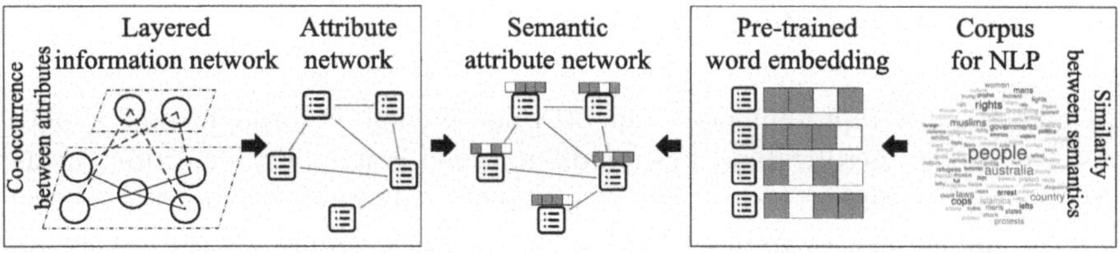

Fig. 4.5 The generation of semantic attribute network

As a commonly used tool, NRL can automatically capture deeply hidden associations, and represent the network into a low-dimensional space. We utilize it to represent the semantic attribute network G_Θ. Then, each attribute θ_v is represented as a vector $\boldsymbol{\theta}_v$ that retains the similarity and co-occurrence associations between attributes.

4.4.1.3 Entity Representation

The performance of subsequent models depends on entity representation. Next, we introduce how to make entity representation with the entity layer and attribute representations.

We first obtain the entity network $G'_\mathscr{V} = (\mathscr{V}, \mathscr{E}, \mathscr{R})$ from the entity layer to reflect the entity co-occurrence associations. In the stage of association representation, the entity network has been formed. Based on the learned attribute representation, every attribute θ_v is transformed into a corresponding vector $\boldsymbol{\theta}_v$, where the attribute representations integrate the co-occurrence and similarity associations between attributes in the stage of attribute representation. Then, we combine the entity network and the layer correspondence association between attributes and entities to link attribute representations for each entity. Therefore, the attributed entity information network can be formally denoted as $G_\mathscr{V} = (\mathscr{V}, \mathscr{E}, \mathscr{R}, \boldsymbol{\Theta})$, where $\boldsymbol{\theta}_v \in \boldsymbol{\Theta}$ is the corresponding representation of θ_v.

The relationship $e \in \mathscr{E}$ reflects the co-occurrence association between entities, and the similarity between attribute representations reflects the co-occurrence and similarity associations between attributes. Similar to the attribute representation stage, we consider the co-occurrence associations between entities and the similarity between attribute representations during learning entity representation. By applying the NRL algorithm on the semantic attribute network G_Θ, we obtain the corresponding entity representation $\mathbf{v}$ for each entity v, where $\mathbf{v}$ retains all associations in data.

Note that the semantic attribute network G_Θ and the attributed entity network $G_\mathscr{V}$ have similar forms, where the attribute set Θ (or the pre-trained vectors $\mathscr{A}$) and the entity set $\mathscr{V}$ (or the attribute representations $\boldsymbol{\Theta}$) play the same role in NRL algorithms. Therefore, the same NRL algorithm can be used in two different stages to integrate different types of information into the corresponding network representation.

4.4.2 Fraud Prediction Based on AI2Vec

The learned entity representations support fraud prediction through some applications, e.g., node classification, link prediction, and community detection. In this work, we further study the similarity of different entity representations to enhance the anti-fraud performance. In OCLFP, anomalous associations are the top concern of most researchers. Therefore, we mainly focus on the associations between loan applications, i.e., the similarity of transaction identifier and each connected entity from the micro perspective, and the similarity of transaction identifier and other indirectly relevant transaction identifiers from the macro perspective. The microscopic similarity reflects whether a loan application is reasonable, e.g., whether an entity is the forged information. The macroscopic similarity reveals the relative relationship between a loan application and others, e.g., whether it involves the fraudulent group to fabricate data in batches.

We apply the Euclidean distance as the similarity between the two entity representations. Other similarity metrics can also be used with performance differences [50]. From the micro view, our method introduces the similarity between the entity v_t and each entity v_d belongs to the first-order neighbor of v_t in the attributed entity network as microscopic features. Given the learned representations $\mathbf{v}_t$ and $\mathbf{v}_d$, we have:

$$sim(\mathbf{v}_t, \mathbf{v}_d) = \sqrt{\sum\nolimits_{i=1}^{D} \left(\mathbf{v}_t^i - \mathbf{v}_d^i\right)^2}, \tag{4.1}$$

where D is the dimension of $\mathbf{v}_t$ and $\mathbf{v}_d$, and $\mathbf{v}^i$ is the i-th dimension of $\mathbf{v}$. We judge the fraud risk of an online loan application through consistency verification.

From the macro perspective, we adopt the similarity between the loan application transactions v_t and v_t' that belongs to the second-order neighbor of v_t in the attributed entity network as macroscopic features. Given the learned representations $\mathbf{v}_t$ and $\mathbf{v}_t'$, we get the similarity $sim(\mathbf{v}_t, \mathbf{v}_t')$. We estimate the fraud risk of a loan application by exploring related transactions. For instance, fraudulent groups use a set of information to apply multiple similar loan applications for more revenue, since it is a high cost to forge information for deceiving a loan.

In reality, the loan applications have different numbers of first-order and second-order neighbors. For microscopic features, the missing entities result in a decrease in first-order neighbors, so we uniformly replace them with the constant term. For macroscopic features, the number of second-order neighbors is uncertain, so we only retain the first s features with the smallest similarity. Generally, it is no need to set up a large s since most loan applications are isolated.

In real-world scenarios, we cannot update the network and learned representations for every loan application transaction because of the huge computing consumption in NRL. Besides, the lending business usually has an audit period, so updating for every transaction is unnecessary. To avoid gang frauds, the platform periodically inspects similar applications. After a large number of similar applications are detected, the referred information will be difficult to apply for loans, so criminals rush to apply for loans before the platform is found to make full use of the information [51].

During network embedding, a lot of unnecessary correlation information will be learned, which will increase the complexity and reduce the efficiency. It is really necessary to eliminate those premature loan application transactions [52], which is helpful for controlling the size of the network to improve efficiency in network embedding. We process the loan applications in batches, which reduces the consumption of repeatedly updating networks and representations.

4.4.3 *The Customized NRL Algorithm*

In this part, we give more details to help explain how we design and deploy our network representation learning algorithm. Inspired by the unsupervised and sampling-based representation learning [23, 24, 38], we design a customized and improved NRL algorithm and take the attributed entity network as an example to introduce the details. The overview of our proposed NRL algorithm is presented in Fig. 4.6.

4.4.3.1 Learning Network Representations

Conceptually, we learn the representations from the following two recognized criteria: (1) Given the observed relationship (v_t, v_d, r) and the unobserved relationship (v_t, v_d, r') in the network, we would like to learn how likely there is a relationship of type r from entity v_t point to v_d. (2) We assume for any two entities, the similarity between the learned entity representations is positively correlated with the similarity between corresponding attribute representations.

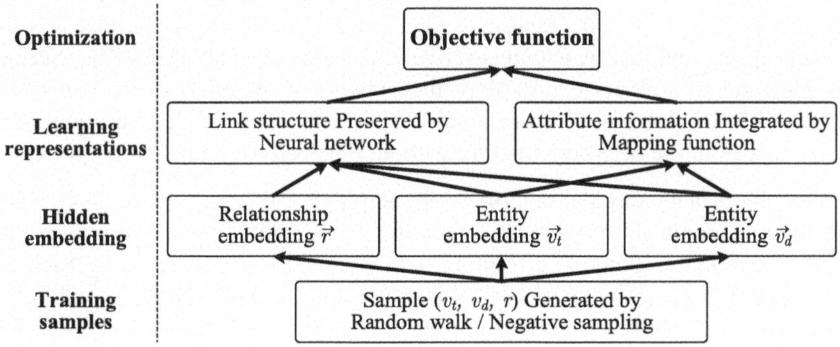

Fig. 4.6 The architecture of our proposed NRL algorithm

For the convenience of describing the goal of network representation leaning, we denote the link structure as $\mathscr{L}$ and the attribute information as $\mathscr{I}$. By using Bayes' rule, the goal can be expressed as the following formula:

$$\begin{aligned} &\arg\max_{\mathbf{x},\mathbf{r}}\ p(\mathbf{x}, \mathbf{r} \mid \mathscr{L}, \mathscr{I}) \\ &= \arg\max_{\mathbf{x},\mathbf{r}}\ p(\mathscr{L}, \mathscr{I} \mid \mathbf{x}, \mathbf{r}) p(\mathbf{x}, \mathbf{r}) / p(\mathscr{L}, \mathscr{I}) \\ &= \arg\max_{\mathbf{x},\mathbf{r}}\ p(\mathscr{L}, \mathscr{I} \mid \mathbf{x}, \mathbf{r}) p(\mathbf{x}, \mathbf{r}) \\ &= \arg\max_{\mathbf{x},\mathbf{r}}\ p(\mathscr{L} \mid \mathbf{x}, \mathbf{r}) p(\mathscr{I} \mid \mathbf{x}), \end{aligned} \tag{4.2}$$

where $\mathbf{x}$ and $\mathbf{r}$ are the target optimal entity and relationship representations, respectively. In Eq. (4.2), the denominator $p(\mathscr{L}, \mathscr{I})$ can be neglected in the process of maximization since it is no involvement to $\mathbf{x}$ and $\mathbf{r}$. To simplify the analysis, we assume that there is no prior knowledge about the optimal representation. We also assume that the link structure is independent of the attribute information ($\mathscr{L} \perp \mathscr{I}$) and the relationship representation is also independent of the attribute information given the entity representation ($\mathbf{r} \perp \mathscr{I} \mid \mathbf{x}$).

Next, we present how to learn representations based on the observations on link structures $\mathscr{L}$ and attribute information $\mathscr{I}$. To get the representation, we first prepare the training data to feed the NN model in the NRL algorithm. An efficient method can achieve a trade-off between computational efficiency and model quality, so it is very essential to adopt an appropriate method to sample the network.

Random Walk We adopt the random walk to generate the observed training samples, which can be run rapidly and incrementally. The random walk has been widely applied to extracting local structures from the network. Note that a few studies [34] try to capture different relationships by meta-path, a special random walk with bias. However, it highly depends on a user-guided way to determine a user-given meta-path set and the weight of each meta-path for representation learning. We apply the random walk to get rid of the dependency on user-given meta-path and leave the task of capturing relationships to the model in an automatically learned way. Besides extracting network structure, there are three other advantages of using the random walk. First, the random walk has relatively low time and space complexity, which is suitable for large-scale networks [23]. Second, multiple random walkers can explore the network in parallel, which can run in different threads, processes, or machines [38]. Third, it is possible to accommodate small changes in the network structure without the need for global recomputation.

We generate a noncircular random walk sample $(v_{t_1}, v_{d_1}, v_{t_2}, v_{d_2}, v_{t_3}, \cdots)$, where v_t is the loan application transaction identifier in $\mathscr{V}_t$ and v_d is an entity in $\mathscr{V}_d$. For a random walk, we generate a training data (v_t, v_d, r) or (v_d, v_t, r) by each sub walk (v_t, v_d) or (v_d, v_t), where $r \in \mathscr{R}$ is the relationship of entities v_d and v_t. We denote the set of all the observed samples as $\mathscr{S}_p$. Since two transaction identifiers can be connected by multiple different entities, this leads to a large probability of the circular walk by applying random walk directly to the network. The circular walk seriously affects the quality of the training data. For instance, when v_{t_1} and v_{t_2} have eight co-linked entities, whereas v_{t_2} and v_{t_3} have only one co-linked entity, the target random walk is to include all three entities, e.g., $(v_{t_1}, v_{d_1}, v_{t_2}, v_{d_2}, v_{t_3}, \cdots)$, rather than a cycle walk $(v_{t_1}, v_{d_1}, v_{t_2}, v_{d_2}, v_{t_1}, \cdots)$. Therefore, we abandon the observed sample with the same entity by checking each entity in a random walk.

Negative Sampling In addition to sampling observed data, the model also needs unobserved data to learn representation. Inspired by Word2Vec [49], we generate the unobserved training samples by negative sampling. For each observed sample (v_t, v_d, r), we generate negative samples by randomly replacing one of the three values with other v_t', v_d', r', where v_t', v_d', and r' are randomly selected from $\mathscr{V}_t$, $\mathscr{V}_d$, and $\mathscr{R}$, respectively. We further filter out the bad negative samples which can be obtained during the generation of observed samples. We denote the set of all the unobserved samples as $\mathscr{S}_n$, and the set of all the training samples as $\mathscr{S}$ with $\mathscr{S} = \mathscr{S}_p \cup \mathscr{S}_n$ and $\mathscr{S}_p \cap \mathscr{S}_n = \emptyset$.

Preserving Link Structure To learn the representations for link structure, we need to ensure that the connected entities are closer together in the embedding space and to distinguish the relationship between the connected entities. Thus, we employ a classification model to predict the targeted relationships between the pair of entities. We simplify the targeted relationships into whether the predicted relationship is correct or not since a strict distinction between each relationship leads to significant computational overhead. Specifically, we employ a neural network (NN) model as shown in Fig. 4.7 to predict the targeted relationships between the pair of entities and simplify the targeted relationships into whether the predicted relationship is correct or not. For each $(v_t, v_d, r) \in \mathscr{S}$, we use $p(v_t, v_d, r)$ to represent the probability that the NN model predicts correctly. We formulate the following likelihood function with respect to link structures:

$$p(\mathscr{L} \mid \mathbf{x}, \mathbf{r}) = \prod\nolimits_{(v_t, v_d, r) \in \mathscr{S}} p(v_t, v_d, r). \tag{4.3}$$

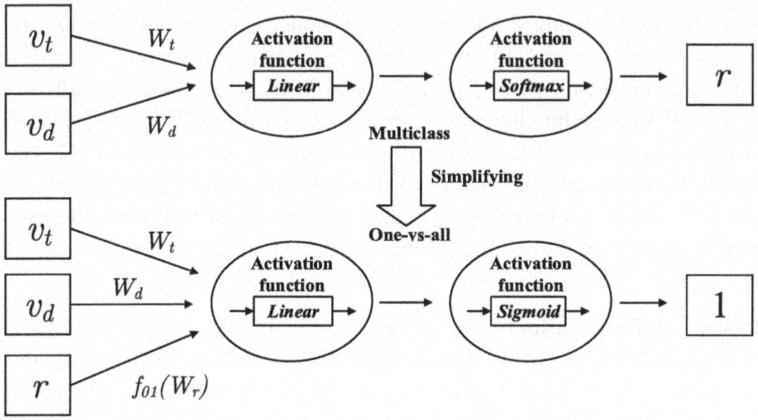

Fig. 4.7 The adopted NN model

Note that samples in $\mathscr{S}_p$ and $\mathscr{S}_n$ have different effects on the model. For a training sample in $\mathscr{S}_p$, $p(v_t, v_d, r)$ approaching 1 means that the model correctly predict r belonging to the pair of v_t and v_d. Otherwise, for a training sample in $\mathscr{S}_n$, $p(v_t, v_d, r)$, approaching 0 means that the nonexistent relationship r between v_t and v_d is correctly identified by the model. Thus, we derive as below:

$$p(v_t, v_d, r) = \begin{cases} o(v_t, v_d, r), & if\ (v_t, v_d, r) \in \mathscr{S}_p \\ 1 - o(v_t, v_d, r), & if\ (v_t, v_d, r) \in \mathscr{S}_n \end{cases}, \tag{4.4}$$

where $o(v_t, v_d, r) = \text{sigmoid}\left(\sum W_t'\mathbf{v}_t \odot W_d'\mathbf{v}_d \odot f_{01}(W_r'\mathbf{r})\right)$ and W_t', W_d', and W_r' are the corresponding transpose matrices in the NN model.

For a training sample (v_t, v_d, r), we take the probabilities of correctly predicting r between v_t and v_d as the output. In the model, the input layer takes in three one-hot vectors, $\mathbf{v}_t$, $\mathbf{v}_d$, and $\mathbf{r}$, representing v_t, v_d, and r. The entity of network consists of two parts, $\mathscr{V}_t$ and $\mathscr{V}_d$, and the length of $\mathbf{v}_t$ and $\mathbf{v}_d$ is equal to $|\mathscr{V}|$. The length of $\mathbf{r}$ is equal to $|\mathscr{R}|$. In the latent layer, the input one-hot vector is converted to a D-dimensional latent vector where D is the dimensionality of the hidden layer. We represent the corresponding hidden layer vectors as $W_t'\mathbf{v}_t$, $W_d'\mathbf{v}_d$, and $f_{01}(W_r'\mathbf{r})$, where W_t and W_d are two $|\mathscr{V}| \times D$ matrices and W_r is a $|\mathscr{R}| \times D$ matrix. W_t', W_d', and W_r' are the corresponding transpose matrices in the NN model. Note that we adopt a regularization function $f_{01}(\cdot)$ to treat r differently. Since r plays a different role from the entity, we define that the function $f_{01}(\cdot)$ restricts the values of $W_r'\mathbf{r}$ into the range from 0 to 1. In the output layer, the model outputs a probability that predicts whether the relationship r exists between v_t and v_d. We adopt the Hadamard function, denoted as $W_t'\mathbf{v}_t \odot W_d'\mathbf{v}_d \odot f_{01}(W_r'\mathbf{r})$, to aggregate three latent vectors in the latent layer. Finally, we take the Sigmoid function as the activation function of the summation, representing as sigmoid($\sum W_t'\mathbf{v}_t \odot W_d'\mathbf{v}_d \odot f_{01}(W_r'\mathbf{r})$). Since the operator $\odot$ satisfies the law of commutation, we do the same operation with the input vectors $\mathbf{v}_d$, $\mathbf{v}_t$, and $\mathbf{r}$. Since the samples (v_t, v_d, r) and (v_d, v_t, r) apply to the same operation, we will only discuss the former to simplify the description.

We treat the relationship r and entities (v_t and v_d) differently for two reasons: (1) It prevents unobserved training samples from being adversely affected by the relationship. Given a pair of v_t and v_d, the sign of corresponding value $W_t'\mathbf{v}_t \odot W_d'\mathbf{v}_d \odot f_{01}(W_r'\mathbf{r})$ may become opposite to that of value $W_t'\mathbf{v}_t \odot W_d'\mathbf{v}_d \odot W_r'\mathbf{r}$ on the optimization process. But, if the entities v_t and v_d are related in $\mathscr{S}_p$, the signs of values in these two entities vectors should be the same. (2) It ensures that the proximity of two entities takes precedence over the relationship which ensures the effectiveness of learned embedding. A big $W_r'\mathbf{r}$ reduces the importance of values of $W_t'\mathbf{v}_t$ and $W_d'\mathbf{v}_d$.

Integrating Attribute Information We assume for any pair of entities (m, n), in the attributed entity network, the similarity between their entity representations $(\mathbf{m}, \mathbf{n})$ is positively correlated with the similarity between their pair of attribute representations $(\boldsymbol{\theta}_m^i, \boldsymbol{\theta}_n^i)$. For each type of attribute, we design a linear bijective mapping function $\boldsymbol{\theta}_n^{i} = f_i(\mathbf{n})$ to bridge the attribute representation $\boldsymbol{\theta}_n^i$ and the entity representation $\mathbf{n}$. If the attribute representations $\boldsymbol{\theta}_m^i \approx \boldsymbol{\theta}_n^i$, the entity representation $\mathbf{m}$ is close to $\mathbf{n}$. We adopt a distribution to quantitatively calculate $p(\mathscr{I} \mid \mathbf{x})$, where the distribution determines the sensitivity of similarity between attribute representations in NRL. To facilitate the optimization, we assume a multivariate normal distribution $\mathscr{N}(\boldsymbol{\theta}^i \mid \mathbf{u}, \frac{1}{2\alpha}I)$ for each type of attribute representation $\boldsymbol{\theta}^i$, where $\mathbf{u}$ is the mean vector and $\frac{1}{2\alpha}I$ is the covariance matrix. The matrix I is an identity matrix. The parameter α can adjust the degree to which attribute information affects the link structure. Then, we have:

$$
\begin{aligned}
p(\mathscr{I} \mid \mathbf{x}) &= \prod_{n \in \mathscr{V}} \prod_{i \in |\Theta|'} \mathscr{N}\left(\boldsymbol{\theta}^i \mid \mathbf{u}, \frac{1}{2\alpha} I\right) \\
&= \prod_{n \in \mathscr{V}} \prod_{i \in |\Theta|'} \left(\frac{\pi}{\alpha}\right)^{-\frac{1}{2}} \exp\left(-\alpha \cdot \beta_n^i \cdot ||\boldsymbol{\theta}_n^i - f_i(\mathbf{n})||_2^2\right),
\end{aligned}
\tag{4.5}
$$

where $|\Theta|'$ is the set of all attribute types in Θ, and β_n^i is set to 0 or 1. If β_n^i is set 1, it means that the i-th attribute of entity n has the corresponding semantic information; otherwise, it means that the attribute does not have the corresponding semantic information. Meanwhile, we set default values for the corresponding word vector $\boldsymbol{\theta}_n^i$ for attributes that have no semantic information. Since β_n^i is 0 in this case, the default value does not affect our learning process. In our work, we adopt the linear bijective mapping function $f_i(\mathbf{n}) = \boldsymbol{K}\mathbf{n} + \boldsymbol{b}$, where the matrices $\boldsymbol{K}$ and $\boldsymbol{b}$ are the parameters in the function $f_i(\cdot)$. Any bijective functions $f_i(\cdot)$ are allowed, and the naive linear mapping function alleviates the computational complexity in tuning potential hyperparameters.

Objective Optimization Combining Eqs. (4.3) and (4.5), we calculate the negative logarithm of $p(\mathscr{L} \mid \mathbf{x}, \mathbf{r})p(\mathscr{I} \mid \mathbf{x})$ to simplify the computation in the optimization process. Thus, our objective function becomes:

$$
\begin{aligned}
\arg\min_{\mathbf{x},\mathbf{r}} \mathit{Object} &= -\log p(\mathscr{L} \mid \mathbf{x}, \mathbf{r}) p(\mathscr{I} \mid \mathbf{x}) \\
&= \sum\nolimits_{(v_t, v_d, r) \in \mathscr{S}} -\log p(v_t, v_d, r) \\
&\quad + \alpha \sum\nolimits_{n \in \mathscr{V}} \sum\nolimits_{i \in |\Theta|'} \beta_n^i \cdot ||\boldsymbol{\theta}_n^i - f_i(\mathbf{n})||_2^2.
\end{aligned}
\tag{4.6}
$$

In Eq. (4.6), we remove the constant terms to simplify the optimization process. We only need to adjust the parameter α manually; the other parameters are learned automatically.

In this work, we adopt the stochastic gradient descent (SGD) to efficiently maximize the objective function *Object*, where SGD optimizes a parameter z using the repetition of following update rule:

$$z \leftarrow z - \eta \frac{\partial \, Object}{\partial \, z} \tag{4.7}$$

For each sample, our model goes backward to adjust the weights $W_t'\mathbf{v}_t$, $W_d'\mathbf{v}_d$, $W_r'\mathbf{r}$ of the NN model, respectively. For each entity, we update the parameters $\boldsymbol{K}$ and $\boldsymbol{b}$ of $f_i(\cdot)$ based on the gradients of *Object*, respectively. Here, we optimize the parameters by using the following updating rules:

$$\begin{aligned}
W_t'\mathbf{v}_t :=& W_t'\mathbf{v}_t - \frac{\partial \, \log p(v_t, v_d, r)}{\partial \, W_t'\mathbf{v}_t} \\
& - 2\alpha \sum\nolimits_{i \in |\Theta|'} \left(\boldsymbol{\theta}_t^i - f_i(W_t'\mathbf{v}_t)\right) \frac{\partial \, f_i}{\partial \, W_t'}, \\
W_d'\mathbf{v}_d :=& W_d'\mathbf{v}_d - \frac{\partial \, \log p(v_t, v_d, r)}{\partial \, W_d'\mathbf{v}_d} \\
& - 2\alpha \sum\nolimits_{i \in |\Theta|'} \left(\boldsymbol{\theta}_v^i - f_i(W_d'\mathbf{v}_d)\right) \frac{\partial \, f_i}{\partial \, W_d'}, \\
W_r'\mathbf{r} :=& W_r'\mathbf{r} - \frac{\partial \, \log p(v_t, v_d, r)}{\partial \, W_r'\mathbf{r}}, \\
\boldsymbol{K} :=& \boldsymbol{K} - 2\alpha \sum\nolimits_{n \in \mathscr{V}} \sum\nolimits_{i \in |\Theta|'} \left(\boldsymbol{\theta}_n^i - f_i(\mathbf{n})\right) \frac{\partial \, f_i}{\partial \, \boldsymbol{K}}, \\
\boldsymbol{b} :=& \boldsymbol{b} - 2\alpha \sum\nolimits_{n \in \mathscr{V}} \sum\nolimits_{i \in |\Theta|'} \left(\boldsymbol{\theta}_n^i - f_i(\mathbf{n})\right) \frac{\partial \, f_i}{\partial \, \boldsymbol{b}}.
\end{aligned}$$

4.4.3.2 Model Updating Scheme and Analysis

In real-world application scenarios, it is really difficult to update the learned representations for any little change with the huge amount of global recomputation cost in NRL. We find that each added or deleted relationship affects a series of random walks which will update the previously learned network representation. In our model, we design an *incremental* updating algorithm to maintain little change in the learned representations. For the newly arrived edges, we generate new observed and unobserved samples by using random walk, where the random walk starts from the nodes of arrived edges. For the deleted edges, we change the related observed samples into unobserved samples in order to reduce the relationship between these two entities of deleted relationships. After updating the samples, we continue to train the learned embedding model with the updated samples set. We describe the procedure of updating the representations by incrementally sampling the changed region in Algorithm 3.

Algorithm 3: Updating the embedding model

Input : The sample set $\mathcal{S} = \mathcal{S}_p \cup \mathcal{S}_n$, the updated relationship set $\mathcal{E}_{updated}$, the deleted relationship set $\mathcal{E}_{deleted}$, the saved embedding model.
Output: The updated sample set $\mathcal{S}$, the updated embedding model.

// new samples generation;
$\mathcal{S}' = \emptyset$;
forall the $(v_t, v_d, r) \in \mathcal{E}_{deleted}$ **do**
  $\mathcal{S}_p = \mathcal{S}_p - \{(v_t, v_d, r)\}$;
  $\mathcal{S}_n = \mathcal{S}_n \cup \{(v_t, v_d, r)\}$;
  $\mathcal{S}'_n = \mathcal{S}'_n \cup \{(v_t, v_d, r)\}$;
end
forall the $(v_t, v_d, r) \in \mathcal{E}_{updated}$ **do**
  Generate samples (including observed sample s_p and unobserved sample s_n) based on random walks starting from entity v_t and v_d;
  $\mathcal{S}_p = \mathcal{S}_p \cup \{s_p\}$;
  $\mathcal{S}'_p = \mathcal{S}'_p \cup \{s_p\}$;
  $\mathcal{S}_n = \mathcal{S}_n \cup \{s_n\}$;
  $\mathcal{S}'_n = \mathcal{S}'_n \cup \{s_n\}$;
end
// model update;
Load the saved embedding model;
Train the model with the samples $\mathcal{S}'$ with a small set from $\mathcal{S}$;
Return the updated sample set $\mathcal{S}$ and save the updated embedding model.

Our model is computationally efficient in terms of both space and time complexity. Given the length-L random walk and N negative samples for each positive sample, we have $(N+1)(L-1)|\mathcal{V}|$ samples to optimize our parameters. Therefore, the overall time complexity is $O(|\mathcal{V}|)$, since N and L are much smaller than $|\mathcal{V}|$. Our model enjoys the time complexity that is linear to the number of node $|\mathcal{V}|$, satisfying the requirement of scalability. The training samples contain $(N+1)L|\mathcal{V}|$ nodes, $(N+1)(L-1)|\mathcal{V}|$ relationships, and $K|\mathcal{V}|$ attributes, where each node has K attributes. As $K \ll |\mathcal{V}|$, our model possesses the space complexity of $O(|\mathcal{V}|)$, which is efficient for practical applications. Besides, the optimization is parallelizable and executed asynchronously, which further facilitates the overall scalability in a large network.

4.5 Experiments

4.5.1 Experimental Setup

4.5.1.1 Datasets

To evaluate our proposed scheme, the experiments are implemented on a real-world online loan dataset from an online lending platform (OLP, a prestigious insurance group), which contains consecutive months of loan application data. We use the dataset from January to June 2017 as the training samples, and set the dataset of July 2017 as the testing samples, where the data of June are selected as the validation set to adjust the parameters. The OLP can obtain the status of whether the user has successfully repaid the loan when the application is due. In this way, it can do some things to prevent overdue repayments, and decide whether to lend money to the user. In this work, the labels of data are provided by the OLP who regards applications for on-time repayment as normal and applications for successive overdue repayment as fraudulent. In addition, some applications are not labeled because of intermittent overdue repayments or the payment dates are not due at the time the data are collected, so we only use them to generate features and do not consider them during the training and testing phase. The data contain a total of 330,903 loan application transactions, where 68,470 transactions (about 22.77% in all transactions) are labeled and 255,557 transactions are not labeled, which reflects that it is difficult to obtain data with labels. For instance, if an application started in Feb., but its repayment was on time in the first 2 months, it became overdue in the third month and repaid on time again in the fourth month. Suppose that the loan will be labeled as fraud when it does not repay for four consecutive months, the application's label can only be determined within 8 months, i.e., Oct., at the earliest. At the same time, the platform starts investigation. During investigation, the fraud label is removed if the borrower pays back all overdue debt and is judged to have no fraudulent intent. To simulate the real application scenario, we do not obtain the labels that are available after Jul. in this work. In the labeled transactions, there are 68,470 normal transactions about 10 times as many as 6,876 fraudulent transactions. The detailed statistics are summarized in Table 4.1. Especially, the unusual Feb. is caused by our data source, where Feb. contains a holiday resulting in a significant reduction of applications. In other months, the distribution of applications does not exhibit time effects.

Table 4.1 The application transaction information

Month	Jan.	Feb.	Mar.	Apr.	May.	Jun.	Jul.
Normal	10,161	5,910	10,771	9,476	10,583	11,198	10,371
Fraudulent	894	608	1,046	886	1,044	1,131	1,267
Unlabeled	36,901	24,111	44,400	35,470	37,500	39,508	37,667

Table 4.2 The selected entities in application transactions

Entity	Attribute	Description
Identifier	Name, Time	The transaction's identifier. We extract the applicant's name and the transaction's time as attributes
ADDR	Province, City, District	The detailed address. We extract three levels of administrative areas as attribute description
CO NAME	None	The name of the company where the online loan applicant works
IDNO	Name	The identity card number. Using holder's name as an attribute
DL	None	The driver's license number of the applicant
VIN	None	The vehicle identification number of the applicant
ENGIN	None	The engine number of the applicant
MP	Name	The mobile number. Using holder's name as an attribute
TEL	Name	The telephone number. Using holder's name as an attribute

Table 4.3 The extracted relations in loan transactions

Relationship	Description
R_ADDR	The loan applicant's address
R_CO	The mobile number of the applicant's colleagues
R_CO_ADDR	The address of the applicant's company
R_CO_NAME	The loan applicant's company name
R_CO_TEL	The telephone number of the applicant's company
R_CPADDR	The loan applicant's residential address
R_CRMP	The mobile number of the applicant's relatives
R_FCMP	The mobile number of the applicant's friends
R_IDNO	The loan applicant's license plate number
R_DL	The driver's license number of loan applicant
R_VIN	The vehicle identification number of loan applicant's car
R_ENGIN	The engine number of loan applicant's car
R_LRMP	The mobile number of the applicant's family members
R_MATEIDNO	The identity card number of the applicant's spouse
R_MATEMP	The mobile number of the applicant's spouse
R_MP	The loan applicant's mobile number
R_OCMP	The mobile number of the applicant's other contacts
R_TEL	The loan applicant's telephone number
R_REGADDR	The loan applicant's native place

In the original dataset, each application transaction is characterized by 64 fields. However, most of the fields are missing or have the same meaning, like multiple different contacts and their corresponding details. We finally choose 9 fields in all as the entity in the layered information network, and extract 19 relationships, which are shown in Tables 4.2 and 4.3, respectively.

Note that there are more types of relationships than entities, so one entity may be connected to different relationships. Restricted by the privacy protection policy, our adopted data have been desensitized by the platform. And applicants are informed by statement subject to local privacy policies before data collection. Therefore, in this work, the selected entities, such as MP and IDNO, have already satisfied the principle of indivisibility.

4.5.1.2 Testbed

Based on the dataset introduced above, we finally generate *eight* microscopic features and *ten* macroscopic features for fraud prediction. We adopt half a year of historical data to build the layered information network and update it when the application transactions are accumulated within a qualified processing latency. To simulate the real application scenario, we build the network under a sliding window strategy. We sort the testing samples by time and treat them as batches. In each batch, we built the layered information network using a batch of the testing samples and its historical data from the previous 6 months. During the fraud prediction, we build the classification model by feeding the microscopic and macroscopic features for fraud prediction. The experiments are conducted under Windows Server 2012 operating system with 40 Intel Xeon E5-2640 v4 CPU and 128 GB of RAM.

4.5.1.3 Metrics

Generally, according to the real-world business requirements of online credit loan fraud prediction, the KS value is a widely used and reliable metric [12] to measure the risk differentiation of the model in financial scenarios. From the de facto standards, a KS value that is not much less than 0.5 with an average processing latency of 10 seconds can be regarded as a qualified performance. We focus on verifying the application transactions already labeled by the OLP. Thus, we mainly adopt the KS value to measure the performance of our scheme. The calculation method of KS value is max $(TPR - FPR)$ under different thresholds, where TPR is the true positive rate and FPR is the false positive rate. We also record TPR and FPR when obtaining the KS value to comprehensively evaluate our model.

4.5.2 Contrast Experiments

We present the performance comparison between the proposed OCLFP-MS and baselines for the fraud prediction task. The following state-of-the-art methods are used to highlight the effectiveness of our framework OCLFP-MS on the online credit loan application dataset, including DT [32], E.T.-RNN [18], crDNN [19], semiGNN [12], and Manual method. The so-called Manual method refers to the features

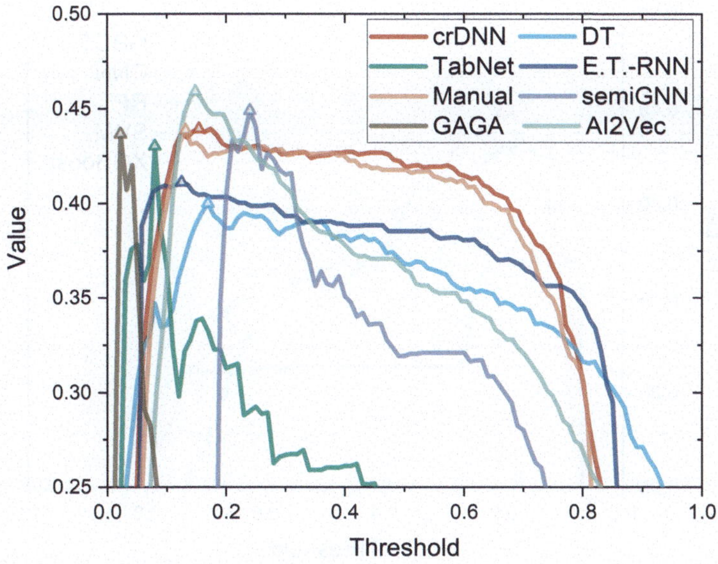

Fig. 4.8 The *KS curves* of different models

that are manually designed by experts for fraud prediction, and is adopted by the platform, i.e., the source of our OCLFP data. We run our AI2Vec-based OCLFP-MS with LMNE algorithm and XGBoost classifier. Note that the semiGNN acts as an end-to-end fraud prediction method, not a data representation tool in this work. We conduct exhaustive grid search to find the optimal parameters for achieving the best fraud detection performance in terms of KS values. The KS curves are presented in Fig. 4.8. The Manual method is designed with manual features designed by platform's experts. We use notation "Δ" to label the KS value. We observe that our scheme achieves better performance than other methods, and the best KS value is over 0.46. The semiGNN is second only to our method in performance. The performance of crDNN and Manual method is close, lower than to above methods. The E.T.-RNN has a gap compared with the previous methods. The DT performs worst, about 0.06 below the best KS value. The performance is proven to be effective when applying our AI2Vec scheme to gradually boost the information density of data.

4.5.3 Module Comparison

4.5.3.1 Performance Comparison of Classifiers

We examine the performance of our AI2Vec-based OCLFP-MS with LMNE and different machine learning models for fraud prediction. We consider five represen-

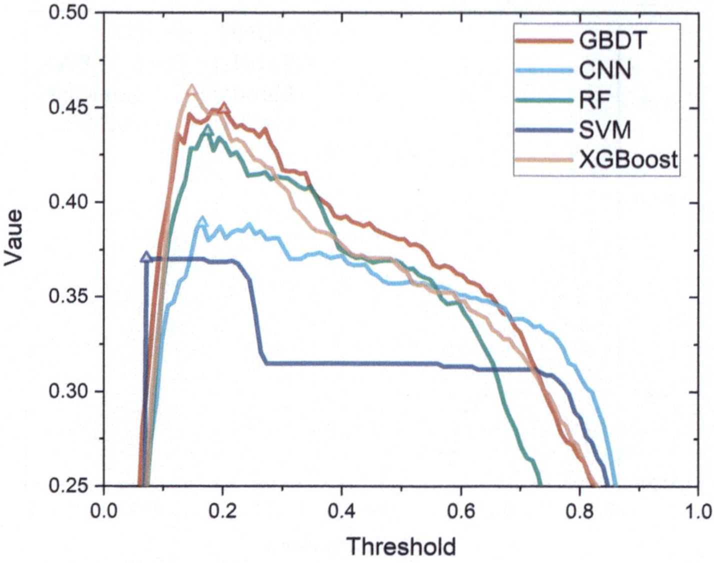

Fig. 4.9 The *KS curves* of our AI2Vec-based OCLFP-MS with the LMNE algorithm and different commonly used machine learning classifiers

tative classifiers, i.e., gradient boosting decision tree (GBDT), convolutional neural networks (CNNs), random forest (RF), support vector machine (SVM), and eXtreme Gradient Boosting (XGBoost), where the classifier parameters are set by default. The experimental results are shown in Fig. 4.9, where we use notation "Δ" to label the KS value. We observe that our framework with XGBoost achieves the best performance with a KS value of 0.46.

The frameworks with GBDT and RF also achieve good results, slightly less about 0.01 and 0.02 in terms of KS values than that with XGBoost. The SVM and CNN classifiers do not work well under our framework, whose KS values are lower than 0.4. The experimental results show that our framework can achieve better performance with tree-based classifiers in the supervised models.

4.5.3.2 Performance Comparison of NRL Algorithms

In this part of experiments, we compare our algorithm LMNE with several state-of-the-art NRL algorithms in the AI2Vec-based OCLFP-MS. For succinctness of the experimental description, we choose the XGBoost classifier as an effective representative, without including the results of all classifiers. According to the goals of NRL algorithms, they can be roughly divided into two groups: network structure and additional information. In general, network structure is the fundamental factor for NRL. Meanwhile, additional information, such as node information and edge information, is also helpful to enable the learned vectors to work well in real

Table 4.4 A summary of different algorithms

Algorithms	[25]	[23]	[36]	[24]	[37]	LMNE
Structure preserving	√	√	√	√	√	√
Edge information		√		√		√
Node information			√	√	√	√
Matrix factorization				√		
Random walk	√	√				√
Graph neural networks			√		√	

application scenarios. Furthermore, to transform a network from the original network space to the representation space, the commonly used models include matrix factorization [24], random walk [25], and graph neural networks [37]. The baseline methods in this work include node2vec [25], HIN2Vec [23], ANRL [36], MARINE [24], and ASNE [37]. A summary of these methods can be found in Table 4.4.

The details of NRL algorithms are given below:

- **node2vec** [25] learns node representations by capturing node pairs within w-hop neighborhood via *random walks* in the network. As a variant of *DeepWalk*, it considers both DFS and BFS neighborhoods.
- **HIN2Vec** [23] learns the representation of *nodes and relationships* in a network at the same time. It can jointly embed the rich information of different relationships and the overall network structure into node representations.
- **ANRL** [36] integrates *network structural proximity and node attribute affinity* into low-dimensional spaces. A neighbor enhancement autoencoder is designed to retain better similarity between data samples. It captures the node attributes as well as network structure information.
- **MARINE** [24] is a unified embedding framework for both *homogeneous and multi-relational networks* to preserve both the proximity and relationship information.
- **ASNE** [37] is a graph embedding algorithm which learns an embedding of nodes and fuses the node representations with *node attributes*. The model adopts a multilayer neural network to take advantage of strong representation and generalization ability of deep learning.

Note that node2vec and HIN2Vec cannot utilize additional node information because they only retain the network structure. This leads to the degeneracy of our framework with these two algorithms to only retain the co-occurrence association between entities. For the above algorithms, we use the same parameters: vector dimension $D = 128$, walk length $L = 20$, walk number $I = 3$, negative number $N = 5$, and max window length $W = 3$.

Figure 4.10 illustrates the performance of different NRL algorithms under the framework of AI2Vec-based OCLFP-MS. Let notation "+" label the version cooperating with manual features from experts; let "Δ" label the KS value. It

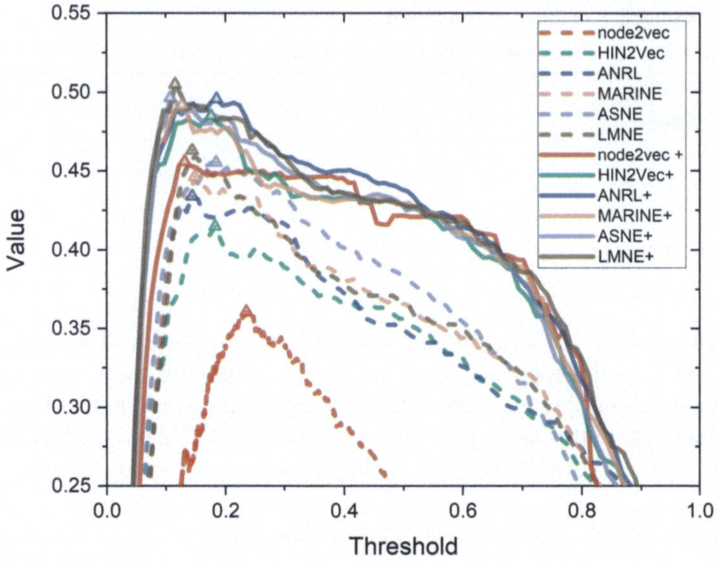

Fig. 4.10 The *KS curves* of OCLFP-MS with XGBoost and different NRL algorithms

shows that our LMNE outperforms all other NRL algorithms for fraud prediction. Notably, the pure structure-based methods, i.e., node2vec and HIN2Vec, perform significantly worse than others. It demonstrates the efficacy of the associations other than entity co-occurrence associations in fraud predicting, as well as the effectiveness of our model in leveraging more information for learning better entity representation. For the NRL algorithms that consider the attribute information, we find that the algorithms incorporating attribute information and link structure play a pivotal role in the improvement of performance. Firstly, MARINE markedly improves over the pure structure-based methods by integrating additional attribute information. Nevertheless, there is a gap between the performance of ASNE and LMNE, which indicates that learning low-dimensional representation based on matrix factorization is insufficient to fully leverage the rich network information. Secondly, we can see that ASNE and ANRL improve the performance by introducing a deep neural network model to capture more complex information. Although ASNE is better than MARINE, ASNE and ANRL neglect the side information (i.e., edge types in heterogeneous networks) during obtaining the low-dimensional representation. Thus, there is still potential for improvement. Finally, the LMNE consistently outperforms MARINE, ASNE, and ANRL by improving the process of representation learning and incorporating more completed information of the network, i.e., network structure, edge information, and node information.

For online credit loan fraud prediction, the running time is indeed an important performance metric of NRL algorithms. We evaluate the running time of obtaining the representations in the attributed entity network based on half-year loan application data. Figure 4.11 shows a comparison of the running time and fraud

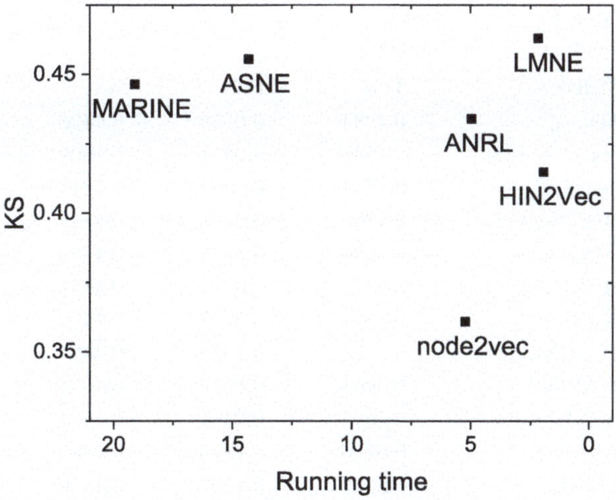

Fig. 4.11 The running time and prediction performance under our OCLFP-MS with different NRL algorithms

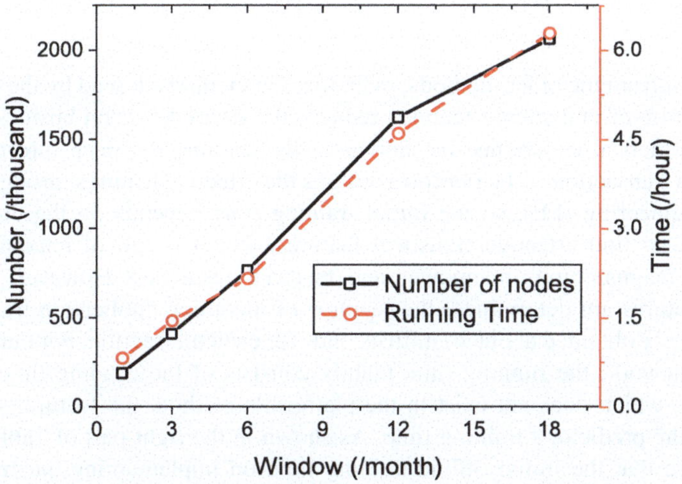

Fig. 4.12 The network scale and the running time of LMNE under different data quantity

prediction performance for different algorithms; our LMNE is superior to all other algorithms when running time and performance are considered simultaneously. Although ASNE and MARINE algorithms are similar to our algorithm in terms of prediction performance, they are really too time-consuming to efficiently obtain the learned representations. Intuitively, the running time of representation learning probably increases accordingly with the expansion of the network scale. We choose LMNE as a representative to evaluate the correlations between the running time of NRL algorithms and the sizes of networks. Based on the historical data, we divide the network into several sizes by month. Figure 4.12 depicts that the running time is approximately linear to the network scale.

Table 4.5 The performance of different NRL algorithms and their combinations with manual features designed by experts

Methods	TPR	FPR	KS	IT (\h)	APL (\s)
Manual	0.4984	0.0558	0.4426	Manual FE	0.04
node2vec [25]	0.4541	0.0931	0.3609	5.221	2.19
node2vec+	0.5504	0.0961	0.4543	Manual FE	
HIN2Vec [23]	0.4896	0.0743	0.4149	1.954	0.84
HIN2Vec+	0.5886	0.1096	0.4790	Manual FE	
ANRL [36]	0.5273	0.0931	0.4341	4.951	17,802
ANRL+	0.5397	0.0417	0.4980	Manual FE	
MARINE [24]	0.5523	0.1059	0.4464	19.132	68,952
MARINE+	0.5963	0.1029	0.4934	Manual FE	
ASNE [37]	0.5045	0.0491	0.4554	14.333	51,703
ASNE+	0.6116	0.1130	0.4986	Manual FE	
LMNE	0.5831	0.1201	0.4630	2.17	0.95
LMNE+	0.6284	0.1216	0.5067	Manual FE	

Different from the other methods, including the method adopted by the platform, the generation of embedding features reduces the dependency on business knowledge and practical experience by automatically learning the deep information in application transactions. The platform designs the effective features through manual feature engineering (FE), so the initial training time depends on the duration of manual FE to find effective statistical features, which is full of uncertainty and consumes magnanimous manpower and material resources. However, once the manual features are determined, the features of incoming application transactions can be quickly obtained about 40 milliseconds for each transaction. For our AI2Vec-based framework, the running time mainly consists of the running time of NRL algorithms, which does not exist in the platform's method, the feature generation time, and the predictor's training time. As shown in the right part of Table 4.5, we can observe that the initial offline training time on implementing the framework requires several hours of model preparation. Especially, our LMNE with 2.17 hours is superior to almost all benchmarks in terms of running time, except for HIN2Vec, whose performance is far less than ours. When a new transaction arrives, we need to update the network and the learned representation accordingly, and then generate the embedding feature for new transactions. In this work, we adopt an incremental algorithm to greatly reduce the repetitive calculations. Table 4.5 also presents the advantages of our incremental approach in terms of average processing delay compared with other algorithms. We observe that node2vec, HIN2Vec, and LMNE with the incremental approach only have a short average processing latency (less than 10 seconds) for each upcoming transaction during online testing. Otherwise, the other algorithms spend a lot of time doing global recalculation for a transaction. Compared to the platform's method, our AI2Vec-based framework sacrifices a part of processing latency to reduce the dependence on manual feature engineering and turns to automatic feature generation.

4.5.4 Algorithm Complementarity

To further exploit our framework, we deliberately implement our AI2Vec-based OCLFP-MS framework by combining the Manual method, i.e., the manually statistical features generated by the experts of platform. We apply XGBoost classifier to judge loan applications, where the parameters are determined by the grid search on the validation set. The results are shown in Table 4.5 and Fig. 4.10. The notation "+" labels the version cooperating with manual features. IT means the initial training time for offline training, and APL means the average processing latency for each upcoming transaction during online testing.

Note that the method being adopted by the platform achieves a KS value of 0.4426, which is inferior to the performance of our framework with MARINE, ASNE, and LMNE. The framework with HIN2Vec and node2vec has a lower performance than the platform's method due to the loss of partial network information. The framework with ANRL has a performance that is closer to the baseline. The embedding features generated by our AI2Vec predict frauds from the perspectives different from those manually acquired by experts. To investigate this difference, we feed the features generated by two different methods into the classification model. We learn that the manual features can bring a gain to our AI2Vec-based OCLFP-MS with all NRL algorithms. The KS value can achieve 0.5067 under AI2Vec, while the *FPR* barely increases. The result fully shows that our framework can effectively explore the potential information neglected by experts.

Since different NRL algorithms are compatible in AI2Vec, we further investigate whether the performance could be improved by combining multiple AI2Vec with different NRL algorithms. We use two different embedding features generated by different AI2Vec to observe the complementarity between them. We select three representative experiments to investigate the effects of embedding features on each other. The results are shown in Fig. 4.13. It is worth noting that the performance of the combinations with LMNE decrease slightly than that of LMNE. Although we get that the performance is improved by integrating the features obtained by node2vec and HIN2Vec, the combination is not better than the single LMNE version. It implies that the combination of different AI2Vec versions does not necessarily have a gain, and even does interfere with the single excellent AI2Vec.

4.5.5 Parameter Study

The LMNE algorithm involves a number of parameters that may affect its performance. We vary some key parameters to determine their impacts on fraud prediction. Except for the parameters being tested, all other parameters are the same as the previous experiment. The results are shown in Fig. 4.14. The figure (a), (b), (c), and (d) respectively show the model performance of different parameters, when we set α, D, L, and N at 0.1, 128, 20, and 5, respectively, in the process of testing others.

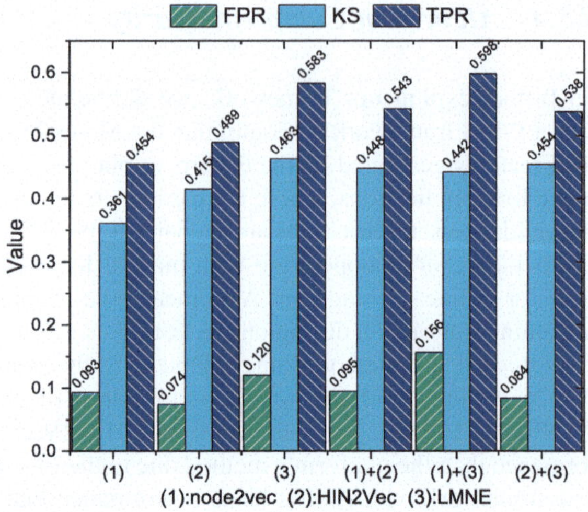

Fig. 4.13 The complementarity between the AI2Vec versions with different NRL algorithms

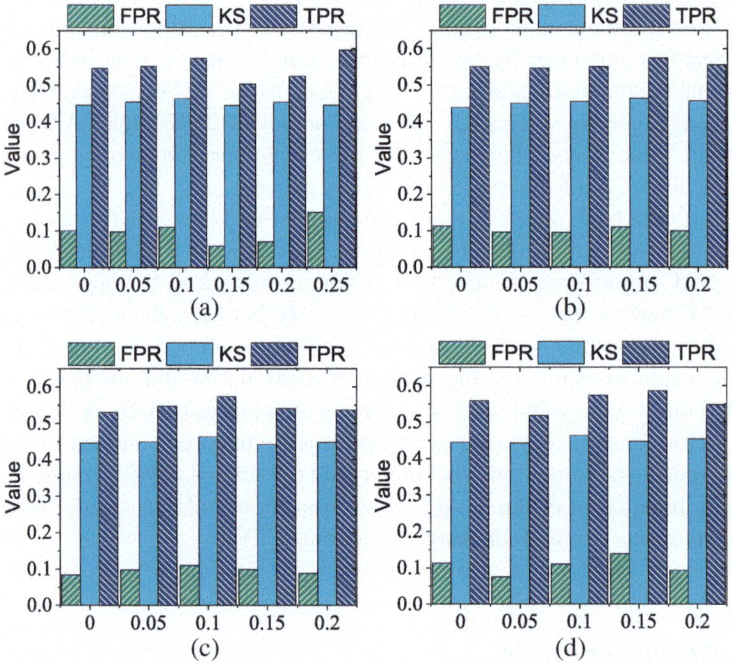

Fig. 4.14 The impacts of different parameters

Importance of Attribute Information First of all, Fig. 4.14a shows the impact of setting different values of α. Generally, a small α represents that the attribute information has less impact than the network structure in the representation space, where α is set in the range 0–1. The best performance is achieved when $\alpha = 0.1$ in this work.

Dimensionality of Vector Space Figure 4.14b shows the performance under different dimensions D. A small D is not sufficient to capture the link information, but a large D may lead to noises and cause overfitting. In our work, the best performance is achieved when D is 128. Generally, a larger network may need a larger D to capture the link information.

Length of Random Walks Figure 4.14c suggests that the performance continues to improve and reaches its maximum when L is 20. In the lending scenario, most applications are isolated, so an excessively long walk will sample the same nodes many times, which cannot reflect the real associations. Meanwhile, the more sample data lead to the more training time. In our network, we set L to be 20, since it achieves a good trade-off between time-consuming and performance.

Number of Negative Samples Figure 4.14d shows that the number of negative samples has a little impact on the performance when it is set between 3 and 7. An appropriate amount of negative samples ensures that the model distinguishes the observed samples from the unobservable samples correctly, and it's not necessary to have too many negative samples. We find that setting the number of N to 5 is a good choice.

4.6 Discussion

Our method achieves effective fraud prediction by boosting the information density of the raw loan application data. If fraudsters plan to deceive the loan, they may use the variable strategy to evade the features detected by the security model. This can cause the data distribution used during model training to be inconsistent with the actual data distribution during deployment, leading to a decrease in model performance.

Training on frequent data updates can mitigate concept drift to some extent, but it is not an ideal solution and cannot completely solve the problem. Although it can make the model better adapt to the new data distribution, the model ignores the features and patterns of the preview data while pursuing the adaptation to new data, which can affect the performance of the model.

Static learning, which trains a model over a training dataset and deploys the trained model for inference with the frozen model parameters, cannot adapt to dynamic environments over time (such as inconsistent data distributions caused by changes in fraud strategies). In online lending services, loan applications are generated in the form of data streams, which require anti-fraud systems to continuously adapt and learn over time. Our model updating scheme only focuses on adapting to the current state of the data, rather than preventing forgetting old knowledge.

Continual learning can learn from a continuous stream of information, preserving inherent knowledge while absorbing new knowledge over time, which has the potential to further alleviate the problem of concept drift. We can consider dividing the training process into continuous stages, where the model updates its learned knowledge during the training process and tries to avoid catastrophic forgetting. By introducing the continual learning scheme, we can make the model more adaptable to new data distributions, thereby reducing the impact of concept drift on model performance.

4.7 Conclusions

In this chapter, to solve one of the most critical problems for data-driven anti-fraud paradigm in online credit loan fraud prediction (OCLFP), i.e., the low density of useful applicant information associations, we have proposed a multistage data representation method, called AI2Vec. It gradually boosts the information density of associations from both aspects of reducing the information carriers and increasing useful information. Besides, we have designed a customized NRL algorithm for AI2Vec, called LMNE, to deal with the special requirements of OCLFP. Based on AI2Vec, we have proposed a new stage-to-stage online credit loan fraud prediction framework, called OCLFP-MS, whose input is the loan application and output is the judgment of fraud application. The performance of our AI2Vec-based framework OCLFP-MS is validated by implementing in a real-world dataset from an online lending platform. The experimental results show that our method outperforms the state-of-the-art methods.

There are some issues left to study: First of all, it is crucial for OCLFP to further investigate the so-called model fairness of machine learning methods. It refers to unjustified distinctions in decisions against individuals based on their membership in a certain group, e.g., regional discrimination and racism. Second, we will seek the potential of temporal information, e.g., designing NRL algorithms that update dynamically as the network evolves, to further improve the performance. Third, it is worth considering how to modify our method for use in unsupervised or weakly supervised learning models. Finally, the versatility of our method can be demonstrated by further extension to other application scenarios, e.g., online payment fraud, insurance fraudulent claims, and money laundering, where the fraud features are hidden in data associations.

References

1. J. Wang, R. Wen, C. Wu, Y. Huang, J. Xion, in *Proceedings of the WWW 2019*, San Francisco, CA, May 13–17 (2019), pp. 310–316
2. Y. Yang, Y. Xu, Y. Sun, Y. Dong, F. Wu, Y. Zhuang, IEEE Trans. Knowl. Data Eng. **33**(1), 169 (2021)
3. K. Ren, A. Malik, in *Proceedings of the ACM WSDM 2019*, Melbourne, VIC, February 11–15 (2019), pp. 510–518

4. Y. Wu, Z. Xie, S. Ji, Z. Liu, X. Zhang, C. Lin, S. Deng, J. Zhou, T. Wang, R. Beyah, IEEE Trans. Dependable Secure Comput. **20**(2), 1169 (2023)
5. T. Liang, G. Zeng, Q. Zhong, J. Chi, J. Feng, X. Ao, J. Tang, in *Proceedings of the ACM WSDM 2021*, Virtual Event, March 8–12 (2021), pp. 229–237
6. B. Hu, Z. Zhang, J. Zhou, J. Fang, Q. Jia, Y. Fang, Q. Yu, Y. Qi, in *Proceedings of the ACM CIKM 2020*, Virtual Event, October 19–23 (2020), pp. 2525–2532
7. D. Cheng, Z. Niu, Y. Tu, L. Zhang, in *Proceedings of the ICPR 2018*, Beijing, August 20–24 (2018), pp. 361–366
8. Z. ul Abi Din, H. Venugopalan, J. Park, A. Li, W. Yin, H. Mai, Y.J. Lee, S. Liu, S.T. King, in *Proceedings of the USENIX 2020*, August 12–14 (2020), pp. 1571–1588
9. X. Ruan, Z. Wu, H. Wang, S. Jajodia, IEEE Trans. Inf. Forensics Secur. **11**(1), 176 (2016)
10. Q. Zhong, Y. Liu, X. Ao, B. Hu, J. Feng, J. Tang, Q. He, in *Proceedings of the WWW 2020*, Taipei, April 20–24 (2020), pp. 785–795
11. J. Du, E. Gelenbe, C. Jiang, H. Zhang, Y. Ren, H.V. Poor, IEEE Trans. Inf. Forensics Secur. **14**(6), 1582 (2019)
12. D. Wang, Y. Qi, J. Lin, P. Cui, Q. Jia, Z. Wang, Y. Fang, Q. Yu, J. Zhou, S. Yang, in *Proceedings of the IEEE ICDM 2019*, Beijing, November 8–11 (2019), pp. 598–607
13. A. Beutel, L. Akoglu, C. Faloutsos, in *Proceedings of the ACM SIGKDD 2015*, Sydney, NSW, August 10–13 (2015), pp. 2309–2310
14. Y. Boshmaf, D. Logothetis, G. Siganos, J. Lería, J. Lorenzo, M. Ripeanu, K. Beznosov, in *Proceedings of the NDSS 2015*, San Diego, CA, February 8–11 (2015)
15. D. Cheng, Y. Zhang, F. Yang, Y. Tu, Z. Niu, L. Zhang, in *Proceedings of the ACM CIKM 2019*, Beijing, November 3–7 (2019), pp. 2547–2555
16. K. Yuan, H. Lu, X. Liao, X. Wang, in *Proceedings of the USENIX 2018*, Baltimore, MD, August 15–17 (2018), pp. 1027–1041
17. L. Tong, B. Li, C. Hajaj, C. Xiao, N. Zhang, Y. Vorobeychik, in *Proceedings of the USENIX 2019*, Santa Clara, CA, August 14–16 (2019), pp. 285–302
18. D. Babaev, M. Savchenko, A. Tuzhilin, D. Umerenkov, in *Proceedings of the ACM SIGKDD 2019*, Anchorage, AK, August 4–8 (2019), pp. 2183–2190
19. F. Tan, X. Hou, J. Zhang, Z. Wei, Z. Yan, IEEE Trans. Neural Netw. Learn. Syst. **30**(5), 1565 (2019)
20. Y. Yang, Y. Xu, C. Wang, Y. Sun, F. Wu, Y. Zhuang, M. Gu, in *Proceedings of the ACM CIKM 2019*, Beijing, November 3–7 (2019), pp. 2043–2052
21. Z. Zhang, P. Cui, W. Zhu, IEEE Trans. Knowl. Data Eng. **34**(1), 249 (2022)
22. R. Levy, T.F. Jaeger, in *Proceedings of the NIPS 2006*, Vancouver, BC, December 4–7 (2006), pp. 849–856
23. T. Fu, W. Lee, Z. Lei, in *Proceedings of the ACM CIKM 2017*, Singapore, November 06–10 (2017), pp. 1797–1806
24. M. Feng, C. Hsu, C. Li, M. Yeh, S. Lin, in *Proceedings of the WWW 2019*, San Francisco, CA, May 13–17 (2019), pp. 470–479
25. A. Grover, J. Leskovec, in *Proceedings of the ACM SIGKDD 2016*, San Francisco, CA, August 13–17 (2016), pp. 855–864
26. S. Lee, J. Kim, IEEE Trans. Dependable Secure Comput. **10**(3), 183 (2013)
27. G. Suarez-Tangil, M. Edwards, C. Peersman, G. Stringhini, A. Rashid, M.T. Whitty, IEEE Trans. Inf. Forensics Secur. **15**, 1128 (2020)
28. C. Wang, H. Zhu, IEEE Trans. Dependable Secure Comput. **19**(1), 301 (2022)
29. X. Li, S. Liu, Z. Li, X. Han, C. Shi, B. Hooi, H. Huang, X. Cheng, in *Proceedings of the AAAI 2020*, New York, NY, February 7–12 (2020), pp. 4731–4738
30. C. Liang, Z. Liu, B. Liu, J. Zhou, X. Li, S. Yang, Y. Qi, in *Proceedings of the ACM SIGIR 2019*, Paris, July 21–25 (2019), pp. 1181–1184
31. H. Lin, G. Liu, J. Wu, Y. Zuo, X. Wan, H. Li, IEEE Trans. Knowl. Data Eng. **32**(10), 1936 (2020)
32. Z. Boz, D. Günneç, S.I. Birbil, M.K. Öztürk, Appl. Artif. Intel. **32**(9–10), 939 (2018)

33. D. Wang, P. Cui, W. Zhu, in *Proceedings of the ACM SIGKDD 2016*, San Francisco, CA, August 13–17 (2016), pp. 1225–1234
34. Y. Dong, N.V. Chawla, A. Swami, in *Proceedings of the ACM SIGKDD 2017*, Halifax, NS, August 13–17 (2017), pp. 135–144
35. C. Yang, Z. Liu, D. Zhao, M. Sun, E.Y. Chang, in *Proceedings of the IJCAI 2015*, Buenos Aires, July 25–31 (2015), pp. 2111–2117
36. Z. Zhang, H. Yang, J. Bu, S. Zhou, P. Yu, J. Zhang, M. Ester, C. Wang, in *Proceedings of the IJCAI 2018*, July 13–19, 2018, Stockholm (2018), pp. 3155–3161
37. L. Liao, X. He, H. Zhang, T. Chua, IEEE Trans. Knowl. Data Eng. **30**(12), 2257 (2018)
38. B. Perozzi, R. Al-Rfou, S. Skiena, in *Proceedings of the ACM SIGKDD 2014*, New York, NY, August 24–27 (2014), pp. 701–710
39. C. Wang, Y. Dou, M. Chen, J. Chen, Z. Liu, P.S. Yu, in *Proceedings of the IEEE Big Data 2021, Orlando, FL*, December 15–18 (2021), pp. 5470–5473
40. Y. Zhu, X. Luo, Y. Li, B. Bu, K. Zhou, W. Zhang, M. Lu, in *Proceedings of the IEEE ICDM 2020*, Sorrento, November 17–20 (2020), pp. 891–899
41. C.W. Munyendo, Y. Acar, A.J. Aviv, in *Proceedings of the IEEE Symposium on Security and Privacy 2022*, San Francisco, CA, May 22–26 (2022), pp. 2304–2319
42. S. Adams, How Tala Mobile Is Using Phone Data To RevolutionizeMicrofinance. https://www.forbes.com/sites/forbestreptalks/2016/08/29/how-tala-mobile-is-using-phone-data-to-revolutionize-microfinance/. Forbes.com. Accessed 29 Aug 2016
43. G.D. Reddy, S. Saxena, E.S. Tinggi, K.R. Isabels, G. Rathnakar, U. Turar, in *2022 4th International Conference on Inventive Research in Computing Applications (ICIRCA)* (2022), pp. 1165–1171
44. P. Mukherjee, Y. Badr, in *Proceedings of the IEEE COINS 2022*, Barcelona, August 1–3 (2022), pp. 1–5
45. H. Nilforoshan, N. Shah, in *Proceedings of the IEEE DSAA 2019*, Washington, DC, October 5–8 (2019), pp. 351–363
46. B. Hu, Z. Zhang, C. Shi, J. Zhou, X. Li, Y. Qi, in *Proceedings of the AAAI 2019*, Honolulu, HI, January 27–February 1 (2019), pp. 946–953
47. Y. He, C. Wang, C. Jiang, IEEE Trans. Knowl. Data Eng. **30**(3), 460 (2018). https://doi.org/10.1109/TKDE.2017.2767599
48. Z. Liu, Y. Dou, P.S. Yu, Y. Deng, H. Peng, in *Proceedings of the ACM SIGIR 2020*, Virtual Event, July 25–30 (2020)
49. T. Mikolov, I. Sutskever, K. Chen, G.S. Corrado, J. Dean, in *Proceedings of the NIPS 2013*, Lake Tahoe, NV, December 5–8 (2013), pp. 3111–3119
50. A. van der Heijden, L. Allodi, in *Proceedings of the USENIX 2019*, Santa Clara, CA, August 14–16 (2019), pp. 1309–1326
51. D.J.N.J. Soemers, T. Brys, K. Driessens, M.H.M. Winands, A. Nowé, in *Proceedings of the AAAI 2018*, New Orleans, LA, February 2–7 (2018), pp. 7831–7836
52. J. Guo, G. Liu, Y. Zuo, J. Wu, in *Proceedings of the IEEE ICDM 2018*, Singapore, November 17–20 (2018), pp. 127–136

Chapter 5
Detecting Behavioral Anomalies for Cybersecurity

Abstract The *behavioral anomaly detection* (BAD) is expected to solve effectively a variety of security issues by detecting the deviances from normal behavioral patterns of protected agents. We propose a new graph-based behavioral modeling paradigm for BAD problem, named *behavioral identification graph* (BIG). Under BIG, the behavioral properties and their co-occurrence associations in behavioral data are modeled as the entities and relationships of graph, respectively; furthermore, both behavioral properties and events are vectorized by a devised event-property composite model, and the behavioral patterns of agents are finally represented as a multidimensional spatial distribution of behavioral properties. Consequently, for a behavior, the intensity of its behavioral anomaly can be transformed into the spatial decentrality of its behavioral agent and properties. To the best of our knowledge, this is the first work to improve behavioral modeling for anomaly detection by integrating *inter* (event-level)- and *intra* (property-level)-associations of behaviors into a unified graph and space. Our method is validated by four representative security issues, i.e., *fraud detection* in online payment services, *intrusion detection* in network communication services, *insider threat detection* in organizational information systems, and *compromise detection* in social networking services.

5.1 Introduction to Behavioral Anomaly Detection

The emerging technologies, such as 5G, mobile Internet, and Internet of Things, provide real-time recording and collecting methods of behavioral data. These behavioral data are the potential source for intelligent information services, such as recommendation services by building behavioral profiles of users [1]. Recently, the *behavioral anomaly detection* (BAD) by behavior modeling has become a promising paradigm for different cybersecurity issues, e.g., online payment anti-fraud, network system intrusion detection, advanced persistent threat, and social network compromised account detection.

The so-called BAD refers to detecting the behaviors that do not match the normal behavior pattern of agents. Actually, it generally has stricter requirements

C. Wang, H. Zhu, *Universal Behavior Computing for Security and Safety*,
https://doi.org/10.1007/978-981-97-9014-2_5

on behavioral data than those of other network services where behavior data are required to meet two conditions [2]: (1) *Behavioral distinctness* for reflecting the personality of agents (2) *Behavioral stability* for detecting the abnormality of behaviors

Unfortunately, in reality, the behavioral data available are usually of limited quality, sparse, and fragmented due to some irresistible limitations [3, 4], such as acquisition technologies, privacy protections, and business characteristics. They are indeed difficult to directly satisfy the behavioral stability and distinctness. A feasible data enhancement way is the most important prerequisite for the effectiveness of behavioral data, especially for that of BAD that needs to build high-resolution behavioral models by using low-quality behavioral data.

To enhance behavioral data, there are generally two ways, i.e., the internal and external methods [5, 6]. In the external methods, the key problem is how to utilize external knowledge from business characteristics and expert rules. In the internal methods, the key idea is to explore the associations hidden within data. In fact, external knowledge is usually limited by the utility and credibility of experience. Therefore, existing methods strive to avoid this limitation by exploiting the internal associations, e.g., learning feature interactions for user modeling [7, 8], modeling sequence interactions for fraud detection [9, 10]. This is also the key problem of this work.

To realize data self-enhancement, an inevitable stage is to extract the essential information and discard the irrelevant information in high-dimensional data. Some studies, e.g., multilayer perceptrons, aim to obtain a powerful feature extraction ability in learning the effective information hidden inside data [11] and have achieved success in many fields, such as image recognition and natural language processing [10]. Most of the above works mainly focus on extracting the deep associations between labels and behavioral features. They form the highly generalized features by iteratively combining the properties of behavioral events. Although these methods have achieved success in behavioral data enhancement, they mainly concentrate on utilizing the association of features between labels without considering other information that can further bring significant improvement.

To mine more information to enhance behavioral data, some researchers are making efforts on utilizing interaction information among behavioral events, e.g., sequence information of behavioral events [9, 12], to build more effective behavioral models. For instance, many studies exploit the users' historical behavior sequences based on Markov chains [13], convolutional neural networks [14], and recurrent neural networks [15]. The low quality of behavioral data probably leads to the failure to build a high-quality behavior model over the data enhanced by local sequential information. Further, graph-based models, such as graph neural networks [16], are proposed to capture global and local information to ensure that nodes aggregate enough information [5, 17, 18]. In general, these methods all boil down to extract the rich interaction information among behavioral events. Nevertheless, they obtain the comprehensive information at event level without considering the internal information of each behavioral event.

Different from the coarse-grained perspective on behavioral events (i.e., the inter-associations between events), a feasible way is to analyze behavioral properties from the fine-grained perspective (i.e., the intra-associations between properties) [19–21]. Along this line, we find that mining co-occurrence associations among behavioral properties can provide rich hidden information for generating high-performance behavioral models. Accordingly, we propose a new paradigm called *behavioral identification graph* (BIG) for BAD problem. In BIG, we introduce a property graph to represent the fine-grained co-occurrence associations among properties, where the nodes, edges, and structures contain property information, sequence information, and topology information in behavioral events, respectively. Cooperating with the property graph, we design an event-property composite model to leverage the co-occurrence information (represented by the property graph) and label information (attached on behavioral events) to learn latent representations of nodes/events for BAD problem. Specially, our event-property composite model includes three important aspects: (1) At property-level, the fine-grained associations are extracted by network representation learning algorithms. (2) At event-level, the coarse-grained associations are utilized to improve the representations. (3) Between property-level and event-level, the fine-grained and coarse-grained associations can be automatically balanced to improve the practical generality of the model.

To summarize, the core technique of BIG is to integrate *inter* (event-level)- and *intra* (property-level)-associations of behaviors into a unified graph and space to improve behavioral modeling for anomaly detection. The main technical contributions are as follows:

- BIG utilizes the information from multiple perspectives to leverage sufficient hidden information simultaneously, such as deep feature information, sequence context information, and global aggregation information.
- BIG represents behavioral events as a fine-grained property graph by modeling the co-occurrence associations of behavioral properties and exploits the label information in behavioral events to improve the representations.
- BIG provides an interface in the property graph to absorb external knowledge for assisting in assigning different weights to the complex internal associations of behavioral properties.

We extensively validate our BIG on various datasets in four representative scenarios with cybersecurity issues, i.e., transaction dataset in online payment services (for *fraud detection*), traffic dataset in network communication services (for *intrusion detection*), system operation dataset in organizational information systems (for *insider threat detection*), and trajectory dataset in social networking services (for *compromise detection*). The results show that our method significantly outperforms three state-of-the-art competing methods. Besides, we demonstrate BIG is compatible with various existing network representation learning algorithms.

5.2 Related Work

5.2.1 *Anomaly Detection*

Many researchers have studied anomaly detection tasks for a long time. Early explorations focused on rule-based methods [22]. But the advent of the Internet has opened the data-driven era, which makes the rule-based methods relying on expert experience relatively weak [23]. Recently, many studies reduce anomaly detection as a classification problem. Various classifier-based models are designed to achieve better anomaly detection performance. Ambusaidi et al. [24] chose special features and fed them into classifiers to achieve better accuracy and lower computational cost. Marteau [25] designed a semi-supervised approach based on random partitioning trees to solve the intrusion detection problem. Meanwhile, with the excellent performance of deep neural networks, recent works try to apply them to anomaly detection. Zong et al. [26] designed a novel deep learning framework that avoids the model falling into local optimization by combining deep autoencoding and Gaussian mixture model for anomaly detection.

5.2.2 *Behavior Modeling for Anomaly Detection*

The behavior-based method is one of the common means for anomaly detection [27], where the historical behavior records of behavioral agents are used to model the normal behavioral patterns. In general, most behavior-based methods mine the associations between behavioral features and labels, where the effective features are extracted to detect the deceptive behaviors [28–30]. For example, Pang et al. [31] proposed an anomaly detection framework to directly learn anomaly scores in an end-to-end fashion rather than the feature representations. However, the performances of these methods for anomaly detection are limited since they usually consider only the behavioral event independently.

The sequence information of historical behaviors can further help differentiate anomaly behaviors from regular ones [9, 12]. Some researchers utilized the sequence information of events to capture the different temporal influences of historical events. Somepalli et al. [32] proposed a specialized architecture by leveraging self-attention and intersample attention to enhance the data sample by relating it to other data samples. Tang et al. [14] introduced the convolutional neural network on the embedding sequence to capture the short-term contexts. The above methods capture the relationship between different events to a certain extent, but they still take no account of the interactions between events in a complete way.

Considering the powerful representation ability of graphs [19, 33], some researchers focused on using graphs for spotting anomaly behaviors. Fan et al. [34] proposed a novel deep network embedding method to detect the abnormal events that usually rarely occur in a network. Zeng et al. [18] constructed a log-based knowledge graph and then employed an embedding model to infer log semantics to help abstract behaviors. Liu et al. [5] adopted a rule-based method to map relationships among log entries that reflect the behaviors into a graph and introduced a graph embedding model to extract representations. Nevertheless, most of the existing studies pay attention to the coarse-grained information at event-level without exploiting the latent information of each behavior.

Different from the above methods, we make effects on improving the behavioral models by utilizing the fine-grained information among behavioral properties. Along this line, for fraud detection in online payment services, we have made an attempt to explore appropriate network representation learning schemes for different behavioral models to utilize fine-grained associations [20]. A major shortcoming of this preliminary work was that it failed to integrate the interaction information among behavioral events for task-driven optimization. In fact, there is no clear modeling strategy that integrates inter (event-level)- and intra (property-level)-associations of behaviors from multiple perspectives into a unified graph and space. We aim to deal with this problem by adopting a new graph-based behavioral modeling paradigm.

5.2.3 Network Representation Learning

Network representation learning can be used to leverage the rich knowledge by exploring deep associations in the network structure. The common network representation learning algorithms obtain the node/edge representations by using matrix factorization, random walk, neural networks, and their variations [35]. Fu et al. [36] proposed a novel method to learn the node representations by jointly embedding the rich information of different relationships and the overall network structure. Feng et al. [37] designed a three-way tensor factorization to preserve the proximity information for each type of relationship. Velickovic et al. [38] proposed to use the attention mechanism to sum the features of neighboring nodes, which replaced the fixed standardized operation in graph convolutional networks. Yun et al. [39] designed an end-to-end method to learn the graph structure that involves useful meta-paths and multi-hop connections. The existing methods have been proved to be effective for mining network structural information by a large number of studies.

However, the above methods have a certain upper limit on performance, because they are separated from the specific tasks. In anomaly detection, existing methods do not map the event labels to behavioral properties, which makes them difficult for graph neural networks to get the label information of the event effectively. In addition, there are some challenges to combining the network structure with specific tasks. To this end, we aim to design a module based on network representation learning that not only utilizes the event-level label information but also explores the property-level structure information.

5.3 Overview

5.3.1 Motivating Examples

Consider the following two representative behaviors, i.e., transaction behaviors in online payment services and traffic behaviors in network communication services.

A fraudster wants to bypass regulation by stealing funds from victims' accounts through false shopping in online payment services (depicted in Fig. 5.1a). One day, the fraudster accesses victims' accounts or tricks them into transferring funds themselves. They do not provide their personal accounts to receive stolen money. To evade detection, the fraudster tends to mimic normal behaviors in their daily transactions by purchasing similar items, e.g., goods that victims often browse, from fake merchants who act as money launderers, making it difficult to track the money being spent. Note that these fake accounts are often potentially associated, as they are often created by fraudsters who buy large amounts of business information and use that information to sign up. This strategy attempts to disguise the fraud transaction behavior as part of ordinary shopping activity and thus misguide analysts to flag it as normal behavior.

Malicious network traffics are one of the major threats that have jeopardized the viability of the Internet over the years. Threat actors usually persecute individuals and organizations by hiding in undetected cyberspace or compromising computers (depicted in Fig. 5.1b). Especially, when a computer is illegally controlled, threat actors generally lurk daily and ordinary users cannot feel anything abnormal during use. Once threat actors need to launch an illegal behavior, they will call the compromising computers to act, which will become the accomplice of illegal behaviors. Therefore, the traffic flows generated from undetected cyberspace or compromising computers will look the same as the legal traffic flows on the Internet, e.g., source ip, destination ip, source port, destination port, and protocol.

Behavioral Events In real applications, behavioral events that can be recorded enable analysts to gain insight into cybersecurity with data provenance. Each behavioral event should record a lot of information such as source, type, procedure, and occurrence time. As we mentioned above in the two scenarios, a behavior can

(a) Transaction behaviors (b) Traffic behaviors

Fig. 5.1 Illustrations of two representative scenarios

be regarded as an operation record in the real world or a set of state information in a computer system. Specifically, a behavioral event is defined as a set of fields (i.e., the properties of an event) in our work. As long as we can abstract out a series of fields to describe a behavioral event from real applications, e.g., payment transaction and Windows system event, our method is able to mine the property-level associations to enrich the event-level information for detecting abnormal events. Note that the raw data is usually structured or unstructured. For the former, it is easy to select fields, and even the structured data itself is already a series of fields, e.g., banks usually record a payment transaction as a structured record. For the latter, e.g., computer logs, we need to extract the fields based on business knowledge with the help of an existing tool. For example, with the help of natural language processing, we can extract the fields of a computer log, such as subject (user id), operation type (visit/send), object (website), and host (server id).

Existing Solutions To improve anomaly detection, the research community uses behavior-based modeling to mine the association between events. In essence, this improvement comes from the use of rich interactive information, e.g., sequence information, between behavioral events, so as to learn behavior preferences with more contextual semantics. Inspired by the powerful representation ability of graphs [19, 39, 40], many researchers employ graph-based methods to spot anomalies efficiently [16–18]. Most of them try to use the edges in the graph to represent the event-level dependencies, e.g., log-level operation sequence, the common connected data objects. However, existing solutions only pay attention to the coarse-grained information and do not introduce more information in essence. To address this issue, we aim to exploit the co-occurrence associations at property-level and cooperate with event-level information for behavior modeling, which provides a new graph-based behavioral modeling paradigm for anomaly detection problems.

5.3.2 Architecture

The workflow of BIG is shown in Fig. 5.2, which consists of the following components:

BIG: Behavioral Identification Graph Given the behavioral events consisting of a set of properties, the fine-grained associations between these properties provide rich information and act as the basis for generating high-performance behavior models. To represent these associations more properly, our BIG employs a property graph, which presents the fine-grained associations between behavioral properties, and introduces an event space, which expresses the behavioral events in a unified low-dimensional vector space, to reconstruct losslessly behavioral events. The event-property composite model is used to learn node/event representations containing deep association information. BIG takes the behavioral events consisting of a series of properties as input and outputs the efficient representations by constructing the property graph and learning the event-property composite model.

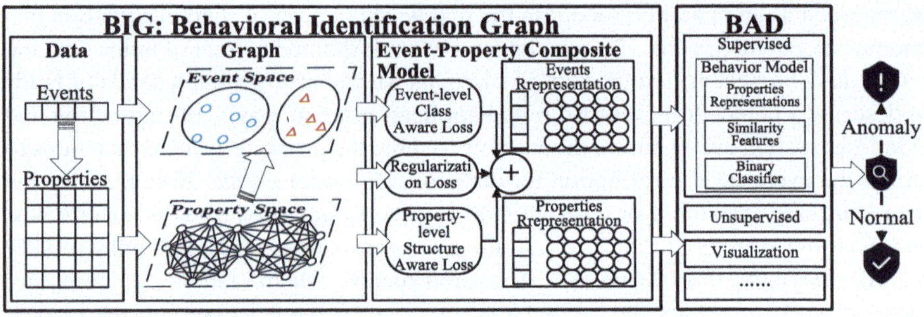

Fig. 5.2 Workflow of our method

BAD transforms each behavioral event into an event representation or a matrix of property representations based on the learned representations and then detects the anomaly behavioral events by adopting the detection models, such as the supervised or unsupervised methods. As shown in Fig. 5.2, the event-property composite model has two inputs: the property space (inter-associations between properties) and event space (intra-associations between events). Our BIG integrates these two associations of behaviors into a unified space, which is indeed a novel perspective to address anomaly detection problems based on network representation learning algorithms.

BAD: Behavioral Anomaly Detection After adopting the event-property composite model, a behavioral event can be transformed into an event representation or a matrix of property representations based on the learned node/event representations. Next, the supervised/unsupervised models can be used to detect the anomaly behaviors [1, 41]. In this work, we implement a supervised model as an example.

5.4 Behavioral Identification Graph

5.4.1 Problem Statement

We first present some related introduction of basic concepts.

Anomaly detection can identify whether an event is abnormal or not, where the result is usually a binary value. Given a set of behavioral events $B := [b_1, b_2, \cdots, b_{|B|}]$, where $|B|$ is the size of the behavioral event set. Each behavioral event b has n properties, e.g., type, place, and participant. The behavioral event is denoted as $b_i := [b_i^1, b_i^2, \cdots, b_i^n | t]$, where b_i^j is the value of the j-th property and t represents the occurrence time of the event. Our task is to detect whether a coming event b is abnormal or not according to the event's properties $[b_i^1, b_i^2, \cdots, b_i^n | t]$.

Property graph is denoted as $G = (V, E, R)$, where V is the set of nodes, each $v \in V$ representing a property value in the event; E is the set of edges between nodes, each $e \in E$ representing a relationship between two property values; and

R is the set of relationships, each $r \in R$ representing a type of relationship. Each edge is an unordered pair, denoted as $e = (u, v, r)$, and is associated with a weight $w_{u,v} > 0$, which indicates the closeness of the two nodes u and v.

Hawkes process is a typical temporal point process that models the discrete sequential events by assuming that historical events can influence the occurrence of the current event. The conditional intensity function of the Hawkes process characterizes the arrival rate of sequential events, as shown in Eq. (5.1):

$$p(t|B(t)) = p_b(t) + \int_{-\infty}^{t} \kappa(t-s)\mathrm{d}N(s), \tag{5.1}$$

where $B(t)$ is the set of historical events before current time t, $p_b(t)$ is the base probability of a specific event at time t, $\kappa(\cdot)$ is a kernel function that models the time decay effect of historical events on the current time t, and $N(\cdot)$ is the number of events occurring before a specific time.

Network representation learning can learn effective node representations for anomaly detection, where the commonly used models mainly include graph neural networks and network embedding. Given a property graph $G = (V, E, R)$, network representation learning algorithms aim to represent each node $v \in V$ into a low-dimensional space Ψ^d, i.e., learn a function $\psi(v)$: $V \rightarrow \Psi^d$, where d is the dimensionality of node representations and $d \ll |V|$. In the representation space Ψ^d, the similarity between the learned node representations is correlated with the structural closeness, e.g., the high-order proximity, and the node information in the property graph.

5.4.2 Property Graph Constructing: From Data to Graph

First of all, we need to transform the original data, i.e., all behavioral events, in B into a property graph G. In general, the values of properties in events can be simply divided into discrete and continuous values; we discretize these continuous values and then build a native graph based on the unique identifier of behavioral event. Specifically, we set the unique identifier u and each property value v of a behavioral event as nodes and define node pairs (u, v) appearing in a behavioral event as edges in the native graph. In the native graph, different behavioral event identifiers become indirect neighbors through jointly connected property value nodes when two behavioral events have the same values of properties.

In the property graph, we aim to reconstruct the behavioral patterns losslessly through the co-occurrence relationships. Hence, we design the property graph as the multiple sets of the co-occurrence relationships in the behavioral events, where the co-occurrence relationship consists of two properties and their edge in the graph. Besides, we consider all the co-occurrences between any two properties, rather than some of them, in an event and remove the unique identifier nodes to obtain the property graph. We provide an example of the property graph in online

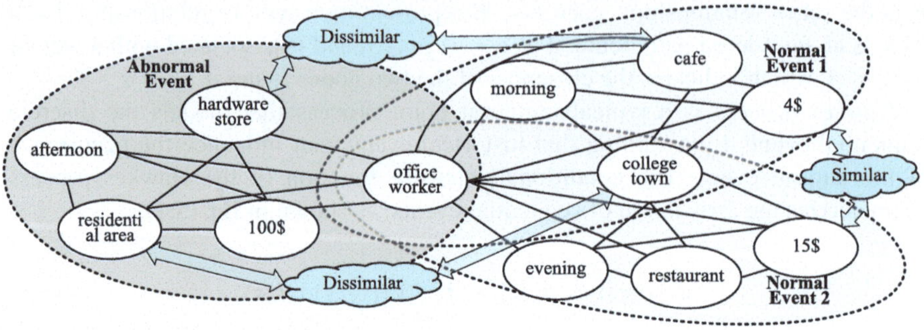

Fig. 5.3 An example of property graph in online payment services

payment services in Fig. 5.3. Our method uses the co-occurrence relationships among properties in a behavior event to build the property graph. In any behavioral event, there are edges between the nodes corresponding to any two property values. Nodes generated in frequent common behavioral events should be similar (e.g., two normal behaviors have similar payment amounts), while nodes generated in occasional anomalies should be dissimilar to those in normal behavioral events (e.g., the abnormal and normal behaviors occur in dissimilar merchant and place).

In the property graph, the relationship between any two property values is not only directly affected by the behavior events that contain these property values but also indirectly affected by other behavior events that do not contain them through the neighbor association of property nodes. We divide each behavioral event into multiple property values, and then analyze each behavioral event from the property-level, where one behavioral event can be regarded as a set of associations of multiple property values in the hidden space.

At the same time, to ensure that these scattered properties can restore behavioral events after losing the unique identifier nodes, we introduce an event space corresponding to the property space to store the unique identifier nodes. Here, the behavioral patterns, presented in the form of a graph, are fed into the subsequent module, and the subsequent event-property composite model is responsible for distinguishing the importance of different co-occurrence relationships.

When a node pair co-occurs in multiple events, we set the sum of all its initial weights in these events as its weight in the property graph, denoted by $w_{u,v}$. That is,

$$w_{u,v} = \sum_{b \in B} \sum_{(u,v) \subset b} \lambda_{u,v}, \tag{5.2}$$

where $\lambda_{u,v}$ denotes the initial weight of node pair (u, v). When determining the weight of an edge, the relationship r does not affect the value of the weight, so we simplify $e := (u, v)$ to represent it here. A larger weight of an edge means that the two nodes (corresponding to two property values) are more closely relevant in the property graph.

5.4.2.1 Initial Weight

It is worth noting that different node pairs should have different initial weights, such as the weight between "2 pm" and "New York" should be different from the weight between "2 pm" and "100 dollars." Therefore, we need to set the appropriate initial weights between any two nodes, which is obviously very time-consuming and depends on external expert knowledge. To this end, we adopt the widely used pre-trained vectors to generate initial weights, which can be used as an alternative solution in the absence of expert knowledge. The pre-trained word vectors, e.g., *fasttext*,[1] can be obtained in many common corpus data with strong generality and then be fine-tuned or directly applied.

For any nodes u and v, we first query their corresponding vectors $\mathbf{u}$ and $\mathbf{v}$ in pre-trained word vectors, and then we can calculate the initial weight $\lambda_{u,v}$ according to Eq. (5.3):

$$\lambda_{u,v} = \frac{\sum_{s=1}^{|\mathbf{u}|}(\mathbf{u}_s \times \mathbf{v}_s)}{\sqrt{\sum_{s=1}^{|\mathbf{u}|} \mathbf{u}_s^2} \times \sqrt{\sum_{s=1}^{|\mathbf{v}|} \mathbf{v}_s^2}}, \tag{5.3}$$

where $\mathbf{u}_s$ and $\mathbf{v}_s$ represent the value of the s-th dimension of the vector $\mathbf{u}$ and $\mathbf{v}$, respectively.

5.4.2.2 Weight Smooth

In this work, the weight of an edge is strongly related to the number of historical events, and the summation in a large dataset might generate a marked disproportion among different edges. This leads to one edge having a small weight between rare values in the dataset but another edge having a large weight between common values since the latter can occur frequently in many events. For instance, in the online payment scenario, the number of occurrences of the edge between a credit card number and a specific location will be significantly less than that between a payment type and the location. This large deviations in edges' weights are not conducive to reflect the real relationships between different nodes in reality. It makes the subsequent model easier to neglect those seemingly unimportant relationships, i.e., edges with small weights in the property graph. To treat each relationship fairly, we introduce a smooth function to reduce the gap in the weights of edges and transform the weight $w_{u,v}$ to an interval [0, 1]. The smooth function $S(\cdot)$ is shown in Eq. (5.4):

$$S(w_{u,v}) = \frac{1}{1 + \exp(-\ln(\alpha \times w_{u,v}) + \theta)}, \tag{5.4}$$

[1] The pre-trained word vector is available via https://fasttext.cc/docs/en/english-vectors.

where the parameters α and θ jointly affect the smoothness of weights. Parameter α mainly controls the changing degree of weights when the value of $w_{u,v}$ is small, and parameter θ plays a more important role when w is relatively large. In addition, parameter α is positively correlated with the transformed weight $S(w_{u,v})$, while parameter θ has a negative correlation with the transformed weight $S(w_{u,v})$. In general, we need to set α to be a small value to ensure that the ratio of two edges' weights becomes smaller and set θ to be a big value for ensuring that the ratio of two edges' weight keeps as constant as possible when $w_{u,v}$ is small.

Our smooth function encourages the tremendous weight values to quickly decay to reduce their influence and retains the effect caused by subtle weight differences to distinguish truly unimportant relationships in the property graph as far as possible. In this work, we set α to 1.8 and θ to 5, and the adjustment procedures will be introduced later in Sect. 5.5.4. This strategy encourages the gap to be reduced moderately when the weight $w_{u,v}$ is tremendous and the gap to be changed as little as possible when it is a small value.

5.4.2.3 Time Effects of Hawkes Process

In reality, the property graph is formed by adding nodes and edges sequentially, i.e., the structure of the property graph constantly changes as behavioral events occur. During this process, an edge is composed of node pairs in different events, which are not formed simultaneously. Obviously, the sequential graph generation process will bring more information than taking the network as an observed snapshot, which is the accumulation of all same edges. Therefore, in order to learn the rich information in the sequential graph generation process, we are inspired by the Hawkes process and design a function to treat edges generated in different periods differently.

Similarly, we assume that discrete sequential events, i.e., historical events before current time t, can influence the graph structure at the current moment. The formation sequence of an edge in the property graph is indeed a counting process in the historical events. Thus, we naturally apply the Hawkes process to model the formation sequence of the edge $e = (u, v, r)$, and the weight of e in time t can be formulated as Eq. (5.5):

$$w_{u,v} = \lambda_{u,v} + \sum_{t_h < t} \lambda_{u,v} \cdot \kappa(t - t_h), \tag{5.5}$$

where $\kappa(t - t_h) := \exp(-\beta(t - t_h))$, t_h is the occur time of the edge (u, v) prior to time t in the graph formation sequence. The kernel function $\kappa(\cdot)$ denotes the time decay effect and is often be expressed in the form of an exponential function. It is worth noting that parameter β can influence the time decay effect with different intensities.

The time effect function shows that the structure of property graph does not only depend on the count of node pairs in all events but is also influenced by the events with the time decay effect. That is, the weight of an edge is influenced by the recent events at a higher intensity, while the earlier events occurring in history have less impact on the weight.

To sum up, the final weight of an edge in the property graph is calculated by:

$$w_{u,v} = \lambda_{u,v} + S\left(\sum\nolimits_{t_h<t} \lambda_{u,v} \cdot \kappa(t - t_h)\right).$$

The construction of the property graph acts as the initialization of our method and has a limited impact on enhancing the behavioral data. The property graph provides a way to model the relationships between property values by representing the interdependencies with edges. Given a set of behavioral events, the existing methods are insufficient to mine the behavioral pattern deeply. For example, when some users have low-frequency behavioral events, their existing behavioral data is possibly too sparse to effectively serve as a data basis for behavior modeling. Next, we utilize and improve network representation learning algorithms to extract more information.

5.4.3 Event-Property Composite Model

Network representation learning (NRL) algorithms have been widely adopted to mine deeper information from complex networks [42]. In practice, although NRL algorithms can obtain the representations of nodes or edges through a variety of methods, e.g., random walk [40], matrix factorization [37], and graph neural networks [39], the topological structure of a network is the vital task commonly concerned by all above algorithms [35, 42]. We propose an event-property composite model $EPC(G, B; \phi_g)$ to mine deep information, where G is the property graph, B is the set of behavioral events, and ϕ_g denotes the parameters of the event-property composite model. It could not only learn the deep information contained in the network structure but also introduce event-level information to improve the performance of anomaly detection.

We have the objective function L for the event-property composite model as shown in Eq. (5.6):

$$L = L_s + L_c + L_{reg}, \tag{5.6}$$

where L has three components: the structure-aware loss L_s at property-level, the class-aware loss L_c at event-level, and the regularization loss L_{reg}. The detailed learning process of representation is illustrated in Fig. 5.4. Finally, we parameterize the final objective function for efficient representations as follows:

$$\phi_g^* = \text{argmin}(L), \tag{5.7}$$

where ϕ_g^* refers to the optimal parameters of the event-property composite model.

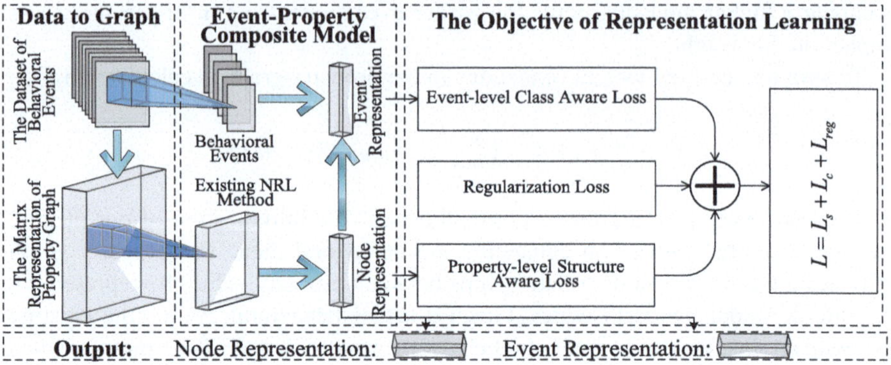

Fig. 5.4 The structure of BIG

5.4.3.1 Property-Level Structure-Aware Loss

We aim to extract the fine-grained associations in behavioral property values. In this part, we introduce existing NRL algorithms to learn node representations. An effective representation preserving comprehensive and deep relationships can act as the important means of relational data enhancement [20] to make up for low-quality behavioral data. Our BIG is compatible with most existing NRL algorithms. We take two representative cases to illustrate this point.

For a representative network embedding method, i.e., MARINE [37], the negative logarithm of its objective function with respect to the network structure can be formulated as:

$$L_s = L_{marine} = \sum\nolimits_{e \in E, e' \notin E} -\log \sigma(s(e) - s(e')), \tag{5.8}$$

where $e = (u, v, r)$ is an observed edge and $e' = (u', v', r')$ is a non-existing edge in the property graph G, and r is the relationship of edge; to incorporate both proximity and relationship information between two connected nodes, $s(\cdot)$ is a customized scoring function, and $\sigma(\cdot)$ is the sigmoid function that translates the score into a probability between 0 and 1.

For a classical graph neural network algorithm, i.e., GNN [43], the optimization process can be formulated as:

$$L_s = L_{gnn} = \sum\nolimits_{u \in V_l} (x_u - o_u), \tag{5.9}$$

where V_l represents the set of all supervised nodes in the graph, x_u is the label of the node u, and o_u is the output label of node u predicted by GNN. In detail, GNN learns a local output function $\Theta(\cdot)$ to get the set of output labels $\mathbf{O} = \Theta(\mathbf{H}, \mathbf{X}_N)$, where $\mathbf{H}$ is the set of nodes' representations and $\mathbf{X}_N$ is the set of nodes' attribute representations. In our property graph, a node (corresponding to a property value of the behavioral event) might simultaneously appear in normal or abnormal events, so we cannot simply assign the labels of behavioral events to nodes. One of the main goals of the existing NRL algorithms is to preserve the network structure, so we view node types as the labels of nodes, with the expectation that algorithms can more effectively distinguish the different relationship types in the property graph. It is worth noting that some GNN-based models do not consider the weight of edges when learning the graph structure, so a part of information in the behavioral pattern may be lost, which will be further discussed in Sect. 5.5.3.

5.4.3.2 Event-Level Class-Aware Loss

We aim to extract the coarse-grained associations in behavioral events. In this part, we extract event representations from node representations corresponding to the property values in a behavioral event. Some studies [7] found that property interactions are very useful for behavior modeling. So we design a naive event-level extractor to capture the first-order property interactions in the property space. At the same time, event representations in the event space can be fed into NRL algorithms to capture more effective node representations synchronously. We formulate the representation of node u as $\psi(u)$, where $\psi(\cdot)$ is the mapping function from nodes in the property graph into the low-dimensional representation space. In a behavioral event b_i, if node u corresponds to the j-th property value b_i^j, then the representation of b_i^j can be denoted as $\mathbf{b}_i^j = \psi(u)$.

The event-level extractor is formulated as:

$$\mathbf{b}_i = \frac{1}{n} \sum\nolimits_{j=0}^{n} \varepsilon_j \cdot \mathbf{b}_i^j, \tag{5.10}$$

where $\mathbf{b}_i$ denotes the event representation of the behavioral event $b_i \in B$ and ε_j is the weight for j-th property value in the behavioral event b_i. At the event-level, we aim to classify behavioral events as far apart as possible in the low-dimensional representation space. For example, in anomaly detection tasks, two groups formed by normal and abnormal behaviors should be as far apart as possible. To quantify the distance between two event classes, we introduce the cluster center $\mathbf{c}$ to represent all events in the special class c and denote it as the representatives of the class. Hence, the cluster center $\mathbf{c}$ can be formulated as:

$$\mathbf{c} = \frac{1}{N_c} \sum\nolimits_{b_i \in B_c} \mathbf{b}_i, \tag{5.11}$$

where B_c is the set of all events belonging to class c and N_c is the count of all events in B_c.

Give any two event classes c_i and c_j, we calculate their distance as:

$$D(c_i, c_j) = \|\mathbf{c}_i - \mathbf{c}_j\| \tag{5.12}$$

by using the Euclidean distance which is sensitive to numerical differences and is suitable for distinguishing different event classes. We further extend the case to multiple event classes and then get the event-level class-aware loss in the property graph as shown in Eq. (5.13):

$$L_c = \frac{-2}{|C| \times (|C| - 1)} \sum\nolimits_{c_i \in C} \sum\nolimits_{c_j \in C \setminus \{c_i\}} D(c_i, c_j), \tag{5.13}$$

where C is the set of all event classes and $| \cdot |$ is the number of elements in a set.

5.4.3.3 Regularization Loss

Since the event-level class-aware loss L_c is negatively correlated with weights $\varepsilon := [\varepsilon_1, \varepsilon_2, \cdots, \varepsilon_n]$, ε will tend to be updated to a set of larger values to minimize the overall loss. Therefore, we consider a regularization loss L_{res} to penalize weights with lasso regularization, which is shown in Eq. (5.14):

$$L_{reg} = \sum\nolimits_{i=1}^{n} |\varepsilon_i|. \tag{5.14}$$

In general, lasso regularization can control the complexity of the model and reduce the risk of overfitting. In addition, it plays a key role in our model for balancing the event-level and property-level losses. Note that different NRL algorithms have different loss functions, e.g., probability and distance. The above loss functions might not be on the same order of magnitude. By learning ε automatically, our model makes L_s and L_c in the same order of magnitude as far as possible, to avoid the model only focusing on one-level loss.

5.4.4 Behavioral Anomaly Detection

Node representations learned by NRL algorithms are beneficial for downstream anomaly detection tasks. Given the mapping function of node representations $\psi(\cdot)$, we can obtain the corresponding representation of any property values in a behavioral event, e.g., $\mathbf{b}_i^j = \psi(u)$. So a behavioral event, $b_i = [b_i^1, b_i^2, \cdots, b_i^n | t]$, can be transformed to a matrix $\mathbf{b}_i$ of size $|\mathbf{b}_i^j| \times n$. Time t is not involved because time information has been included in node representations. Next, we can feed the above matrix $\mathbf{b}_i := [\mathbf{b}_i^1, \mathbf{b}_i^2, \cdots, \mathbf{b}_i^n]$ into the subsequent task to realize anomaly detection, or example, feeding it into a classifier to achieve binary classification, applying the clustering algorithm to detect abnormal gang nodes, or providing certain explainability by the visualization technology.

In this part, we focus on utilizing learned representations to improve the performance of supervised anomaly detection tasks. We simplify the anomaly detection task to a binary classification problem to find abnormal behavioral events from the normal ones in supervised models. Abnormal behaviors tend to deviate from normal behaviors on some property values, so the association of any two property values in a behavioral event can sensitively identify behavioral deviations. The similarity between two learned representations is one of the good means to evaluate the degree of the association between two property values. So we calculate the similarities of any two representations as new features based on the matrix $\mathbf{b}_i$. Specifically, we can get $n(n-1)/2$ similarities as the constructed features f_i to reflect the association of any two property values in a behavioral event b_i.

The similarity is formalized as follows: given a behavioral event's matrix $\mathbf{b}_i = [\mathbf{b}_i^1, \mathbf{b}_i^2, \cdots, \mathbf{b}_i^n]$, for any two representations $\mathbf{b}_i^j$ and $\mathbf{b}_i^k$, we get:

$$sim(\mathbf{b}_i^j, \mathbf{b}_i^k) = \frac{\mathbf{b}_i^j \cdot \mathbf{b}_i^k}{||\mathbf{b}_i^j|| \times ||\mathbf{b}_i^k||} \tag{5.15}$$

by using the cosine similarity. It is worthy to note that our model does not establish an independent model for each behavioral agent, so we do not emphasize numerical differences to distinguish behavioral agents. We adopt the cosine similarity to pay more attention to the difference between two representations in direction, rather than in numerical value.

To reflect the internal association of a behavioral event more fully, we further calculate the average and variance of the constructed features f_i and add them into f_i. Let $sim\_avg(\mathbf{b}_i)$ and $sim\_var(\mathbf{b}_i)$ denote the average and variance of a behavioral event b_i. They can be calculated as follows:

$$\begin{aligned} sim\_avg(\mathbf{b}_i) &= \frac{2}{n(n-1)} \sum\nolimits_{j=1}^{n-1} \sum\nolimits_{k=j+1}^{n} sim(\mathbf{b}_i^j, \mathbf{b}_i^k), \\ sim\_var(\mathbf{b}_i) &= \frac{2}{n(n-1)} \sum\nolimits_{j=1}^{n-1} \sum\nolimits_{k=j+1}^{n} v(j,k), \end{aligned} \tag{5.16}$$

where $v(j,k) = \left(sim(\mathbf{b}_i^j, \mathbf{b}_i^k) - sim\_avg(\mathbf{b}_i)\right)^2$.

Taking users' privacy or the difficulty of data collection into consideration, some behavioral events might have missing properties in practical applications. For this case, we do not consider node representations corresponding to those missing properties. Further, when calculating the average and variance, we set n to be $n-k$ and set k to be the number of missing properties so that these missing items are omitted.

In the real scenario, we choose a time point t, construct our BIG model by using behavioral events that occurred before t, and train a classifier model based on the learned representations. For an incoming behavioral event after t, we calculate its similarities and feed them into the trained classifier model to get a discriminant result corresponding to the behavioral event. The process of building an anomaly detection model is described in Algorithm 4.

Algorithm 4: The procedure of anomaly detection

Input: The set of behavioral events B, the time point t
Output: The set of judgment results R
1 Divide B into training set B_{train} and test set B_{test} according to the time point t;
2 Build the property graph based on B_{train};
3 Obtain the mapping function $\psi(\cdot)$ by applying the event-property composite model $EPC(G, B_{train})$;
4 **foreach** $b_i \in B_{train}$ **do**
5   Initialize constructed features f_i to Empty;
6   Transform b_i into $\mathbf{b}_i$ based on $\psi(\cdot)$;
7   **foreach** $\mathbf{b}_i^j \in \mathbf{b}_i$ **do**
8   **foreach** $\mathbf{b}_i^k \in \mathbf{b}_i$ **do**
9   Get $sim(\mathbf{b}_i^j, \mathbf{b}_i^k)$ using Eq. (5.14);
10             $f_i = f_i.append(sim(\mathbf{b}_i^j, \mathbf{b}_i^k))$;
11   **end**
12   **end**
13   Get $sim\_avg(\mathbf{b}_i)$ and $sim\_var(\mathbf{b}_i)$ using Eq. (5.15);
14     $f_i = f_i.append(sim\_avg(\mathbf{b}_i))$;
15     $f_i = f_i.append(sim\_var(\mathbf{b}_i))$;
16 **end**
17 Train a classifier model $CLF(f_i; \phi_c)$ based on all constructed features f_i and extract validation samples from training set to determine the optimal parameters ϕ_c^*;
18 Initialize judgment results R to Empty;
19 **foreach** $b_i \in B_{test}$ **do**
20   Repeat step $5 - 13$;
21   Get the classification result $r_i = CLF(f_i; \phi_c^*)$;
22     $R = R.append(r_i)$;
23 **end**
24 Return judgment results R.

5.5 Experiments

We validate the proposed method by empirically evaluating real-world data. First, we describe the setup of our experiments. Then, we demonstrate the effectiveness of our BIG on supervised anomaly detection tasks of three aspects:

Evaluating BIG shows that our BIG can outperform baseline methods in the standardized partial AUC and true positive rate of anomaly detection.

Validating BIG illustrates how much gain our BIG can bring in terms of data enhancement and how excellent performance our BIG can achieve by cooperating with various existing models.

Tuning BIG discusses how to tune parameters of the property graph to affect the graph generation and how to select components of BIG to improve the overall performance.

5.5.1 Experimental Setup

5.5.1.1 Datasets

To evaluate the performance of our detection methods, we select four representative security issues that can be solved by behavioral anomaly detection: fraud detection in online payment services, intrusion detection in network communication services, insider threat detection in organizational information systems, and compromise detection in social networking services. Specifically, we use the transaction behavior dataset (a real-world online banking payment transaction dataset from one of the biggest commercial banks in China [20]), the traffic behavior datasets (a network traffic dataset[2] from Kyoto university's honeypot [44] and a darknet traffic dataset[3] generated by the Canadian Institute for Cybersecurity [45]), the system behavior dataset (an insider threat test dataset[4] from Carnegie Mellon University [46]), and the trajectory behavior dataset (an online social geographic check-ins dataset[5] from a popular location-based social network, Gowalla [47]). Different from the above four datasets, the initial Gowalla dataset only contains normal check-in records of users, that is, there are no abnormal behaviors. In order to simulate abnormal events in records, we first treat three consecutive check-in records of a user as an event. Then we randomly extract a part of the events and randomly choose a value of "user" property to replace the normal value. Thus, several abnormal behavioral events can be constructed.

5.5.1.2 Implementation

In our model, we build the property graph to represent these original data and then apply node representations obtained by the event-property composite model to detection models.

Data processing is a critical step in data mining. The real-world data from different platforms are typically noisy, redundant, and inconsistent. In this work, one of the main challenges is how to transform the raw data into the property graph in our behavioral identification graph (BIG), especially how to transform those properties

---

[2] **Kyoto**: http://www.takakura.com/Kyoto_data.

[3] **Darknet**: https://www.unb.ca/cic/datasets/darknet2020.html.

[4] **CERT**: https://kilthub.cmu.edu/articles/dataset/Insider_Threat_Test_Dataset/12841247.

[5] **Gowalla**: http://snap.stanford.edu/data/loc-Gowalla.html.

of continuous values into nodes. For the discrete properties, we naturally retain this discreteness, treat each independent value as a node in BIG, and then mine the associations between them. Due to the difference in meaning and scale, it is difficult to propose a specific method suitable for all continuous property values. Next, we will describe how to deal with different continuous property values in each dataset.

For the Banking dataset, two kinds of properties need to be transformed, i.e., "time" and "amount." We discretize the value of "time" inspired by Zhao et al. [48]. The time of a day can be divided into four intervals. We set four intervals of an hour: $[0, 3)$, $[6, 11)$, $[15, 24)$, and $[3, 6) \cup [11, 15)$, according to the time distribution of the dataset. We further divide the value of "time" into eight unique values by distinguishing whether it is a weekday. For the "amount" property, we discretize it into four different values according to the following intervals: $[0, 60)$, $[60, 300)$, $[300, 3600)$, and $[3600, +\infty)$.

For the Kyoto, Darknet, and CICDDoS datasets, the continuous values of properties mainly consist of two types, i.e., the count number and the rate. For the former, e.g., "destination bytes," "count," and "dst host srv count," we discretize them with a unit of count. For the latter, we discretize them with an interval of 1%.

For the Gowalla dataset, the properties to be processed include "check-in time," "longitude," and "latitude." For the former, we divide the value of "check-in time" into 24 unique values, spaced one hour apart. For the geographic information, we adopt the geohash [49] to convert two-dimensional longitude and latitude into a string that uniquely corresponds to a rectangular area on the map. In general, the length of geohash code represents the regions of different sizes, where the longer the code is, the smaller the rectangle it represents.

For the CERT dataset, the properties "weekday" and "hour" are extracted from the "date" in the dataset. Different operations are recorded in different files; we generate the property "type" according to the file division. Each operation will lead to different active states, and we regard them as the property "state," such as login, open pdf, connect, and link URL.

It is noteworthy that the k-fold cross validation may cause the time-crossing problem, so it is not suitable for the experiments in this work. Time-crossing means that the information that has not yet occurred is utilized when the samples are tested. Different from the k-fold cross validation, we select a specific time point on the time sequence, divide the data before the selected time point into the training set, and divide the data after the selected time point into the testing set. The statistics of dataset are listed in Table 5.1. N# means Number, and R# means Ratio. Dividing the dataset in time sequence avoids the time-crossing problem and is more in line with the real application scenario than randomly selecting the verification samples.

We further count the number of nodes and edges in BIG, which are detailed in Table 5.1. Note that the number of node types in Gowalla is more than the number of its properties because we treat three consecutive check-in records of a user as one event.

The experiments are conducted under Windows 10 Pro operating system with Intel i7-9700K @ 3.6 GHz CPU, 16 GB RAM, and GeForce RTX 2080 Ti Graphics card.

Table 5.1 The characteristic information of five datasets

Dataset	Training			Testing			Property graph		
	N# of events	N# of positive	R# of positive	N# of events	N# of positive	R# of positive	N# of nodes	N# of edges	N# of node types
Banking	2,434,210	40,393	1.66%	1,028,437	24,898	2.42%	94,848	2,371,314	8
Kyoto	514,631	287,425	55.85%	134,665	90,408	67.14%	1880	56,358	8
Darknet	99,071	17,116	17.28%	42,459	7195	16.95%	8899	237,053	9
CERT	7,835,885	4298	0.055	917,805	2001	0.218%	2471	622,024	6
Gowalla	6234	622	9.98%	4996	442	8.85%	5333	116,775	7

5.5.1.3 Evaluation Metric

In binary anomaly detection tasks, the AUC (area under the *ROC curve*) is a widely used metric [50]. However, according to the industry requirement, the detection models should pay more attention to improving the TPR *(true positive rate)* under the low FPR *(false positive rate)* to avoid disturbing the normal behaviors as few as possible. For the insidious anomalies, the attackers can bypass the regulatory means by generating similar anomalies in batches, which will cause a lot of losses. So an achieved TPR with an FPR higher than a special threshold, maxfpr, makes no sense in this work. We only focus on the meaningful part of the *ROC curve* without considering the whole AUC. Hence, the standardized partial AUC ($\text{SPAUC}_{\text{FPR}\leq\text{maxfpr}}$) is adopted to evaluate our model [9] and denotes the area of the head of the *ROC curve* when FPR $\leq$ maxfpr. The $\text{SPAUC}_{\text{FPR}\leq\text{maxfpr}}$ is calculated as follows:

$$\text{SPAUC}_{\text{FPR}\leq\text{maxfpr}} = \frac{1}{2}\left(1 + \frac{\text{AUC}_{\text{FPR}\leq\text{maxfpr}} - \text{minarea}}{\text{maxarea} - \text{minarea}}\right),$$

where maxarea = maxfpr and minarea = $(\text{maxfpr}^2)/2$.

Similar to AUC, the larger $\text{SPAUC}_{\text{FPR}\leq\text{maxfpr}}$ achieved by the model shows that the model has the better classification performance. In different real-world scenarios, we should pay attention to $\text{SPAUC}_{\text{FPR}\leq\text{maxfpr}}$ under different maxfpr according to the de facto standards. While the maxfpr is very low, $\text{SPAUC}_{\text{FPR}\leq\text{maxfpr}}$ has a small range of variation since all methods are difficult to detect the anomalies. Hence, in this work, we use $\text{SPAUC}_{\text{FPR}\leq 0.002}$, $\text{SPAUC}_{\text{FPR}\leq 0.01}$, $\text{SPAUC}_{\text{FPR}\leq 0.05}$, $\text{SPAUC}_{\text{FPR}\leq 0.2}$, and $\text{SPAUC}_{\text{FPR}\leq 0.3}$ for the Banking, Kyoto, Darknet, CERT, and Gowalla datasets, respectively. In addition, we also use precision, recall (TPR), disturbance (FPR), and F1-score to comprehensively evaluate our model.

5.5.2 Contrast Experiments

We compare the performance of our proposed method with four representative classifiers, including logistic regression (**LR**), random forest (**RF**), gradient boosting decision tree (**GBDT**), and extreme gradient boosting (**XGB**). In addition, we also choose three popular anomaly detection methods as baselines, including the following:

- **DevNet** [31] synthesizes neural networks, Gaussian prior, and zscore-based deviation loss to perform data-efficient and effective learning of the anomaly scores.
- **DAGMM** [26] joints autoencoders and Gaussian mixture model to generate an efficient representation and reconstruction error for anomaly detection on high-dimensional data.

- **SAINT** [32] projects all features into a dense vector space and design self-attention, intersample attention, and contrastive pre-training to boost performance in the semi-supervised way.

The SPAUC and TPR performance of our method and competing methods are shown in Table 5.2. Here, TPR is obtained when FPR is set to 0.002, 0.01, 0.05, 0.2, and 0.3, respectively. Our method performs best on five datasets in the respective SPAUC and TPR performance. It achieves satisfactory performance (more than 94% TPR under the given FPR) in all datasets, which shows that our method can effectively identify anomalies. On Kyoto, although most all methods achieve high and close TPR, our method has higher SPAUC, which means that it will achieve better performance at lower FPR. On Darknet, it has 1.725% higher SPAUC and 2.412% higher TPR than the next best baseline. Similarly, on Gowalla, it has 0.860% higher SPAUC and 5.803% higher TPR than the next best baseline. Compared with the popular baselines, our method achieves significant improvement in most cases. The result shows that BIG efficiently leverages the hidden information from multiple perspectives, resulting in high-performance anomaly detection, i.e., substantially high SPAUC and TPR.

To intuitively show the detection performance of our method, we sampled five testing users who had abnormal operations in CERT and report the detection results in Table 5.3. #D is the number of detected anomalies and #T is the number of total anomalies. Although our method does not detect all anomalies, it achieves superior performance, that is, more than 80% of anomalies can be captured with less disturbance. Experiments show that our method can still effectively identify abnormal events by mining the property-level associations in behavioral data in the face of concealed behaviors.

Table 5.2 Performance of our method and competing methods

Method	Banking		Kyoto		Darknet		CERT		Gowalla	
	SPAUC FPR≤0.002	TPR	SPAUC FPR≤0.01	TPR	SPAUC FPR≤0.05	TPR	SPAUC FPR≤0.2	TPR	SPAUC FPR≤0.3	TPR
LR	0.520	0.081	0.641	0.466	0.561	0.283	0.685	0.590	0.485	0.269
RF	0.929	0.957	0.973	0.999	0.828	0.862	0.750	0.818	0.609	0.688
GBDT	0.935	0.961	0.983	0.999	0.868	0.917	0.788	0.883	0.655	0.781
XGB	0.927	0.956	0.953	0.999	0.766	0.770	0.755	0.743	0.593	0.681
DevNet	0.503	0.022	0.843	0.950	0.640	0.508	0.806	0.933	0.634	0.655
DAGMM	0.528	0.072	0.805	0.860	0.686	0.648	0.571	0.292	0.556	0.524
SAINT	0.671	0.629	0.890	0.998	0.778	0.903	0.788	0.769	0.824	0.935
Ours	**0.942**	**0.965**	**0.987**	**0.999**	**0.885**	**0.941**	**0.823**	**0.944**	**0.833**	**0.993**

Table 5.3 The detailed results on five selected users (e.g., RGG0064) in CERT

User id	#D/#T	TPR	FPR
RGG0064	12/12	1.0	0.02809
IKR0401	23/28	0.82142	0.05907
IUB0565	155/190	0.81578	0.09870
KRL0501	156/168	0.92857	0.07224
AKR0057	192/238	0.80672	0.10366

5.5.3 *Module Comparison*

5.5.3.1 Gains of Data Enhancement

Note that our BIG is a general method that can adopt various advanced NRL algorithms. To fully show the gain of our BIG, we adopt two typical types of NRL algorithms, i.e., network embedding algorithms (HIN2Vec, Marine) and graph neural network algorithms (GAT, GTN), respectively.

In view of the above algorithms, we uniformly set the dimension of learned representations to be 128 and adopt the parameters suggested by their authors in the codes for other parameters. For succinctness of the experimental description, we only choose extreme gradient boosting (XGB) classifier as a representative detection model. Meanwhile, we feed the original properties into the same detection model to prove that our BIG is equipped to deliver huge gains of data enhancement. The *ROC curves* of experiments are depicted in Fig. 5.5. "ORI" means the version that feeds original properties into extreme gradient boosting classifier to detect abnormal events. From the results, we have the following findings.

Firstly, we find that our BIG has brought significant gains compared with the single detection model based on original properties in all tasks. In addition, we also observe that all NRL algorithms play a good role and have close performances on the same dataset. These results show that our BIG can work well with various advanced existing algorithms.

Secondly, on the Banking (Darknet) dataset, our BIG brings significant gains to the detection model when the FPR is greater than 0.025% (1.25%). In terms of TPR, our BIG has 6% (15%) higher TPR than the single detection model when the FPR is greater than 0.2% (5%). It is noting that the samples are unbalanced and the abnormal samples are only about 2% (17%). The imbalance leads to the single detection model cannot effectively distinguish the abnormal and normal behaviors. Thus, our BIG shows effectiveness in mining deeply the property-level associations to enhance behavioral data.

Thirdly, on the Kyoto dataset, there is a similar improvement to that on the Banking dataset. It differs in that the gain of BIG decreases with the increase of FPR. This phenomenon results from the equilibrium ratio, about 1, of normal and abnormal samples in the Kyoto dataset. Therefore, with the increase of FPR, even the detection model based on original properties can still effectively detect abnormal behaviors. Although our BIG can get the maximum gain at a very low FPR, e.g.,

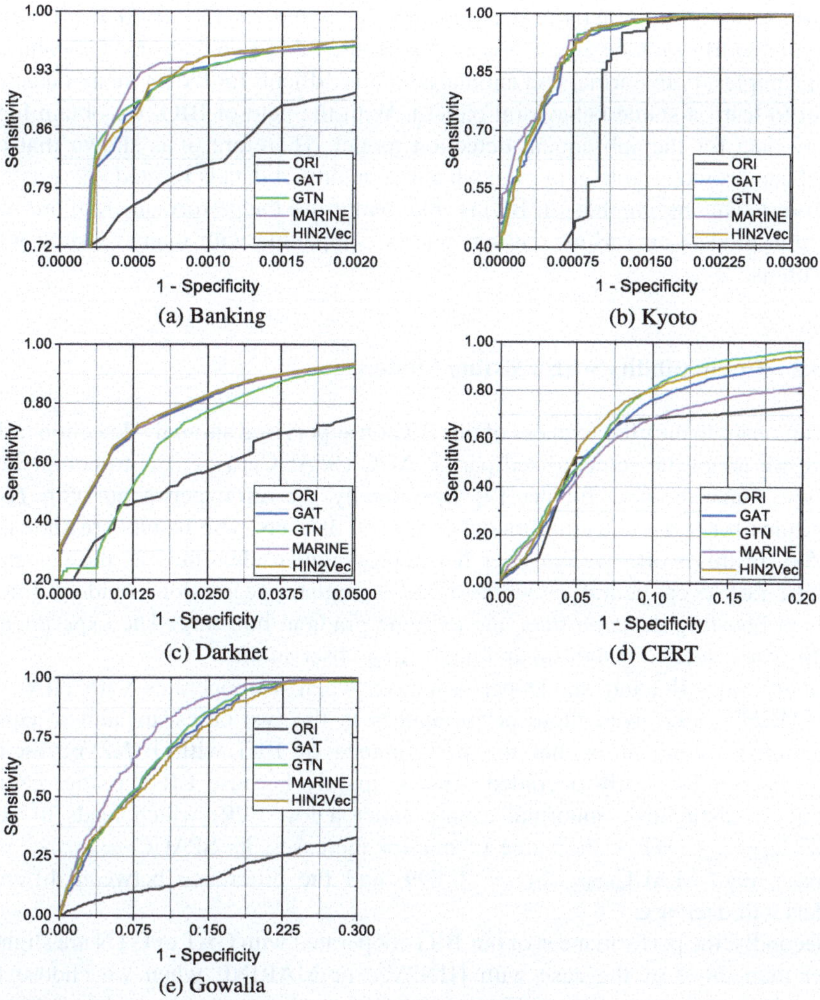

Fig. 5.5 The gains of BIG to the detection model based on original properties on 5 (datasets) × 5 (algorithms) = 25 tasks

0.01%, this phenomenon is not representative. It is worth noting that the overall detection ability of the model is weak when FPR is a very small value.

Fourthly, we can observe that the rising trend of TPR tends to be stable when FPR exceeds 0.2. With the increase of maxfpr, our method will achieve greater gain on SPAUC. Furthermore, we calculate the special case when maxfpr is 1. The version based on GAT, GTN, MARINE, and HIN2Vec achieves the AUC performance of 0.86626, 0.93236, 0.87778, and 0.92323, respectively. The results are higher than the AUC performance of "ORI" (0.80844).

Fifthly, we notice that the detection model based on the original properties performs poorly on Gowalla. Compared with the other two datasets, Gowalla has fewer samples, both normal and anomaly, so it is difficult for the anomaly detection model to learn a stable behavioral pattern. With the help of BIG, we obtain high-quality data for the subsequent detection model. Therefore, it is shown that our model can provide stable gain even when the quality of data is limited.

Overall, we observe that our BIG is able to improve the performances of ordinary detection models on various datasets and is compatible with many existing NRL algorithms.

5.5.3.2 Compatibility with Existing Models

We demonstrate the effectiveness of our BIG on supervised anomaly detection tasks, where we adopt the standardized partial AUC ($\text{SPAUC}_{\text{FPR}\leq\text{maxfpr}}$) to comprehensively evaluate the performance [9]. Specifically, we run experiments with $four$ algorithms for $five$ detection models on $five$ datasets. The results are shown in Fig. 5.6. In this work, we consider five representative classifiers as the detection models, i.e., deep neural networks (DNN), logistic regression, random forest, gradient boosting decision tree, and extreme gradient boosting. The experimental results reveal several insightful and interesting observations.

Firstly, our BIG has stable performances when it cooperates with HIN2Vec or MARINE. Moreover, these performances in the two cases are almost equal. A significant contrast is that the performance of BIG with HIN2Vec on the Kyoto dataset is greatly degraded. This is mainly because LR classifier cannot effectively distinguish abnormal events under a low FPR, which leads to poor $\text{SPAUC}_{\text{FPR}\leq 0.01}$. When the value of maxfpr increases, its $\text{SPAUC}_{\text{FPR}\leq\text{maxfpr}}$ will increase, e.g., $\text{SPAUC}_{\text{FPR}\leq 0.05} = 0.899$, and the difference between different models will decrease.

Secondly, the performances of our BIG cooperated with GAT or GTN are slightly lower than those of the case with HIN2Vec or MARINE when we choose the ensemble models, e.g., RF, GBDT, and XGB, as the detection model. Different from network embedding algorithms, graph neural networks usually need some labeled training data to run in semi-supervised or supervised learning mode. In our BIG, a node might appear in different behavioral events, including both normal and abnormal events, which leads that a label (normal or abnormal) cannot be given to this node. In addition, graph neural networks neglect the weight of edges when learning the graph structure, which leads to the loss of some information.

Finally, from the comprehensive results, almost all integration models achieved better results, with XGB performing best of all tasks, and the performance is slightly worse when BIG is combined with DNN or LR. We can conclude that our BIG can achieve good anomaly detection performances and can widely cooperate with most existing models as the key component.

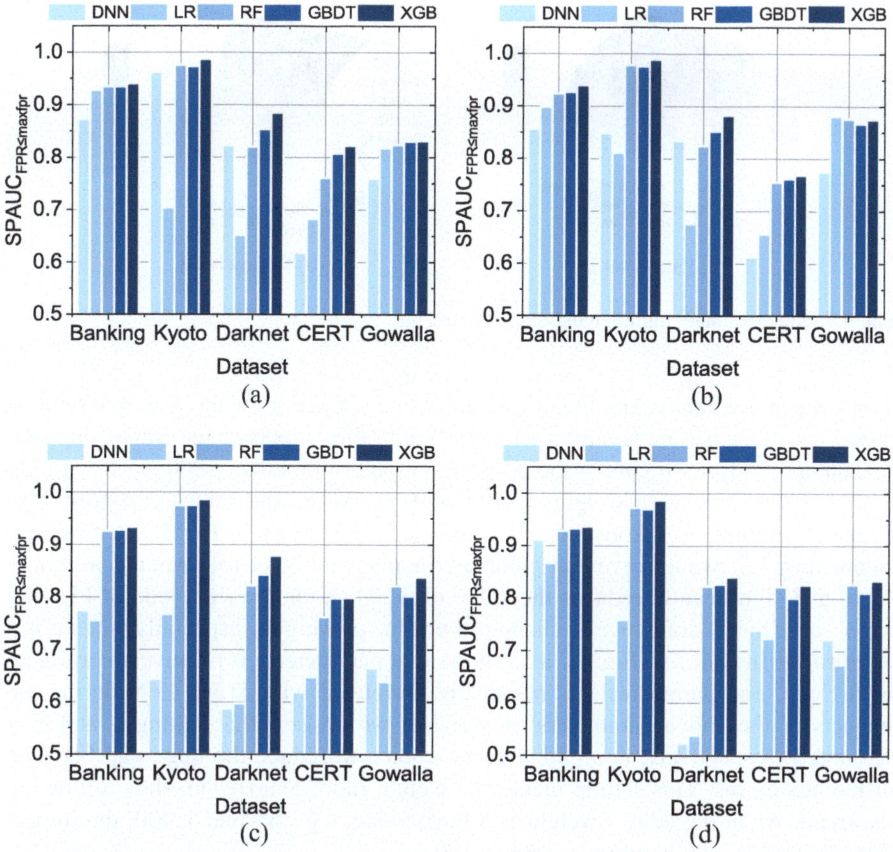

Fig. 5.6 Supervised anomaly detection $\text{SPAUC}_{\text{FPR}\leq\text{maxfpr}}$ on 4 (algorithms) $\times$ 5 (models) $\times$ 5 (datasets) $=$ 100 tasks

5.5.4 Parameter Study

The setting of key parameters in Eq. (5.4) has a significant impact on the weight of an edge during building our BIG. In this work, we accumulate the weights of edges that co-occur in multiple behavioral events. It leads to the huge weight difference between different edges. The huge gap is easy to make NRL algorithms neglect those edges with small weights in the process of learning node representations, especially when the ratio of weights is of a great disparity. Therefore, we introduce a transformation fuction $S(w)$ in Eq. (5.4) to convert the weights. We aim at setting a common large weight, closed to the maximum weight, to be a value close to 100, so that the information carried by an edge with a small weight can be utilized.

To determine how the key parameters affect weights in Eq. (5.4), we vary the parameters α and θ to examine their impacts on weights. We set parameters α and

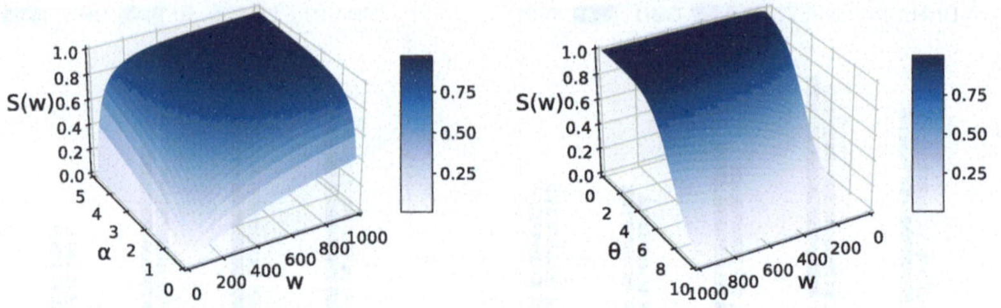

(a) Parameter α (b) Parameter θ

Fig. 5.7 Key parameter tuning in building our BIG

θ at 1.8 and 5 when testing the other, respectively. Except for the parameters to be tested, all other parameters are consistent with other experiments. Based on these five datasets, most weights in the property graph are less than 1000, so we mainly represent the range with weights less than 1000. We set the values of α from 0 to 5, and the values of θ from 0 to 10. The results are shown in Fig. 5.7. On the left side of Fig. 5.7, we observe that parameter α plays a bigger role when a weight is small and significantly reduces the value of weights. On the right side of Fig. 5.7, we observe that θ dramatically changes the value of weights, especially when it is a huge value. In addition, Fig. 5.7 also shows that parameter α is positively correlated with $S(w)$, and parameter θ is negatively correlated with $S(w)$. Considering the impacts of the above parameters on weights, we observe that setting α and θ at 1.8 and 5, respectively, is an appropriate choice to reduce the huge gap between different weights. This setting makes the weight ratio, $S(w)/S(1)$, show following the trend: when the edge's weight is a huge value, e.g., beyond 1,000, the impact decays rapidly, and the ratio is close to 100.

5.5.5 Ablation Study

To demonstrate how each component of our model contributes to the overall performances of anomaly detection, we conduct the ablation experiments. We not only compare the impact of the existence of some key components but also compare the differences of some replaceable components. Therefore, we divide the ablation experiments into two parts: (1) Omitting some tips in building our BIG, e.g., the time impact, the pre-trained word embedding, and the event-layer information (2) Replacing some key components with the replaceable ones, e.g., the construction of new features, NRL algorithms, and the classification models In term of constructing new features, we introduce the Euclidean distance ("dis") and the integration of node representations ("vec1order" and "vec2order") for comparison. The "vec1order" means that we simply represent properties in a behavioral event b with n properties as node representations and sum the node representations by weights, as shown below:

$$\mathbf{b}_{1order} = \sum_{i=0}^{n} \varepsilon_i \cdot \mathbf{b}^i.$$

The "vec2order" means that we further add Hadamard products of any two node representations corresponding to the properties in an event record on the basis of $\mathbf{b}_{1order}$. The calculation of "vec2order" is shown below:

$$\mathbf{b}_{2order} = \mathbf{b}_{1order} + \sum_{i=0}^{n} \sum_{j=0}^{n} \varepsilon_i \cdot \varepsilon_j \cdot \mathbf{b}^i \odot \mathbf{b}^j.$$

Regarding NRL algorithms, we choose MARINE [37], GAT [38], and GTN [39] as the replaceable ones. For classification models, we choose some representative classifiers, i.e., deep neural networks (DNN), logistic regression (LR), random forest (RF), and gradient boosting decision tree (GBDT) as the replaceable ones.

We conduct the ablation experiments on the Banking dataset. All experiments are compared with the version based on cosine to construct new features, HIN2vec algorithm to learning node representations, and XGB classifier to detect anomaly events. The results are shown in Tables 5.4 and 5.5. From the results, we can make four interesting observations.

Firstly, the performance of "ours" is more outstanding than "without time," "without pre-trained," and "without event" in most cases, which proves these tips for building BIG are useful. The "without time" outperforms "ours" is a phenomenon that only appears in the extremely low FPR. We further calculate the $\text{SPAUC}_{\text{FPR} \leq 0.002}$ of "without time" which is 0.934. The difference of

Table 5.4 Results (TPR, precision, F1-score) of the ablation experiment on the dataset Banking

	FPR=0.05%			FPR=0.10%		
Ablation	TPR	Prec	F1-score	TPR	Pre	F1-score
Ours	87.646%	97.751%	0.9242	94.674%	95.919%	0.9529
Without time	88.702%	97.782%	0.9302	92.164%	95.812%	0.9395
Without pre-trained	86.272%	97.721%	0.9164	90.983%	95.760%	0.9331
Without event	83.392%	97.653%	0.8996	93.248%	95.855%	0.9453
dis	86.348%	97.723%	0.9168	90.473%	95.737%	0.9303
vec1order	85.826%	97.700%	0.9138	89.156%	95.677%	0.9230
vec2order	84.288%	97.668%	0.9049	89.248%	95.681%	0.9235
MARINE	91.666%	97.890%	0.9468	94.642%	95.921%	0.9528
GAT	88.236%	97.775%	0.9276	93.534%	95.871%	0.9469
GTN	87.485%	97.747%	0.9233	93.188%	95.856%	0.9450
DNN	72.966%	97.327%	0.8340	86.907%	95.570%	0.9103
LR	85.216%	97.684%	0.9102	89.256%	95.677%	0.9236
GBDT	86.035%	97.719%	0.9151	91.232%	95.763%	0.9344
RF	87.035%	97.740%	0.9208	91.919%	95.8011%	0.9382

Table 5.5 Results (TPR, precision, F1-score) of the ablation experiment on the dataset Banking

	FPR=0.15%			FPR=0.20%		
Ablation	TPR	Precision	F1-score	TPR	Precision	F1-score
Ours	87.646%	97.751%	0.9242	94.674%	95.919%	0.9529
Without time	88.702%	97.782%	0.9302	92.164%	95.812%	0.9395
Without pre-trained	86.272%	97.721%	0.9164	90.983%	95.760%	0.9331
Without event	83.392%	97.653%	0.8996	93.248%	95.855%	0.9453
dis	86.348%	97.723%	0.9168	90.473%	95.737%	0.9303
vec1order	85.826%	97.700%	0.9138	89.156%	95.677%	0.9230
vec2order	84.288%	97.668%	0.9049	89.248%	95.681%	0.9235
MARINE	95.313%	94.095%	0.9468	96.437%	92.403%	0.9438
GAT	95.369%	94.040%	0.9470	96.012%	92.255%	0.9418
GTN	94.602%	94.005%	0.9430	96.020%	92.255%	0.9410
DNN	89.577%	93.675%	0.9158	90.385%	91.812%	0.9109
LR	92.068%	93.851%	0.9295	94.168%	92.115%	0.9313
GBDT	93.184%	93.912%	0.9355	95.281%	92.200%	0.9371
RF	93.726%	93.957%	0.9384	95.196%	92.193%	0.9367

$\text{SPAUC}_{\text{FPR}\leq 0.002}$ also reflects that the consideration of time delay is effective in our model.

Secondly, comparing "ours" with "dis," "vec1order," and "vec2order," we could find that "ours" is more effective. A significant difference in the above four versions is that "ours" and "dis" compute the similarity between any two representations, while "vec1order" and "vec2order" merge different representations by vector computation. The results show "ours" and "dis" have better performances than the other two versions. Compared with "vec1order," "vec2order" utilizes the interaction information between any two representations, so its performance is better than "vec1order." Between "ours" and "dis," the cosine similarity pays more attention to the difference between two representations in direction, while Euclidean distance is sensitive in numerical value. The model fits well with the cosine similarity since it is easier to distinguish the tendency of anomaly behaviors.

Thirdly, comparing different NRL algorithms, the cooperation with different algorithm has a light impact on the performance. The differences in performances mainly come from the representations learned by different NRL algorithms. Since most deployed algorithms can capture fine-grained associations between the original properties, this component can be replaced when more advanced algorithms are proposed.

Finally, comparing different classification models, XGB achieves the best performance than others. Next to it, RF and GBDT have similar performances, and LR and DNN have relatively bad performances. In addition, the differences generated by different classification models are greater than those generated by different NRL algorithms in the overall performances of anomaly detection.

5.6 Discussion

5.6.1 Stability Analysis

Because of concerns that not all anomalies are immutable with temporal changes, e.g., a deceiver may attempt to obfuscate a behavior by intentionally introducing new anomalies, we further investigate the effectiveness of our method in different anomaly modes, that is, testing our method with different types of anomalies from those in the training dataset.

In this way, we evaluate the performance of our method in a distributed denial of service attack dataset (CICDDoS[6]) which consists of different types of attacks. In particular, we select two different attacks as anomalies, NetBIOS (denoted as DDoS_A) and LDAP (denoted as DDoS_B). Table 5.6 refers to the part statistics of CICDDoS, where N# means Number, and R# means Ratio.

We implement our BIG with four representative network representation learning algorithms and random forest to detect anomalies in this part.

Table 5.7 shows the performance comparison between different types of attacks. **DDoS_A/B** means that we use the data generated by DDoS_A for training and the data generated by DDoS_B for testing. Notably, we only report SPAUC under two different maxfpr rather than SPAUC and F1-score when the TPR is equal to maxfpr, since the ratio of positive samples is close to 1. As can be seen, our method can achieve significant improvements over the implementations based on original features. After the attack mode changes, our method can still capture useful information from the previous attack mode to improve the detection ability in most cases, although the improvement is reduced compared with the implementation of the same attack mode.

5.6.2 Complexity Analysis

In order to facilitate to analyze the complexity of implementing our method, we give a discussion on the cost of time and memory.

Table 5.6 The characteristic information of the CICDDoS dataset

	DDoS_A			DDoS_B		
Dataset	N# of events	N# of positive	R# of positive	N# of events	N# of positive	R# of positive
Training	4,094,986	4,093,279	99.96%	2,181,542	2,179,930	99.93%
Testing	3,455,899	3,454,333	99.95%	2,113,234	2,111,667	99.92%

[6] **CICDDoS**: https://www.unb.ca/cic/datasets/ddos-2019.html.

Table 5.7 Performance of our method under different types of attacks

	DDoS_A/A		DDoS_B/A	
	SPAUC	SPAUC	SPAUC	SPAUC
Method	FPR≤0.005	FPR≤0.01	FPR≤0.005	FPR≤0.01
Ori	0.80922	0.90424	0.80674	0.90031
HIN2Vec	0.99812	0.99828	0.96905	0.97888
MARINE	0.99975	0.99975	0.94795	0.95818
GAT	0.98789	0.98789	0.94755	0.94891
GTN	0.99974	0.99980	0.94961	0.95003
	DDoS_B/B		DDoS_A/B	
	SPAUC	SPAUC	SPAUC	SPAUC
Method	FPR≤0.005	FPR≤0.01	FPR≤0.005	FPR≤0.01
Ori	0.74519	0.87192	0.80426	0.90912
HIN2Vec	0.98005	0.98473	0.79699	0.84267
MARINE	0.94292	0.95132	0.87453	0.91093
GAT	0.94631	0.94794	0.87058	0.89005
GTN	0.94906	0.94970	0.79759	0.85394

Before learning the representations from the property-level and event-level information, we need to build the property graph on the training dataset B, where B includes $|B|$ behavioral events and each behavioral event consists of n properties. Because an edge's weight needs to be calculated during building the property graph, it requires $n(n-1)|B|/2$ calculations for initial weights and then obtains the final weights through $|B|$ calculations. So building the property graph takes $O(|B|n^2)$ time. Assuming that each property of the behavioral event has m values on average, we generate a property graph with nm nodes. It takes $O(m)$ memory to store the nodes and $O(m^2)$ memory to store the edges since $n \ll m$ in most cases. Note that the running time largely depends on that of existing algorithms that cooperate with our method, when calling the event-property composite model and anomaly detection model.

After obtaining the learned representations, generating the constructed features takes only $O(n^2)$ time (e.g., set n as 10) for an incoming behavioral event. Therefore, our method has the ability to make decisions for an event in a fast response time under a modest condition, e.g., a desktop computer with 3 GHz CPU.

5.6.3 Adaptability Analysis

Our method may be further improved, although it outperforms competitors in most cases. Modularity is one of our guiding principles in design. Each module can be freely swapped out for more effective solutions or application-specific trade-offs in terms of performance. So an intuitive way for improving the existing version is to adopt more efficient network representation learning algorithms and behavior anomaly detection algorithms than those implemented in our current experiments. In addition, this work can be further conducted in two potential directions, i.e., choosing parameters automatically and processing anomalies in an end-to-end way.

For the former, most of the parameters to be set in our method belong to existing network representation learning algorithms and behavior anomaly detection algorithms, e.g., the dimension of representation and the learning rate. Choosing the appropriate parameters will largely depend on the prior knowledge of the algorithm used together. Certainly, it increases the workload of analysts to a certain extent. So a mechanism that can automatically select the optimal parameters for our method is very meaningful for the analysis of workload reduction.

For the latter, our behavioral identification graph (BIG) acts as an enhancer to downstream solutions, which is separate from anomaly detection methods. Although the straightforward adoption of more advanced methods can obtain some performance improvement, it may yield the representations that are suboptimal or even incompatible with respect to the specific anomaly detection methods in the two-stage structure. We address this problem by incorporating the event-level information into the objective of representation learning, which substantially improves the expressiveness of the representations. As a matter of fact, our BIG is an indirect optimization of the events' label information. To further improve our method, the end-to-end anomaly detection is an interesting solution to achieve the efficient use of data.

5.7 Conclusions

For behavioral anomaly detection in cybersecurity, we propose a new graph-based behavioral modeling paradigm, named BIG. Different from the existing works, we further utilize the fine-grained information among behavioral properties to improve the behavioral models. The core technique of BIG is to integrate inter (event-level)- and intra (property-level)-associations of behaviors into a unified graph and space to improve behavioral modeling for anomaly detection. To analyze the graph efficiently, we design an event-property composite model to leverage both the properties information from the fine-grained level and the events information from the coarse-grained level to learn deeply latent representations. We conduct extensive experiments on four representative cybersecurity issues to validate the effectiveness of our proposed method. On the whole, our BIG can actually serve as a new behavioral modeling paradigm for behavioral anomaly detection problems.

There are some interesting issues left to study: (1) Under BIG, it is beneficial to study the cooperation between BIG and its downstream detection models, such as weakly supervised classification and unsupervised clustering models. (2) It is anticipated to provide a visual system based on our BIG to assist decision-makers, like topological data analysis (TDA). For example, an interesting future direction is to explore the quality of explainability by introducing the attention mechanism at the event-level.

References

1. P. Sun, J. Li, M.Z.A. Bhuiyan, L. Wang, B. Li, Inf. Sci. **479**, 456 (2019)
2. H. Mazzawi, G. Dalaly, D. Rozenblatz, L. Ein-Dor, M. Ninio, O. Lavi, in *Proceeding of the IEEE ICDE 2017, San Diego, CA, USA, April 19–22, 2017* (2017), pp. 1140–1149
3. C. Jing, C. Wang, C. Yan, in *Proceedings of the Financial Cryptography and Data Security 2019, Frigate Bay, St. Kitts and Nevis, February 18–22, 2019*, vol. 11598 (2019), pp. 588–604
4. Z. Jiang, L. Sun, P.S. Yu, H. Li, J. Ma, Y. Shen, in *Proceedings of the IEEE ICDE 2020, Dallas, TX, USA, April 20–24, 2020* (2020), pp. 1862–1865
5. F. Liu, Y. Wen, D. Zhang, X. Jiang, X. Xing, D. Meng, in *Proceedings of the CCS 2019, London, UK, November 11–15, 2019* (2019), pp. 1777–1794
6. Z. Li, J. Zeng, Y. Chen, Z. Liang, CoRR **abs/2111.07093** (2021)
7. J. Lian, X. Zhou, F. Zhang, Z. Chen, X. Xie, G. Sun, in *Proceedings of the ACM SIGKDD 2018, London, UK, August 19–23, 2018* (2018), pp. 1754–1763
8. X. Chen, X. Song, G. Peng, S. Feng, L. Nie, in *Proceedings of the SIGIR 2021 Virtual Event, Canada, July 11–15, 2021* (2021), pp. 1084–1093
9. Y. Zhu, D. Xi, B. Song, F. Zhuang, S. Chen, X. Gu, Q. He, in *Proceedings of the WWW 2020, Taipei, China, April 20–24, 2020* (2020), pp. 928–938
10. L. Sun, B. Cao, J. Wang, W. Srisa-an, P.S. Yu, A.D. Leow, S. Checkoway, IEEE Trans. Mob. Comput. **20**(4), 1465 (2021)
11. Z. Wang, W. Zhang, N. Liu, J. Wang, in *Proceedings of the AAAI 2020, New York, NY, USA, February 7–12, 2020* (2020), pp. 6331–6339
12. C. Wang, C. Wang, H. Zhu, J. Cui, IEEE Trans. Dependable Secure Comput. **18**(5), 2122 (2021)
13. R. He, J.J. McAuley, in *Proceedings of the IEEE ICDM 2016, December 12–15, 2016, Barcelona, Spain* (2016), pp. 191–200
14. J. Tang, K. Wang, in *Proceedings of the WSDM 2018, Marina Del Rey, CA, USA, February 5–9, 2018* (2018), pp. 565–573
15. G. Zhou, X. Zhu, C. Song, Y. Fan, H. Zhu, X. Ma, Y. Yan, J. Jin, H. Li, K. Gai, in *Proceedings of the ACM SIGKDD 2018, London, UK, August 19–23, 2018* (2018), pp. 1059–1068
16. J. Wang, R. Wen, C. Wu, Y. Huang, J. Xion, in *Proceedings of the WWW 2019, San Francisco, CA, USA, May 13–17, 2019* (2019), pp. 310–316
17. Y. Yang, Y. Xu, Y. Sun, Y. Dong, F. Wu, Y. Zhuang, IEEE Trans. Knowl. Data Eng. **33**(1), 169 (2021)
18. J. Zeng, Z.L. Chua, Y. Chen, K. Ji, Z. Liang, J. Mao, in *Proceedings of the NDSS 2021, virtually, February 21–25, 2021* (2021)
19. H. Peng, J. Li, Q. Gong, Y. Song, Y. Ning, K. Lai, P.S. Yu, in *Proceedings of the IJCAI 2019, Macao, China, August 10–16, 2019* (2019), pp. 3238–3245
20. C. Wang, H. Zhu, IEEE Trans. Dependable Secure Comput. **19**(1), 301 (2022)
21. S. Eswaran, M. Sachdeva, V. Vimal, D. Seth, S. Kalpam, S. Agarwal, T. Mukherjee, S. Dattagupta, in *Proceedings of the ACM SIGKDD 2020, Virtual Event, CA, USA, August 23–27, 2020* (2020), pp. 2657–2665
22. W.W. Cohen, in *Proceedings of the ICML 1995, Tahoe City, California, USA, July 9–12, 1995* (1995), pp. 115–123
23. L. Tong, B. Li, C. Hajaj, C. Xiao, N. Zhang, Y. Vorobeychik, in *Proceedings of the USENIX 2019, Santa Clara, CA, USA, August 14–16, 2019* (2019), pp. 285–302
24. M.A. Ambusaidi, X. He, P. Nanda, Z. Tan, IEEE Trans. Comput. **65**(10), 2986 (2016)
25. P. Marteau, IEEE Trans. Inf. Forensics Secur. **16**, 2157 (2021)
26. B. Zong, Q. Song, M.R. Min, W. Cheng, C. Lumezanu, D. Cho, H. Chen, in *Proceedings of the ICLR 2018, Vancouver, BC, Canada, April 30–May 3, 2018*, pp. 1–19 (2018)
27. X. Feng, Q. Li, H. Wang, L. Sun, in *Proceedings of the USENIX Security 2018, Baltimore, MD, USA, August 15–17, 2018* (2018), pp. 327–341

28. M. Abouelenien, V. Pérez-Rosas, R. Mihalcea, M. Burzo, IEEE Trans. Inf. Forensics Secur. **12**(5), 1042 (2017)
29. W. Youyou, M. Kosinski, D. Stillwell, PNAS **112**(4), 1036 (2015)
30. X. Ruan, Z. Wu, H. Wang, S. Jajodia, IEEE Trans. Inf. Forensics Secur. **11**(1), 176 (2016)
31. G. Pang, C. Shen, A. van den Hengel, Proceedings of the ACM SIGKDD 2019, Anchorage, AK, USA, August 4–8, 2019 (2019), pp. 353–362
32. G. Somepalli, M. Goldblum, A. Schwarzschild, C.B. Bruss, T. Goldstein, CoRR **abs/2106.01342** (2021)
33. T. Cui, G. Gou, G. Xiong, Z. Li, M. Cui, C. Liu, in *Proceedings of the USENIX Security 2021, August 11–13, 2021* (2021), pp. 4329–4346
34. S. Fan, C. Shi, X. Wang, in *Proceedings of the ACM CIKM 2018, Torino, Italy, October 22–26, 2018* (2018), pp. 1483–1486
35. C. Yang, M. Sun, Z. Liu, C. Tu, in *Proceedings of the IJCAI 2017, Melbourne, Australia, August 19–25, 2017* (2017), pp. 3894–3900
36. T. Fu, W. Lee, Z. Lei, in *Proceedings of the ACM CIKM 2017, Singapore, November 06–10, 2017* (2017), pp. 1797–1806
37. M. Feng, C. Hsu, C. Li, M. Yeh, S. Lin, in *Proceedings of the WWW 2019, San Francisco, CA, USA, May 13–17, 2019* (2019), pp. 470–479
38. P. Velickovic, G. Cucurull, A. Casanova, A. Romero, P. Liò, Y. Bengio, *Proceedings of the ICLR 2018, Vancouver, BC, Canada, April 30–May 3, 2018, Conference Track Proceedings of theedings* (2018), pp. 1–12
39. S. Yun, M. Jeong, R. Kim, J. Kang, H.J. Kim, in *Proceedings of the NeurIPS 2019, December 8–14, 2019, Vancouver, BC, Canada* (2019), pp. 11960–11970
40. W. Yu, W. Cheng, C.C. Aggarwal, K. Zhang, H. Chen, W. Wang, in *Proceedings of the ACM SIGKDD 2018, London, UK, August 19–23, 2018* (2018), pp. 2672–2681
41. Y. Liu, L. He, B. Cao, P.S. Yu, A.B. Ragin, A.D. Leow, in *Proceedings of the AAAI 2018, New Orleans, Louisiana, USA, February 2–7, 2018* (2018), pp. 117–124
42. P. Cui, X. Wang, J. Pei, W. Zhu, IEEE Trans. Knowl. Data Eng. **31**(5), 833 (2019)
43. Z. Wu, S. Pan, F. Chen, G. Long, C. Zhang, P.S. Yu, IEEE Trans. Neural Networks Learn. Syst. **32**(1), 4 (2021)
44. J. Song, H. Takakura, Y. Okabe, M. Eto, D. Inoue, K. Nakao, in *Proceedings of the BADGERS@EuroSys 2011, Salzburg, Austria, April 10, 2011* (2011), pp. 29–36
45. A.H. Lashkari, G. Kaur, A. Rahali, *Proceedings of the ICCNS 2020, Tokyo, Japan, November 27–29, 2020* (2020), pp. 1–13
46. J. Glasser, B. Lindauer, in *2013 IEEE Security and Privacy Workshops* (2013), pp. 98–104
47. E. Cho, S.A. Myers, J. Leskovec, in *Proceedings of the ACM SIGKDD 2011, San Diego, CA, USA, August 21–24, 2011* (ACM, New York, 2011), pp. 1082–1090
48. S. Zhao, T. Zhao, H. Yang, M.R. Lyu, I. King, in *Proceedings of the AAAI 2016, February 12–17, 2016, Phoenix, Arizona, USA* (2016), pp. 315–322
49. D. Lee, R. Moussalli, S.W. Asaad, M. Srivatsa, in *Proceedings of the FCCM 2016, Washington, DC, USA, May 1–3, 2016* (2016), pp. 176–183
50. T. Fawcett, Pattern Recogn. Lett. **27**(8), 861 (2006)

Chapter 6
Layered Behavior Modeling for Recommendations

Abstract Sequential recommendation, a critical aspect of recommendation systems, focuses on predicting users' next preferred items based on their behavioral history. Existing advanced models, including transformer-based and dynamic graph-based approaches, often prioritize macroscopic event sequences or node relationships, overlooking fine-grained information within user and item attributes. This gap poses challenges in constructing stable models, particularly in sparse data scenarios. To address this, we propose an approach that seamlessly integrates side information while maintaining computational efficiency. Our novel method, named Micro Event Graph (MiEG), employs layered computations to project fine-grained attribute information onto a temporal event graph. Subsequently, final node representations are generated through graph neural network processing. Extensive experiments on real-world datasets showcase MiEG's superiority over the state-of-the-art baselines, underscoring its exceptional information mining capabilities. Degeneration experiments further reveal MiEG's competitiveness even in nonsequential recommendation scenarios.

6.1 Introduction to Layered Information Aggregation

In recent times, sequential recommendation (SR) has emerged as a crucial technology across various web platforms [1, 2]. Diverging from traditional methods, SR emphasizes capturing temporal dynamics and sequential patterns, leading to notable improvements in accuracy [3]. Earlier approaches relied on Markov chain-based methods to model item transitions [4, 5]. The advent of Transformers [6] introduced self-attention mechanisms [7] for capturing correlations, achieving the state-of-the-art results. Additionally, some SR models harnessed the power of graph neural networks (GNNs) [8] in the context of session-based recommendations [9, 10]. Recent advancements in SR focus on two key perspectives to enhance performance.

Side information fusion method (Fig. 6.1a: The side information fusion method encodes attributes into item sequences, where x^i denotes the attribute embedding, and x_t^i represents the embedding that incorporates temporal information post-processing by a sequential model). Intuitively, harnessing highly related information

C. Wang, H. Zhu, *Universal Behavior Computing for Security and Safety*,
https://doi.org/10.1007/978-981-97-9014-2_6

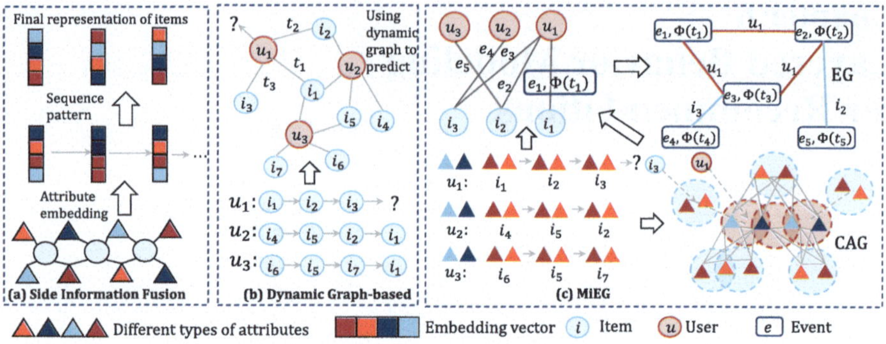

Fig. 6.1 Recent advancements in sequential recommendation

can significantly enhance the quality of recommendations. In previous studies [11–13], the integration of attribute information into the recommendation process involved the utilization of GNNs or self-supervised tasks. These methods were employed to extract attribute representations, which were subsequently combined with item embeddings to serve as input for sequential networks such as GRUs [14] or BERTs [15]. Although these methods exhibited improved performance compared to traditional sequential recommendation approaches, their limitations lay in their ability to capture high-order connectivity between different user sequences and their underutilization of attribute information at a global level (the connectivity between different attributes).

Dynamic Graph-Based Method (Fig. 6.1b: The dynamic graph-based method constructs a graph with interaction time t on the edges). To capture the associative information between users and items in recommendations, certain studies [16, 17] have embraced dynamic graph representation learning methods to model the temporal characteristics of user-item interactions. Some approaches [18] aimed to learn dynamic graph embeddings by encoding timestamps, achieving notable improvements by integrating sequence information with temporal collaboration information. These methods utilized self-attention mechanisms to capture temporal relationships within the sequence [19]. Despite their successes, these methods did not fully exploit the potential benefits of integrating fine-grained side information into the recommendation process.

Although the constraints on utilizing information in the aforementioned methods are not significant in scenarios with ample high-quality data, they become notably evident when dealing with sparse data. This can result in unstable behavior modeling in the context of sequential recommendations. To address these challenges and enhance information utilization, it is essential to merge these two methods, compensating for their respective limitations. However, combining them in a simple, naive way can pose significant computational challenges, such as exponentially increasing costs when incorporating associations among attributes into the dynamic graph.

Table 6.1 Comparison of different methods in terms of information mining

Information degree	Side information fusion method	Dynamic graph-based method	Our method
Temporal information	√	√	√
Collaboration information		√	√
Local attribute information	√		√
Global attribute information			√

Building upon these considerations, we propose a layered projection approach. This involves initially using the attribute layer to learn attribute representations, followed by leveraging the event layer to project the attribute space onto the event nodes of the global graph. This integration accomplishes the fusion of side information and dynamic graph components. Detailed comparisons with other methods can be found in Table 6.1.

In this chapter, we introduce a model, named **Mi**cro **E**vent **G**raph (MiEG, as shown in Fig. 6.1c), that merges information from attributes (side information), time (user interaction history), and space (spatial topology of users and items). Our method constructs a complete attribute graph (CAG) and an event graph (EG) for acquiring attribute representations and event representations, respectively. $\Phi(t)$ represents the embedding of a timestamp. MiEG comprises three layers: (1) The *attribute layer*, which fully connects the attributes contained in each events to generate a heterogeneous graph, and then learns the node embedding of each attribute. (2) The *event layer*, where we transform the user-item interaction bipartite graph into an event graph. We inject attribute embeddings obtained from the attribute layer and timestamp embeddings into the nodes, allowing the use of dynamic graph learning methods to capture temporal and topological information in event node representations. (3) The *agent layer*, which processes the output of the event layer to obtain node embeddings on the graph. User and item embeddings are derived from these node embeddings to complete the recommendation.

This work's contributions can be outlined as follows:

- We propose a layered model, MiEG, for incorporating side information into an event graph with temporal information. The layered architecture effectively addresses the issue of exponentially increasing computational cost when merging diverse types of information into a unified space.
- We design a fine-grained treatment of side information that captures attribute correlations from both the micro and macro perspectives. This enables us to generate attribute representations with enhanced granularity and precision.
- We perform extensive experiments on real-world datasets, demonstrating substantial performance enhancements when compared to the state-of-the-art techniques.

6.2 Related Work

The field of sequential recommendation (SR) has gained significant attention in recent years. Traditional methods focus on modeling temporal relationships between items, while recent approaches incorporate side information and utilize graph structures to capture dynamic user-item interactions.

6.2.1 *Sequential Recommendations*

Compared with the general recommendation, sequential recommendations aim to predict a user's next preference based on their most recent activities and model sequential patterns between successive items to generate accurate recommendations for the user [2, 7, 20, 21]. Neural network models, such as recurrent neural networks (RNNs) [22] and convolutional neural networks (CNNs) [23], are popular choices for sequential problems and have been used in the recommendation field to develop classic models such as GRU4Rec [21] and CASER [24]. However, both of these models suffer from length limitation and are unable to learn the influence of all historical items, thus unable to encode sequences effectively. With the development of the Transformer [6], current research focuses on designing self-attention SR models to predict future item sequences [7, 25, 26]. SASRec [25] and STAMP [27] adopt the self-attention mechanism to encode sequences of items by assigning attention weights to each historical item and aggregating these items to infer the sequence embedding at location t. There are also several sequence recommendations based on graph neural networks [9, 10, 28, 29], where items are typically connected as nodes sequentially in interaction order into a graph for graph representation learning, mainly for session-based recommendations. Failing to consider significant side information, these methods might lead to suboptimal recommendations and decreased accuracy, particularly when dealing with intricate or limited data. Additionally, the neglect of indirect associations presents a challenge for these methods to excel in scenarios involving short sequential data.

6.2.2 *Side Information Fusion for SR*

Numerous researches have focused on integrating side information at different stages of recommendation. Early attempts employed straightforward methods [13, 30] to integrate side information into SR models. Recent innovations have explored more sophisticated approaches. For instance, S^3Rec [12] incorporates self-supervised attribute prediction during the pretraining stage. ICAI-SR [11] integrates side information into item representation by employing a dedicated attribute-to-item aggregation layer before the attention layer, utilizing separate attribute sequential models during training. NOVA-BERT [31] separates the flow of side information and item ID information within the BERT [15] model to maintain item ID independence. DIF-SR [32] introduces a novel decoupled auxiliary information fusion attention mechanism, allowing for high-level attention matrices and adaptive gradients, thereby enhancing the learning of item representations. Despite these advancements, these methods face challenges in effectively leveraging sparse user-item interaction data and extracting global attribute information.

6.2.3 Dynamic Graph-Based Method for SR

Graph neural networks (GNNs) play a crucial role in recommender systems by extracting features from graph-structured data [10, 29, 33, 34]. As the nature of graph data evolves, dynamic graph models [16, 18] have been developed to adapt to these changes. CTDNE [35] introduces a versatile framework for integrating temporal information into network embedding methods. TDIG [36] incorporates LSTM to amalgamate information from user nodes at different periods. TGSRec [19] introduces an innovative collaborative attention mechanism that captures collaborative signals from both users and items while considering the temporal dynamics within sequential patterns. DGSR [16] designs a dynamic graph attention neural network to propagate and aggregate information of different users and their sequences. Another notable approach, PTGCN [17], utilizes a bipartite graph structure to model user-item interactions, effectively integrating temporal dynamics and high-order connectivity. While these recent approaches have demonstrated success in sequence recommendation scenarios, the challenge lies in their effectiveness in incorporating side information into the recommendation process.

To overcome the limitations inherent in existing SR methods, we employ dynamic graph structures to extract both attribute and interaction information. The integration of these dynamic graph structures into our methodology facilitates the robust modeling of intricate relationships among users, items, and attributes. This comprehensive representation not only enhances performance but also effectively addresses the constraints associated with current SR methods.

6.3 Overview

As depicted in Fig. 6.2, we introduce our novel model, named **Mi**cro **E**vent **G**raph (MiEG). The model embraces a layered architecture, consisting of three layers: the attribute layer, the event layer, and the agent layer. We provide detailed descriptions of some important notations in Table 6.2.

In the **attribute layer**, we construct a *complete attribute graph* to generate a heterogeneous graph. Then we utilize *dual-level attention mechanism* to learn the micro-level representation of each attribute. In the **event layer**, we transform

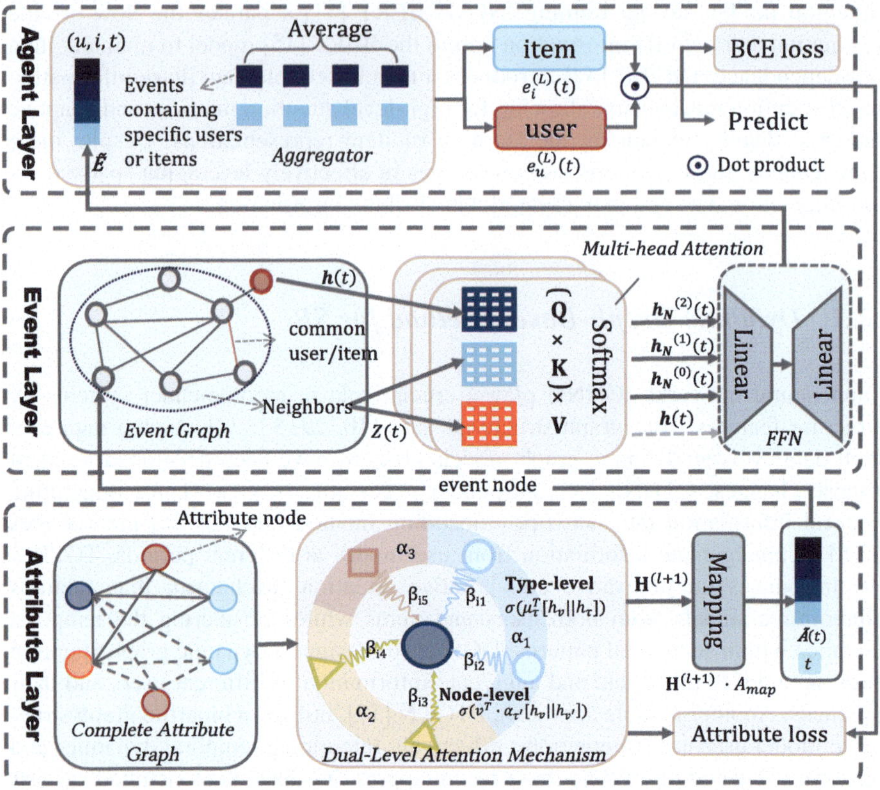

Fig. 6.2 The framework of **MiEG**

Table 6.2 Notations

Notations	Definitions
$\mathscr{U}$	The set of users
$\mathscr{I}$	The set of items
$\mathscr{A}_u$	The set of user attributes
$\mathscr{A}_i$	The set of item attributes
$\mathscr{A}_e$	The set of event attributes
$\mathscr{C}$	The set of attribute classes
α	Type-level attention score
β	Node-level attention score
$\mathbf{h}$	The embedding of hidden layers
$\tilde{\mathbf{A}}$	The embeddings of attribute nodes
$\tilde{\mathbf{E}}$	The embeddings of event nodes
$\mathbf{A}_{map}$	The mapping matrix of event attributes

the user-item bipartite graph into an event graph, followed by applying a self-attention-based dynamic graph approach to model the event nodes within the *event graph*. This facilitates the learning of macro-level correlations among events. The **agent layer** processes the output of the event layer for the node embedding on the graph. In the top-down direction, the agent layer projects events occurring between entities onto the event layer, while the event layer projects attributes contained in event nodes onto the attribute layer. Conversely, in the bottom-up direction, the attribute layer transfers learned attribute representations from the complete attribute graph (CAG) to event nodes. Subsequently, the event layer captures the relational transfers of events from the event graph (EG) to the agent layer, thereby forming representations for each agent. This bidirectional flow ensures a comprehensive modeling of relationships and representations across the attribute, event, and agent layers within the layered structure.

6.3.1 Attribute Layer

Existing approaches [37–39] use one-hot transformations of attributes, stitching them together for embedding. It simplifies attribute expression as vectors, suitable for training in common recommendation models. We use an attribute layer to mine relationships between side information for better attribute utilization.

Suppose we have a set of M users $\mathscr{U} = \{u_1, u_2, \ldots, u_M\}$ and a set of N projects $\mathscr{I} = \{i_1, i_2, \ldots, i_N\}$. Each user u has K_u attributes ($\mathscr{A}_u = \{a_u^1, a_u^2, \ldots, a_u^{K_u}\}$) and each item has K_i attributes ($\mathscr{A}_i = \{a_i^1, a_i^2, \ldots, a_i^{K_i}\}$). One user-item interaction is counted as one event. An event consists of a triple of user, item, and timestamp. For an event $e = (u, i, t)$, it has $\mathscr{A}_u$ of the user, $\mathscr{A}_i$ of the item, the attributes $\mathscr{A}_b$ of the interaction behavior, and the timestamp t, i.e., $\mathscr{A}_e = \{\mathscr{A}_u, \mathscr{A}_i, \mathscr{A}_b, t\}$. We fully connect all the attributes in $\mathscr{A}_e$ except t in the attribute layer to generate an attribute graph.

This process generates a complete attribute graph (CAG) that captures both local and global relationships among various attributes involved in user-item interactions. For *local attribute information*, a fully connected structure is formed among the attributes of a single event, enabling the complete learning of local correlations among attributes. Regarding *global attribute information*, attributes from different events are projected into the same space, sharing embeddings.

6.3.2 Event Layer

We enhance event node representation from a broader perspective by transforming the user-item interaction bipartite graph into an event graph (EG) (the upper part of Fig. 6.1c). For instance, for a trajectory like i_1-u_1-i_2, we convert edges $e_1(i_1, u_1, t_1)$

and $e_2(u_1, i_2, t_2)$ into nodes, connecting these two event nodes through a shared node u_1 in the bipartite graph. The event graph $\mathscr{E}$ can be defined as:

$$\mathscr{E} = \{(e(u_1, i_1, t_1), e(u_2, i_2, t_2)) \mid u_1 = u_2 \text{ or } i_1 = i_2\},$$

where $(e(u_1, i_1, t_1), e(u_2, i_2, t_2))$ denotes an edge between two event nodes if they share a common user $u_1 = u_2$ or item $i_1 = i_2$. This transformation strategically repositions the dynamic bipartite graph problem into the well-established domain of dynamic homogeneous graph problems. It enables the use of widely adopted and computationally efficient methods for computing homogeneous graphs without edge weights, eliminating the necessity for complex techniques commonly associated with dynamic heterogeneous graphs in dynamic bipartite graph computation. Furthermore, this approach alleviates modeling instability issues resulting from edge overlap in bipartite graphs. By adopting this strategy, we incur a higher spatial complexity, but it results in enhancements in model accuracy and computational efficiency.

Our method then leverages the self-attention-based dynamic graph methods to reveal the temporal and topological correlations inherent in inter-event relationships. The output of this layer is embeddings of event nodes, denoted as $\tilde{\mathbf{E}}$.

6.3.3 Agent Layer

Once event representations have been acquired, both users and items emerge as agents within discrete events. We aggregate event representations on a specific agent to generate the definitive representation:

$$e_u(t) = Aggregate(\tilde{\mathbf{E}}_u^{(L)}), \tag{6.1}$$

where $Aggregate(\cdot)$ signifies the averaging aggregation operation. $e_u(t)$ refers to the final embeddings for users, and $\tilde{\mathbf{E}}_u^{(L)}$ is a set of embeddings of events which are relative to user u. Similarly, we obtain the item embedding $e_i(t)$.

Following this, systematic procedures for prediction modeling and optimization are executed. The recommendation score $\mathbf{r}(u, i, t)$ is calculated by taking the dot product of the user and item embeddings at time t, represented as $e_u(t) \cdot e_i(t)$. The dot product of embeddings can signify the distance between two embeddings within the same space. Shorter distances indicate higher similarity, thereby suggesting a higher probability of recommendation. Subsequently, the top N recommended items are ranked based on their scores.

6.4 Micro Event Graph

6.4.1 Complete Attribute Graph

For the event attribute $\mathscr{A}_e$, distinct node types represent each category of attributes, and attribute values are mapped to their respective nodes. This process yields a heterogeneous complete attribute graph with $|\mathscr{C}| = |\mathscr{A}_u| + |\mathscr{A}_i| + |\mathscr{A}_b|$ node classes, covering all nodes in $\mathscr{A}_e$, where $|\cdot|$ denotes the number of elements in a set. The process of building a CAG is described in Algorithm 5.

Algorithm 5: The procedure of building CAG

Input: The behavior sequence $\mathscr{S}_A$, the predefined number of nodes K_n
Output: The adjacency matrix M of CAG

Function: $AMF(\mathscr{S}_A, K_n)$ Initialize $\psi(\cdot)$ as an empty set of mapping rules;
  Initialize $\mathscr{N}_A \in \mathbb{R}^{|\mathscr{C}| \times P}$ with the initial node;
  Sort $\mathscr{S}_A$ in ascending order;
  Divide $\mathscr{S}_A$ into K_n equal-sized partitions:

$$\mathscr{S}_A^i = \begin{cases} \mathscr{S}_A\left[\frac{(i-1)\cdot|\mathscr{S}_A|}{K_n} + 1 : \frac{i\cdot|\mathscr{S}_A|}{K_n}\right], & \text{if } i < K_n \\ \mathscr{S}_A\left[\frac{(K_n-1)\cdot|\mathscr{S}_A|}{K_n} + 1 : |\mathscr{S}_A|\right], & \text{if } i = K_n \end{cases}$$

  where $1 \le i \le K_n$;
  for *each partition* $\mathscr{S}_A^i$ **do**
    for *each attribute value* a_j *in* $\mathscr{S}_A^i$ **do**
      Find the corresponding node n_k in $\mathscr{N}_A$;
      Add a new mapping rule to ψ: $\psi(a_j) \to n_k$;
    end
  end
  return $\psi(\cdot) : \mathscr{S}_A \to \mathscr{N}_A$;

Initial $M \in \mathbb{R}^{|\mathscr{C}| \times P}$ with zeros;
Adaptive mapping function $\psi(\cdot) = \text{AMF}(S_A, K_n)$;
foreach $line \in \mathscr{S}_A$ **do**
  foreach $v_i \in line$ **do**
    foreach $v_j \in line$ *and* $v_i \neq v_j$ **do**
      $\mathbf{v}_{n_i} = \psi(v_i)$; // Transform v_i into $\mathbf{v}_{n_i}$
      $\mathbf{v}_{n_j} = \psi(v_j)$; // Transform v_j into $\mathbf{v}_{n_j}$
      $M[\mathbf{v}_{n_i}][\mathbf{v}_{n_j}] += 1$;
      $M[\mathbf{v}_{n_j}][\mathbf{v}_{n_i}] += 1$;
    end
  end
end
return M;

After performing these procedures, we obtain a global complete attribute graph, where nodes represent the values of variables and edges represent the number of

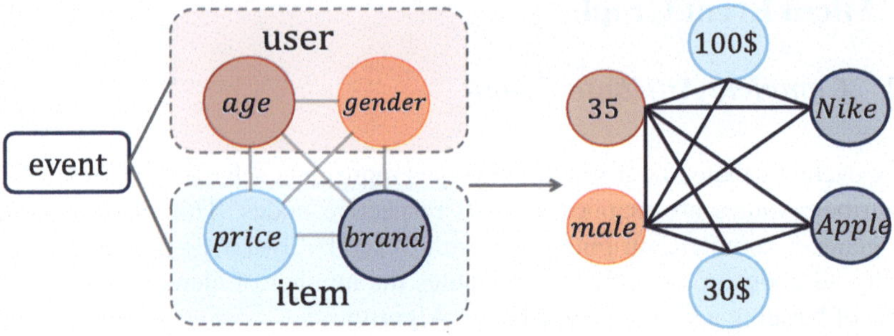

Fig. 6.3 A toy example of the CAG

connections between variable values, indicating the strength of their associations. A toy example of the CAG is given in Fig. 6.3.

The attributes of the event, *age*, *gender*, *price*, and *brand*, are connected as nodes. Suppose a 35-year-old man bought two things: a 100$ Nike pants and a 30$ Apple charger. An event e contains the attribute $\mathscr{A}_e = \{age, gender, price, brand\}$. The first two come from the user and the last two from the item. Connect all four attributes in an event by treating the attribute values as nodes and establishing full connections. Two events are connected to form the attribute graph on the right, linking age/gender and price/brand. By mapping the values of these six variables to different types of nodes, we connect all the nodes within each data row. This process is repeated for all training data.

Our goal is to establish a balanced CAG, preventing attribute values from clustering around a few central nodes. To achieve this, let $\mathscr{S}_A$ represent the set of attribute values, and $\mathscr{N}_A$ denote the set of nodes. We introduce an attribute mapping function (AMF) $\psi(\cdot) : \mathscr{S}_A \rightarrow \mathscr{N}_A$ designed to allocate each attribute value to distinct nodes.

Following this, our approach involves partitioning the sorted attribute values into K_n equally sized subsets, denoted as $\mathscr{S}_A^1, \mathscr{S}_A^2, \ldots, \mathscr{S}_A^{K_n}$, where K_n represents the predefined number of nodes. Each subset $\mathscr{S}_A^i$ contains an identical number of attribute values. Subsequently, we add a mapping rule for each attribute value in each subset $\mathscr{S}_A^i$, assigning it to the corresponding node in $\mathscr{N}_A$. This mapping rule is represented as $\psi(a_j) \rightarrow n_k$, where a_j is an attribute value and n_k is a node in $\mathscr{N}_A$. In doing so, we ensure that each node has an equal number of attribute value mappings. In terms of implementation details, we utilize the matrix $\mathbf{A}_{map}$ to embody the $\psi(\cdot)$. Specifically, the input is a vector of attribute values, and the output is a vector containing the corresponding node indices.

By employing this approach to ensure an equitable distribution of attribute value mappings across nodes, we can construct a balanced CAG, effectively capturing the global associations among attributes without being influenced by a few central nodes.

6.4.2 Dual-Level Attention Mechanism

The model employs a dual-level attention mechanism as described in [40, 41], enabling the calculation of node attention scores from both type-level and node-level perspectives. This dual-level approach effectively captures variations in importance across hierarchical levels.

Neighboring nodes with the same type are aggregated: For type τ, $h_\tau = \sum_{v'} h_{v'}$, where v' belongs to the set $\mathcal{N}_v$ of the neighbors of v and its type is τ. We calculate the type-level attention score $\boldsymbol{\alpha}_\tau$ based on h_v and h_τ:

$$\boldsymbol{\alpha}_\tau = softmax\left(\sigma\left(\mu_\tau^T \cdot [h_v \| h_\tau]\right)\right), \tag{6.2}$$

where σ represents the activation function, μ_τ is the attention vector of the type τ, and $||$ represents the concatenation operation.

The node-level attention score $\beta_{vv'}$ is obtained by incorporating α_τ into the attention mechanism:

$$\boldsymbol{\beta}_{vv'} = softmax\left(\sigma\left(\nu^T \cdot \boldsymbol{\alpha}_{\tau'} [h_v || h_{v'}]\right)\right), \tag{6.3}$$

where ν represents the attention vector and σ represents the activation function. In our work, we employ Leaky ReLU.

The hidden vector propagation rule for attribute nodes is:

$$\mathbf{H}^{(l+1)} = \sigma\left(\sum_{\tau \in \mathcal{A}_e} \mathbf{B}_\tau \cdot \mathbf{H}_\tau^{(l)} \cdot W_\tau^{(l)}\right), \tag{6.4}$$

where $\mathbf{B}_\tau$ is the attention matrix with $\beta_{vv'}$ values.

Attribute embedding $\tilde{\mathbf{A}} = \mathbf{H}^{(l+1)} \cdot \mathbf{A}_{map}$ is obtained after mapping, where $\mathbf{A}_{map} \in \mathbb{R}^{N \times |\mathcal{C}| \times P}$ is the mapping matrix of event attributes, N is the number of events, and P is the number of attribute values, respectively.

6.4.3 Timestamp Embedding

We avoid treating timestamps as attributes due to their weak correlation and difficulty in extracting valuable information. In the event layer, we introduce a specialized network that treats timestamps as a structural element [18, 42], preventing redundancy. We seek a continuous function Φ that maps time to a multidimensional vector space, where Φ is designed to be translation invariant:

$$\Phi(t_1) - \Phi(t_2) = \Phi(t_1 + c) - \Phi(t_2 + c), \tag{6.5}$$

where t_1 and t_2 increase simultaneously and equivalently, ensuring that the difference after the Φ mapping remains constant. To obtain this Φ function, we can define a temporal kernel function $\mathscr{K}$:

$$\mathscr{K}(t_1 + c; t_2 + c) = \Psi(t_1 - t_2) = \Phi(t_1) \cdot \Phi(t_2) = \mathscr{K}(t_1; t_2),$$

where Ψ is a mapping function on the space of real numbers; thus, the span of any pair of timestamps can be inferred from the dot product of time representations. According to Bochner's theorem [43], it is always possible to approximate a periodic function using the Fourier transform. Thus, we can express the function $\Phi(t)$ exactly in a finite-dimensional space as follows:

$$\Phi(t) = \sqrt{\frac{1}{d_T}} \left[\cos(\omega_1 t), \sin(\omega_1 t), \ldots, \cos\left(\omega_{d_T} t\right), \sin\left(\omega_{d_T} t\right)\right]^\top,$$

where $\omega = [\omega_1, \ldots, \omega_{d_T}]^T$ is the learnable parameter.

6.4.4 *Event Graph Learning*

We propose a method for integrating multidimensional information from neighboring nodes. First, we integrate the event node embedding in $\tilde{\mathbf{A}}$ and timestamp embedding Φ:

$$\mathbf{h}^{(l-1)}(t) = \tilde{\mathbf{A}}(t) \| \Phi(t), \tag{6.6}$$

where l represents the layer number.

For event node v_0 at time t, suppose it has neighbors $\mathscr{N}(v_0; t) = \{v_1, v_2, \ldots, v_N\}$, each of which computes $\mathbf{h}_{v_i}(t)$ according to Eq. (6.6), and we stack the information of all neighbors as a matrix:

$$Z(t) = [\mathbf{h}_{v_1}^{(l-1)}(t), \ldots, \mathbf{h}_{v_N}^{(l-1)}(t)]^\top. \tag{6.7}$$

Next, this matrix is utilized in the self-attention network to compute the attention score:

$$\omega = \text{Softmax}\left(\frac{\mathbf{Q}\mathbf{K}^T}{\sqrt{d_k}}\right)\mathbf{V}, \tag{6.8}$$

where **Q**, **K**, and **V** stand for Query, Key, and Value, representing three essential components of the self-attention model. We respectively pass **h**, Z, and Z into **Q**, **K**, **V**.

We can obtain the total embedding $\mathbf{h}^l_{\mathcal{N}}(t)$ by weighting the sum of the neighbor node information:

$$\mathbf{h}^l_{\mathcal{N}}(t) = \sum_{j \in \mathcal{N}_v(t)} \omega^l_{v,j}\, \mathrm{h}^{l-1}_j(t)\,. \tag{6.9}$$

Next, we use Eqs. (6.9) and (6.6) to incorporate neighborhood and query information, respectively, and feed them into a feedforward neural network (FFN) to obtain a nonlinear association between the two embeddings. The equation for computing the updated user embedding is given by:

$$\mathbf{e}^{(l)}(t) = \mathrm{FFN}\left(\mathbf{h}^{(l)}_{\mathcal{N}_v}(t) \| \mathbf{h}^{(l-1)}(t)\right). \tag{6.10}$$

We utilize the multi-head attention mechanism to capture diverse temporal information, enhancing the model's understanding of complex temporal relationships in the events.

6.4.5 *Model Prediction and Optimization*

To compute the attribute loss $\mathcal{L}_{attr}$, we employ the Softmax function to transform the hidden vector $\mathbf{H}$ of output attribute nodes into the predicted attribute value $\hat{y}$. Subsequently, the cross-entropy is employed to quantify the discrepancy between the predicted $\hat{y}$ and the ground truth y. Specifically,

$$\hat{y} = \mathrm{Softmax}(\mathbf{H} \cdot W_{out}), \tag{6.11}$$

$$\mathcal{L}_{attr} = -\sum_{j=1}^{|C|} \sum_{i=1}^{|f_j|} y_i^{(f_j} \log\left(\hat{y}_i^{(f_j)}\right), \tag{6.12}$$

where W_{out} performs a linear transformation on $\mathbf{H}$ to obtain a lower-dimensional representation. This allows for an appropriate input to the softmax function, facilitating the prediction of attribute values. Here, $\hat{y}_i^{(f_j)}$ represents the predicted value for the attribute category f_j and $|f_j|$ indicates the number of this attribute category.

We adopt a pair-wise training approach [44], where we randomly generate 1,000 negative samples for each positive sample. The training set $\mathfrak{D}_T$ consists of tuples (u, i, j, t), where i represents the index of a positive sample and j represents the index of a negative sample, respectively. To optimize the recommendation scores $\mathbf{r}(u, i, t)$ for recommending item i to user u at time t, we utilize binary cross-entropy loss, denoted as $\mathcal{L}_{ev}$, where

$$\mathcal{L}_{ev} = \sum_{(u,i,j,t) \in \mathfrak{D}_T} \log \sigma(r(u, i, t)) + \log \sigma(1 - \mathbf{r}(u, j, t)).$$

Finally, the combined loss function with balance parameter λ can be formulated as:

$$\mathcal{L}_{all} = \lambda \mathcal{L}_{attr} + \mathcal{L}_{ev}. \tag{6.13}$$

6.5 Experiments

- **(RQ1) Performance Comparison**: How does MiEG perform in the sequential recommendation task, compared with the state-of-the-art methods?
- **(RQ2) Ablation Study** : How do different layers (e.g., attribute layer) affect the performance of MiEG?
- **(RQ3) Hyperparameter Study**: How do various hyperparameters affect the performance of MiEG?
- **(RQ4) Degeneration Study**: If we remove the event layer and the model degenerates to a general recommendation model, how does it perform?
- **(RQ5) Efficiency Study**: How does MiEG perform in terms of time and space efficiency?

6.5.1 Experimental Setup

6.5.1.1 Datasets

We investigate the performance of MiEG on four datasets. We conduct our experiments on datasets that are widely used in the state-of-the-art methods.

- **ml100k**[1] [45]: This dataset serves as a standard benchmark for recommendation systems and encompasses ratings from multiple users for numerous movies. Additionally, it incorporates movie metadata and user attribute information.
- **steam**:[2] This dataset contains information on over 40,000 video games from the Steam platform, including user reviews, game genres, and developer information.
- **yelp**:[3] This dataset, a subset of Yelp.com's Challenge 2019 spanning 2014–2016, encompasses reviews of businesses on the Yelp platform. It includes valuable information like user ratings, business categories, and geographical locations.
- **lfm**:[4] The dataset comprises information on social networks, tags, and music artists listened to by a group of 2,000 users. It was collected from the Last.fm API, representing the complete listening habits (till May 5, 2009).

[1] https://grouplens.org/datasets/movielens/.

[2] https://www.kaggle.com/datasets/tamber/steam-video-games.

[3] https://www.yelp.com/dataset.

[4] https://grouplens.org/datasets/hetrec-2011/.

Table 6.3 Statistics of the four datasets

Dataset	ml100k	steam	yelp	lfm
# Users	944	16,288	25,967	1725
# Items	1683	8847	10,753	31,496
# Avg. events/user	105.3	17.1	38.5	579.7
# Avg. events/item	73.6	32.5	92.9	31.7
# Events	100,000	278,288	1,000,000	1,000,000
# Attributes	7	9	28	29
# Sparsity(%)	93.70	99.80	99.98	99.76

We filter out items and users with fewer than five interactions from the datasets. Dataset details are presented in Table 6.3. The ml100k dataset exhibits relatively low sparsity, leading to longer interaction records per user. In contrast, the other datasets have higher sparsity, resulting in shorter average interaction records per user.

To integrate user and item attributes into the model, we merge them into a single table using the event (u, i, t) as the key. We partition the training/validation/testing data based on the interaction timestamps in chronological order, with a ratio of $8 : 1 : 1$ for each dataset.

6.5.1.2 Baselines

- **GRU4Rec** [21] is the first application of RNN (GRU) to the session-based recommendation, with significant improvement over traditional KNN [46] and matrix decomposition.
- **BERT4Rec** [26] is a transformer-based model that uses deep, bidirectional self-attention to model sequences of user behavior.
- **SASRec** [25] is also a transformer-based model that uses a self-attentive mechanism to model historical behavioral information about users.
- **SRGNN** [29] models separated session sequences as graph-structured data and uses graph neural networks to capture complex item transformations.
- **TiSASRec** [47] is a transformer-based model for sequential recommendation with multi-head self-attention and temporal decayed attention.
- **S^{3}Rec** [12] is a self-supervised learning-based model that learns comprehensive user and item representations using four optimization objectives.
- **DIF-SR** [32] enhances the performance of sequential recommendation by decoupling the integration of item embedding and side information in the attention layer and introducing side attribute predictors.
- **TGSRec** [44] is a recent model that takes dynamic temporal information into account, achieving sota in the sequence recommendation problem.
- **PTGCN** [17] combines graph convolution and bipartite graph structures to model user-item interactions while considering temporal dynamics and high-order connectivity.

The baselines used in our experiments can be classified into three categories: traditional sequence neural network-based models such as GRU4Rec, BERT4Rec, and TiSASRec; side information fused models including S^3Rec and DIF-SR; and dynamic graph-based models such as TGSRec and PTGCN.

6.5.1.3 Metrics

Each model generates a list of recommended items ranked for each test case, and its performance is assessed using three commonly used metrics: hit rate, normalized discounted cumulative gain (ndcg), and mean reciprocal rank (mrr):

$$\mathrm{hit}@N = \frac{1}{|\mathcal{U}|}\sum_{u\in\mathcal{U}} \mathbf{1}\left(r_{u,i} \leq N\right), \tag{6.14}$$

$$\mathrm{ndcg}@N = \frac{1}{|\mathcal{U}|}\sum_{u\in\mathcal{U}} \frac{\mathbf{1}\left(r_{u,i} \leq N\right)}{\log_2\left(r_{u,i}+1\right)}, \tag{6.15}$$

$$\mathrm{mrr}\,@N = \frac{1}{|\mathcal{U}|}\sum_{u\in\mathcal{U}} \frac{\mathbf{1}\left(r_{u,i} \leq N\right)}{r_{u,i}}, \tag{6.16}$$

where i represents the ground-truth item for user u, $r_{u,i}$ is the rank generated by a recommendation method for user u and item i, and $\mathbf{1}(\cdot)$ is the indicator function. The indicator function evaluates to 1 if the specified condition is true and 0 otherwise [48]. Evaluation is carried out for the top 10 and top 20 items in the ranking list (with a default of top 10), as these are considered the most relevant to the user.

6.5.1.4 Implementation Details

To ensure a fair comparison, all baseline models are implemented and run within the RecBole framework [49], a PyTorch-based recommender system framework, and evaluated under identical conditions. The Adam optimizer is employed for training all the baselines and our proposed model, with a fixed learning rate of $1e-3$. All models are trained for a maximum of 100 epochs, with early stopping if there is no improvement for three consecutive epochs. The hidden layer dimension is set to $D = 64$, and the batch size is configured as 256, except for BERT4Rec, which utilizes a batch size of 64. In the case of Transformer-based baselines, the number of attention heads is fixed at 2.

Regarding other hyperparameters, we apply a grid search approach to identify the optimal values. The search space includes the embedding size ($D_A \in \{4, 8, 16, 32, 64\}$) of CAG, the embedding size ($D_E \in \{8, 16, 32, 64, 128\}$) of EG, and the number of neighbors $S \in \{1, 3, 5, 10, 15, 20, 25\}$. Furthermore, the balance parameter λ in Eq. (6.13) is set to 0.05. For the remaining hyperparameters in baselines, we follow the suggestions of the original work. All models are trained on four Tesla V100-SXM2-32GB calculate cards.

6.5.2 Contrast Experiments

Experimental results, as presented in Table 6.4, demonstrate the superior performance of MiEG when compared to the baseline models across four datasets, with a marked performance improvement. The best and second-best results are highlighted in **boldface** and underlined, respectively. We have replaced the baseline names in the table with corresponding numbers: [1] **GRU4Rec**, [2] **SRGNN**, [3] **SASRec**, [4] **TiSASRec**, [5] **BERT4Rec**, [6] **S^{3}Rec**, [7] **DIF-SR**, [8] **PTGCN**, and [9] **TGSRec**. Further analysis of these results is provided in the following:

MiEG consistently demonstrates superior performance compared to the baseline methods across a diverse range of settings for all four evaluated datasets. Remarkably, MiEG outperforms other baseline methods in 22 out of 24 dataset and metric combinations. A noteworthy highlight is its exceptional performance gain of over 30% concerning the NDCG metric, an encompassing measure of recommendation quality that takes into account the ranking order of recommended items, compared to the second-best method. These results underscore the effectiveness and robustness of MiEG in enhancing the sequential recommendation task.

Regarding dynamic graph-based methods, TGSRec and PTGCN achieve performance improvements more than twice those of other methods on the ml100k dataset. A comparative analysis leads to the conclusion that on datasets where user-item interactions are frequent and side information is relatively limited, such methods benefit from the dynamic graph architecture, effectively modeling the interaction relationships and temporal information between users and items. Our MiEG approach also incorporates the concept of dynamic graphs, differing in the transformation of bipartite graphs into event graphs, coupled with the incorporation of side information. Consequently, MiEG outperforms dynamic graph methods on such datasets.

For methods that rely on side information, such as S^{3}Rec and DIF-SR, scenarios with abundant side information are more suitable for their effectiveness. Take the steam dataset, for instance, where interaction records are the sparsest, but it integrates rich attributes from both players and games. Consequently, on the steam dataset, DIF-SR outperforms even the latest state-of-the-art methods, whereas its performance on other datasets is not as remarkable. Regarding the exceptional performance of S^{3}Rec on the lfm dataset, our analysis is as follows: S^{3}Rec only learns item attributes while neglecting user attributes, and in lfm, the number of items far exceeds that of users. On a macro level, the model utilizes a higher proportion of information compared to other datasets, thus achieving better performance. This theory can also explain why MiEG outperforms in all datasets. MiEG learns attributes at the event layer, including attributes of items, users, and

Table 6.4 Performance comparison between MiEG and various baselines on the four datasets

Dataset	Metric	[1]	[2]	[3]	[4]	[5]	[6]	[7]	[8]	[9]	MiEG	Imp.
ml100k	ndcg@10	0.0669	0.0650	0.0659	0.0667	0.0345	0.0622	0.0541	0.1705	0.1921	**0.2903**	0.0982
	ndcg@20	0.0922	0.0869	0.0891	0.0891	0.0615	0.0870	0.0847	0.2027	0.1950	**0.2972**	0.0945
	hit@10	0.1230	0.1315	0.1283	0.1400	0.0891	0.1389	0.1488	0.2927	0.2177	**0.3128**	0.0201
	hit@20	0.2344	0.2195	0.2216	0.2301	0.1591	0.2375	0.2600	0.3308	0.2698	**0.3403**	0.0095
	mrr@10	0.0357	0.0352	0.0469	0.0447	0.0209	0.0394	0.0390	0.1740	0.1893	**0.2879**	0.0986
	mrr@20	0.0539	0.0508	0.0531	0.0506	0.0352	0.0461	0.0463	0.2334	0.2010	**0.2997**	0.0663
steam	ndcg@10	0.0770	0.0688	0.0685	0.0785	0.0654	0.1784	0.1999	0.2243	0.1523	**0.3044**	0.0800
	ndcg@20	0.0941	0.0869	0.0958	0.0955	0.0797	0.1912	0.2336	0.2496	0.1630	**0.3237**	0.0740
	hit@10	0.1462	0.1305	0.1269	0.1480	0.1294	0.2151	**0.2561**	0.2148	0.1947	0.2379	−0.018
	hit@20	0.2142	0.2020	0.2161	0.2150	0.1863	0.2994	**0.4864**	0.3978	0.2383	0.4640	−0.022
	mrr@10	0.0526	0.0502	0.0509	0.0576	0.0462	0.0997	0.1515	0.1312	0.1284	**0.1720**	0.0205
	mrr@20	0.0608	0.0550	0.0624	0.0623	0.0501	0.1325	0.1623	0.1441	0.1475	**0.1835**	0.0212
yelp	ndcg@10	0.0521	0.0461	0.0713	0.0728	0.0327	0.0172	0.0427	0.0881	0.0734	**0.1045**	0.0164
	ndcg@20	0.0651	0.0566	0.0847	0.0863	0.0424	0.0354	0.0503	0.0134	0.0913	**0.1574**	0.0661
	hit@10	0.0978	0.0831	0.1212	0.1243	0.0651	0.0344	0.0778	0.0934	0.0847	**0.1256**	0.0013
	hit@20	0.1493	0.1246	0.1744	0.1782	0.1039	0.0463	0.1075	0.1145	0.0865	**0.2144**	0.0362
	mrr@10	0.0227	0.0350	0.0562	0.0571	0.0230	0.0511	0.0405	0.0941	0.0736	**0.1225**	0.0284
	mrr@20	0.0419	0.0378	0.0598	0.0608	0.0256	0.0589	0.0512	0.1147	0.0850	**0.1450**	0.0303
lfm	ndcg@10	0.0189	0.0205	0.0180	0.0199	0.0163	0.0433	0.0176	0.0379	0.0238	**0.0478**	0.0045
	ndcg@20	0.0252	0.0276	0.0236	0.0251	0.0241	0.0539	0.0488	0.0515	0.0418	**0.0550**	0.0011
	hit@10	0.0400	0.0435	0.0348	0.0412	0.0365	0.0741	0.0251	0.0341	0.0178	**0.0798**	0.0057
	hit@20	0.0650	0.0719	0.0574	0.0615	0.0551	0.0971	0.0744	0.0459	0.0200	**0.0976**	0.0005
	mrr@10	0.0126	0.0137	0.0130	0.0136	0.0104	0.0145	0.0112	0.0221	0.0241	**0.0334**	0.0093
	mrr@20	0.0143	0.0156	0.0144	0.0150	0.0122	0.0160	0.0245	0.0270	0.0337	**0.0441**	0.0104

the interaction behaviors themselves, utilizing more information compared to other methods.

Traditional sequence neural network-based models exhibit limitations, likely attributed to their inherent memory constraints, especially when confronted with datasets containing thousands of historical user interactions. In contrast, transformer-based models such as SASRec and TiSASRec demonstrate excellence on both the yelp and ml100k datasets. Their success can be attributed to their proficiency in encoding location data and overcoming sequence length constraints through the transformer architecture. However, BERT4Rec, incorporating the powerful BERT model from NLP, does not perform as well, as pre-trained models are not utilized. Overall, these methods lack a nuanced exploration of fine-grained information and comprehensive consideration of collaborative interactions. They merely treat items as sequences, resulting in suboptimal performance for recommendation tasks.

6.5.3 Ablation Study

We perform ablation experiments to evaluate the importance of the attribute and event layers in MiEG. Two ablated models are introduced for the attribute layer in MiEG: one replaces it with a randomly initialized embedding layer (MiEG-r) and the other utilizes pre-trained GNN parameters as the initial weights for the embedding layer (MiEG-p). The results presented in Tables 6.5 and 6.6 highlight the critical significance of the attribute layer in the MiEG model, demonstrating

Table 6.5 Ablation analysis of MiEG on **ml100k** and **steam** datasets

	ml100k					
Model	ndcg@10	ndcg@20	hit@10	hit@20	mrr@10	mrr@20
MiEG-r	0.2712	0.2834	0.2867	0.2881	0.2706	0.2715
MiEG-p	0.2833	0.2873	0.2761	0.3145	0.2821	0.2891
MiEG-t	0.134	0.159	0.1521	0.1732	0.1394	0.148
MiEG-g	0.1032	0.1396	0.1183	0.211	0.1105	0.1577
MiEG	**0.2903**	**0.2972**	**0.3128**	**0.3403**	**0.2879**	**0.2997**
%Imp.	2.47%	3.45%	9.10%	8.20%	2.06%	3.67%
	steam					
Model	ndcg@10	ndcg@20	hit@10	hit@20	mrr@10	mrr@20
MiEG-r	0.1551	0.2333	0.2083	0.2458	0.151	0.1641
MiEG-p	0.1595	0.2198	0.2152	0.2677	0.1564	0.1638
MiEG-t	0.1217	0.1493	0.1728	0.2361	0.1204	0.1571
MiEG-g	0.0996	0.1106	0.1281	0.2533	0.0804	0.1355
MiEG	**0.3044**	**0.3237**	**0.2379**	**0.464**	**0.172**	**0.1835**
%Imp.	90.85%	38.75%	10.55%	73.33%	9.97%	11.82%

Table 6.6 Ablation analysis of MiEG on **yelp** and **lfm** datasets

	yelp					
Model	ndcg@10	ndcg@20	hit@10	hit@20	mrr@10	mrr@20
MiEG-r	0.0759	0.1033	0.083	0.091	0.0812	0.0856
MiEG-p	<u>0.0858</u>	<u>0.1174</u>	0.0953	0.1198	<u>0.1055</u>	<u>0.112</u>
MiEG-t	0.0789	0.0877	<u>0.1237</u>	<u>0.1779</u>	0.0691	0.0732
MiEG-g	0.0656	0.0701	0.1007	0.167	0.0459	0.0671
MiEG	**0.1045**	**0.1574**	**0.1256**	**0.2144**	**0.1225**	**0.145**
%Imp.	21.79%	34.07%	1.54%	20.52%	16.11%	29.46%
	lfm					
Model	ndcg@10	ndcg@20	hit@10	hit@20	mrr@10	mrr@20
MiEG-r	0.0258	<u>0.043</u>	0.0211	0.0257	0.0242	<u>0.0315</u>
MiEG-p	<u>0.0261</u>	0.0415	0.0239	0.0267	<u>0.024</u>	0.0291
MiEG-t	0.0185	0.0253	0.0396	0.0624	0.0133	0.0149
MiEG-g	0.019	0.0253	<u>0.0404</u>	<u>0.0668</u>	0.0139	0.0147
MiEG	**0.0478**	**0.055**	**0.0798**	**0.0976**	**0.0334**	**0.0441**
%Imp.	83.14%	27.91%	97.52%	46.11%	39.17%	40.00%

improvements in prediction performance and convergence speed. The best and second-best results are highlighted in **boldface** and <u>underlined</u>, respectively.

While the use of pre-trained weights enhances performance, the attribute layer remains indispensable. We also find that removing the attribute layer has minimal impact on performance on the ml100k. However, for steam and lfm, its influence is significant. Our analysis suggests that sparser datasets rely more on the attribute layer to provide additional information, aiding the model in modeling users and items with limited interaction records.

It is worth noting that even in the absence of the attribute layer, the model outperforms the state-of-the-art methods, namely, TGSRec and PTGCN, based on dynamic graphs. Comparative analysis suggests that MiEG's performance enhancement primarily stems from the transformation of the dynamic graph into an EG. In dynamic bipartite graphs, temporal information associated with edges may induce modeling instability due to multiple timestamps on a single edge. In contrast, EG consolidates temporal information within event nodes, mitigating duplicated edges and providing a more stable modeling approach.

In addition, we replace the event layer with two alternative sequential networks: Transformer (MiEG-t) and GRU (MiEG-g). To ensure a fair comparison with the GRU4Rec implementation, we set a maximum length of 200. Experimental results indicate that these replacements do not result in satisfactory performance. One contributing factor is the length limitation of GRU, constraining its ability to consider the entire interaction history. Moreover, these two ablation models cannot capture associative information between users and items, a role precisely fulfilled by the event layer. Nevertheless, the model still outperforms GRU4Rec, emphasizing the significant contribution of the attribute layer.

6.5.4 Parameter Study

We conduct experiments to scrutinize the impact of hyperparameters on our model. These hyperparameters include the number of neighbors (S), the embedding size in CAG learning (D_A), and the embedding size in EG learning (D_E). The outcomes of these experiments are depicted in Fig. 6.4. The plots on the left side show the variation in ndcg@10 (rectangular dots) and hit@10 (triangular dots) on the **ml100k** dataset as the S, D_A, D_E are varied, respectively. The plots on the right side depict the same metrics on the **steam** dataset.

6.5.4.1 Sensitivity of S to MiEG

Our investigation on S reveals interesting findings. Specifically, for the ml100k dataset, we observe that performance improves as S increases. This can be attributed

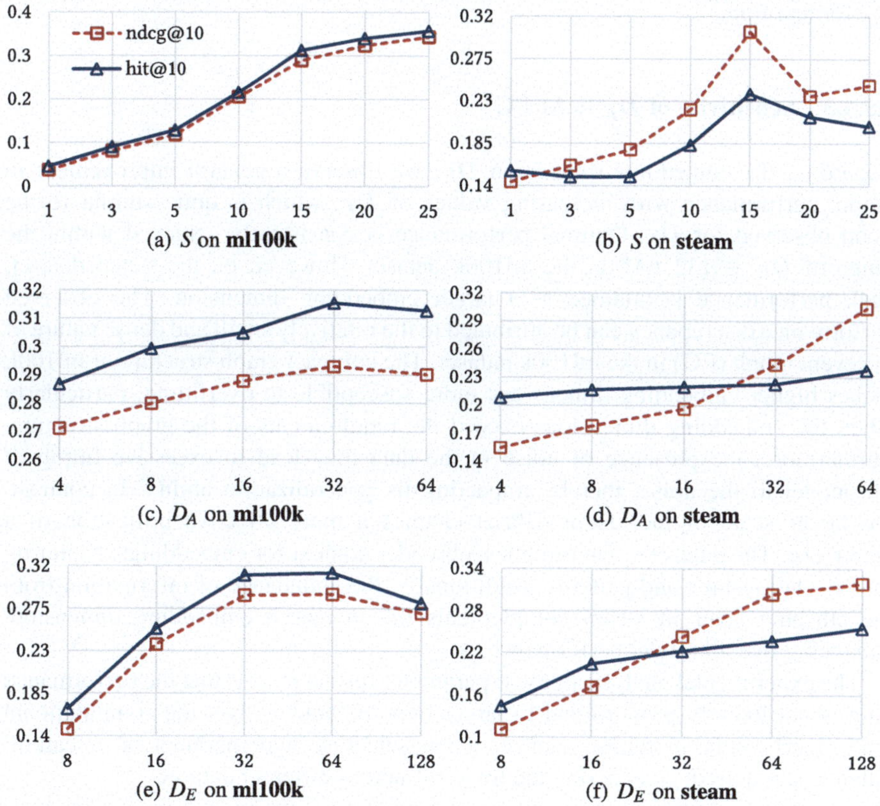

Fig. 6.4 Results of different hyperparameters

to the enhanced capacity to incorporate and integrate information from neighboring nodes. However, in the case of the sparse steam dataset, we observe that the optimal performance is attained when $S = 15$. This can be attributed to the relatively small average user behavior of 17.1 (Table 6.3).

6.5.4.2 Sensitivity of D_A to MiEG

Regarding the representation dimensionality for CAG learning, from the experimental results, we can observe that performance generally increases with higher dimensions. This can be easily explained: a higher embedding dimension can encompass more information, thus more accurately representing attributes, which in turn enhances modeling precision at the event and agent layers. Additionally, it is noteworthy that on the ml100k dataset, the performance saturates at $D_A = 32$, while on the steam dataset, it has not reached the limit even at $D_A = 64$. This is because ml100k contains only seven attributes, while steam includes a much larger number of attributes, resulting in a larger size for CAG and necessitating a larger D_A for learning.

6.5.4.3 Sensitivity of D_E to MiEG

Regarding the embedding dimension D_E, we observe a general improvement in model performance with increasing values of D_E, which is quite similar to the trend observed for D_A. Optimal performance is consistently achieved within the range of $D_E \in [32, 64]$ on the ml100k dataset. However, on the steam dataset, peak performance is attained at a larger embedding dimension. The observed performance discrepancy can be attributed to the relatively small and dense nature of the event graph (EG) in the ml100k dataset. The compact graph structure of ml100k makes higher embedding dimensions more susceptible to overfitting, particularly when the embedding dimensions exceed the requirements of the graph structure. Furthermore, the presence of noise in the data may lead to excessive fitting of the model to the noise, thereby impacting its generalization ability. In contrast, the larger scale of the steam dataset enables a more effective utilization of a larger D_E. The dataset's size provides a broader context for embeddings, capturing richer relationships and patterns. Additionally, the abundance of information from the attribute layer in steam complements the increased embedding dimension, enhancing overall model performance.

The experimental analysis of hyperparameter results reveals that the performance gain is not linearly proportional to larger hyperparameters, yet the computational load significantly increases. Hence, when selecting hyperparameters, a careful balance is required, considering the trade-off across different datasets.

6.5.5 Degeneration Study

In this experiment, we investigate the model's performance in the general recommendation problem and assess its degradation. The main distinction between general and sequential recommendations is that general recommendations do not include temporal relationships in their datasets; thus, the event layer does not contribute in this context. We remove the event layer and modify the model to an attribute layer spliced with the agent layer (MiEG-d) for the general recommendation problem, as temporal relationships are not included in the datasets. LightGCN [34] is a lightweight graph convolutional network designed with the goal of simplicity and effectiveness, aiming to deliver competitive performance. Widely utilized in general recommendation scenarios, it serves as a benchmark for comparison. Therefore, we adopt LightGCN as the benchmark for evaluating the degradation of MiEG in our study. The metrics adopted for comparison are gauc [50], hit, ndcg, and mrr.

In a different problem domain, our model, without the event layer, exhibits competitive performance, as illustrated in Fig. 6.5, trailing the baseline by a margin of less than 2%. This confirms the effectiveness of the attribute layer in acquiring side information, a crucial factor for the proficiency of recommendation systems.

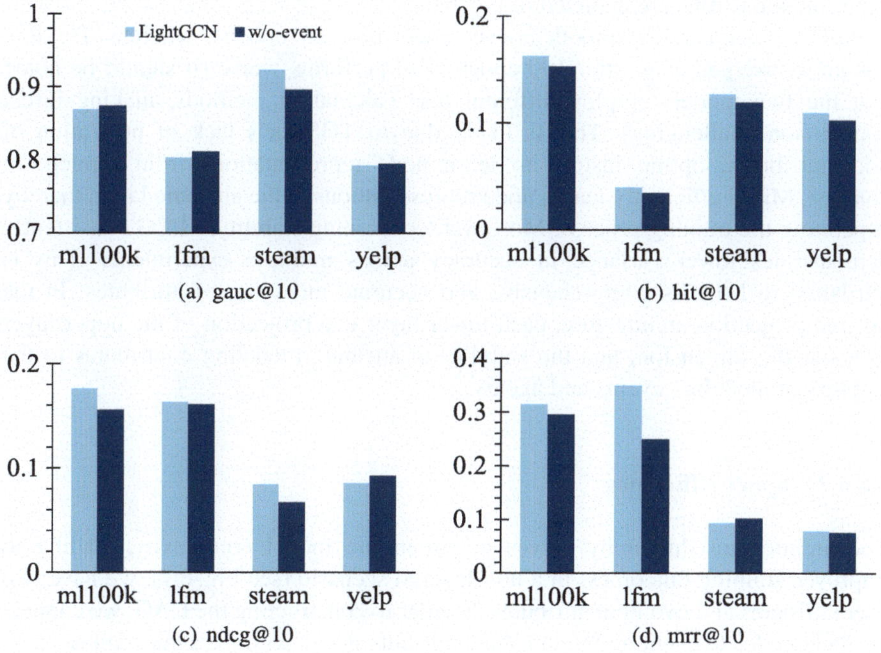

Fig. 6.5 Degenerated MiEG-d vs. LightGCN on four datasets

The observed reduction in recommendation accuracy underscores the indispensability of the event layer in MiEG. This layer plays a crucial role in enabling the model to fully leverage its capacity to amalgamate information across multiple dimensions. In the absence of temporal information from the event layer, the model's performance is inferior to that of other models.

6.5.6 *Efficiency Study*

6.5.6.1 Time Efficiency

For deep learning models, there are variations in training strategies, including batch-wise training, full-graph training, the incorporation of pre-trained models, and two-stage training. Consequently, it is challenging to perform a unified timing comparison. Therefore, we assess the model time efficiency by monitoring its convergence. We choose to use TGSRec as our reference for comparison because it shares similarities with MiEG: a dynamic graph-based method, employing full-graph training, operating as a one-stage model, and lacking pre-training. We disable the early stop option, allowing the model to complete 100 epochs. After each epoch, we perform validation, recording both loss and accuracy for analysis. The experimental results are depicted in Fig. 6.6.

MiEG demonstrates smooth convergence after 10 epochs, whereas TGSRec has not converged even after 100 epochs and performs worse (it should be noted that the two models employ different loss calculation methods, making direct comparison challenging). This is likely due to TGSRec's lack of utilization of node attributes, opting instead to learn node representations from scratch. In contrast, MiEG efficiently learns node representations at the attribute layer, thereby expediting the training process. Moreover, concerning stability, MiEG consistently demonstrates lower variance in accuracy across multiple experiments. This is attributed to MiEG's comprehensive and accurate modeling of attributes. In the layered projection architecture, each lower layer is a projection of the upper layer in a specific dimension, and the stability in attribute modeling contributes to the stability in modeling events and agents.

6.5.6.2 Space Efficiency

The architectural hierarchy serves to streamline model complexity, leading to improved training outcomes. In a non-layered scenario (see Fig. 6.7), we have two user attributes and two item attributes. Similar to constructing the CAG, we connect attribute nodes to event nodes and share attribute nodes with the same values.

Achieving equivalent information utilization, i.e., considering attribute correlations, necessitates extending $|\mathscr{C}|$ attribute nodes from each event node, resulting in a vast heterogeneous graph (naive architecture). In this case, spatial complexity

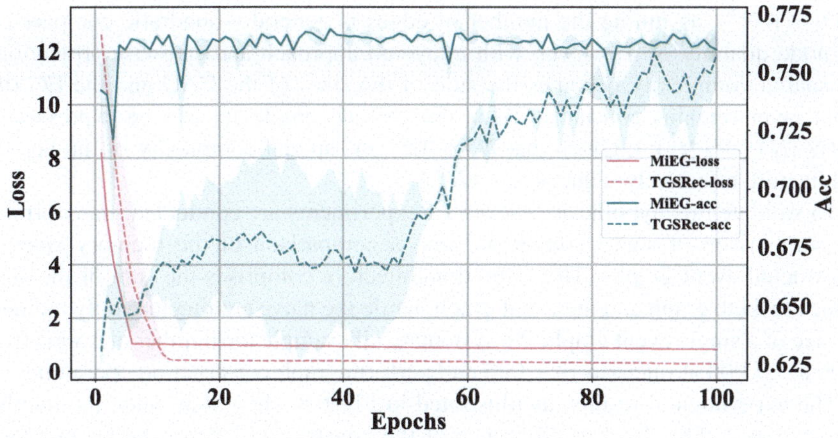

(a) **ml100k**

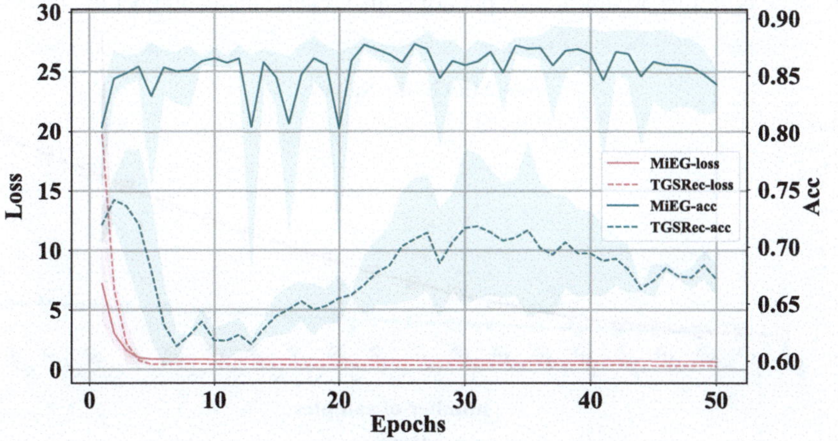

(b) **steam**

Fig. 6.6 The change in loss and accuracy (Acc) with increasing epoch as training progresses

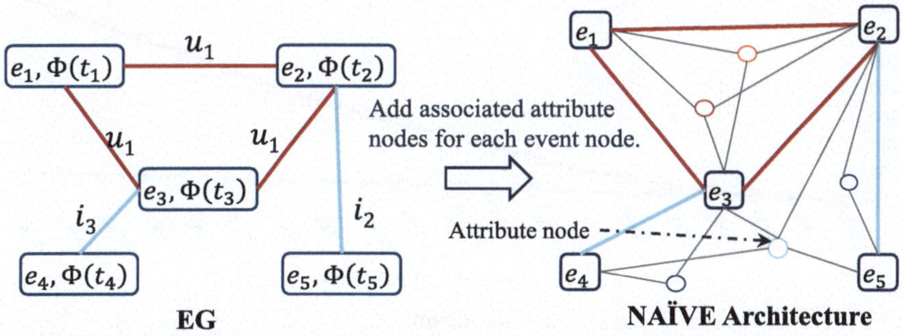

Fig. 6.7 Schematic diagram of naive architecture

is $O(|\mathscr{C}|^2 N^4)$, assuming the number of edges is generally quadratic compared to the node quantity N. However, with a layered approach and stepwise projections, the spatial complexity becomes the sum of the sizes of the CAG and the EG (the agent layer remains constant). Thus, the spatial complexity can be expressed as $O((|\mathscr{C}|N)^2 + (N^2)^2)$. Given that $N \gg |\mathscr{C}|$, the spatial complexity of the layered architecture effectively reduces to $O(N^4)$.

To validate this hypothesis, a series of experiments are conducted, transforming the comparison of space complexity into a comparison of the memory sizes of constructed event graphs. The layered architecture comprises the sum of the sizes of the attribute graph and the event graph, while the naive architecture only includes the size of a single event graph. We systematically sample the number of events from 1000 to 20,000 at intervals of 1,000 and perform graph construction operations.

The experimental results, as illustrated in Fig. 6.8, show that when the number of events is 1,000, the size difference of the constructed heterogeneous graphs is relatively small. However, when the number of events reaches 20,000, the size difference is almost 50 times. This pattern is also related to the number of attributes

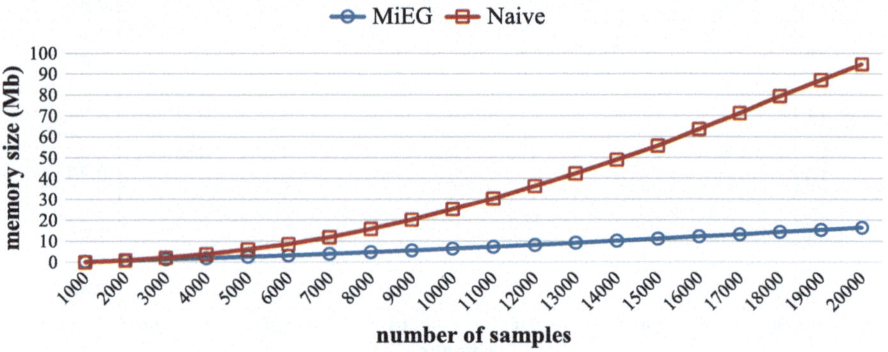

(a) **ml100k**

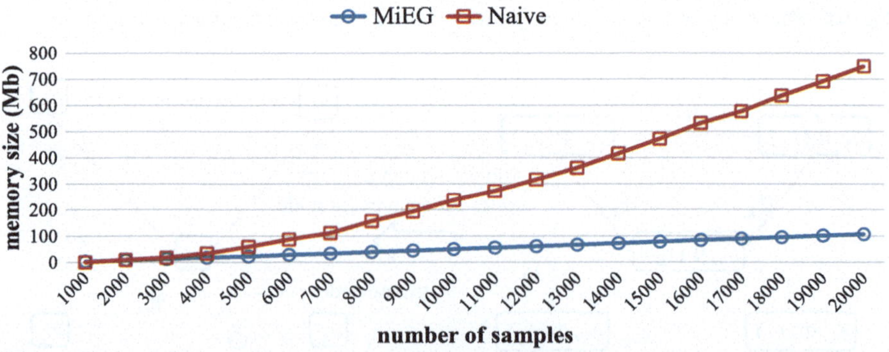

(b) **steam**

Fig. 6.8 The memory size of graph construction under different sampling quantities

K, as datasets with a larger K, such as the steam dataset, exhibit a steeper increase in complexity when using the naive architecture. The experimental results align well with the theoretical analysis.

6.6 Conclusions

We propose a novel model named MiEG, for integrating attribute, temporal, and topological information to enhance the utilization of sparse data for more accurate predictions. Specifically, our proposed scheme leverages the complete attribute graph to extract fine-grained global attribute information. Additionally, we incorporate this information into the event graph, considering the dynamic relationships of events over time and extracting agent representations from event nodes. Furthermore, MiEG can serve as a multi-level information fusion framework, projecting micro-level attribute information onto macro-level event space for agent modeling. Extensive experiments demonstrate that MiEG outperforms the state-of-the-art baselines.

In future research efforts, there is a critical need to explore practical pre-training models aimed at accelerating the training process. An intriguing avenue for further investigation involves the integration of subgraph learning into the model, thereby enhancing its capability for parallel computation.

References

1. R. Ying, R. He, K. Chen, P. Eksombatchai, W.L. Hamilton, J. Leskovec, in *Proceedings of the 24th ACM SIGKDD International Conference on Knowledge Discovery and Data Mining* (2018), pp. 974–983
2. P. Covington, J. Adams, E. Sargin, in *Proceedings of the 10th ACM Conference on Recommender Systems* (2016), pp. 191–198
3. X. He, L. Liao, H. Zhang, L. Nie, X. Hu, T.S. Chua, in *Proceedings of the 26th International Conference on world wide web* (2017), pp. 173–182
4. S. Rendle, C. Freudenthaler, L. Schmidt-Thieme, in *Proceedings of the 19th International Conference on World Wide Web* (2010), pp. 811–820
5. R. He, J. McAuley, in *2016 IEEE 16th International Conference on Data Mining* (IEEE, New York, 2016), pp. 191–200
6. A. Vaswani, N. Shazeer, N. Parmar, J. Uszkoreit, L. Jones, A.N. Gomez, Ł. Kaiser, I. Polosukhin, in *Proceedings of the 31st Conference on Neural Information Processing Systems*, vol. 30 (2017)
7. Z. Yu, J. Lian, A. Mahmoody, G. Liu, X. Xie, in *Proceedings of the 28th International Joint Conference on Artificial Intelligence* (2019), pp. 4213–4219
8. P. Velickovic, G. Cucurull, A. Casanova, A. Romero, P. Lio, Y. Bengio: Stat **1050**, 20 (2017)
9. Z. Wang, W. Wei, G. Cong, X.L. Li, X.L. Mao, M. Qiu, in *Proceedings of the 43rd International ACM SIGIR Conference on Research and Development in Information Retrieval* (2020), pp. 169–178

10. R. Qiu, J. Li, Z. Huang, H. Yin, in *Proceedings of the 28th ACM International Conference on Information and Knowledge Management*, pp. 579–588 (2019)
11. X. Yuan, D. Duan, L. Tong, L. Shi, C. Zhang, in *Proceedings of the 44th International ACM SIGIR Conference on Research and Development in Information Retrieval* (2021), pp. 1687–1691
12. K. Zhou, H. Wang, W.X. Zhao, Y. Zhu, S. Wang, F. Zhang, Z. Wang, J.R. Wen, in *Proceedings of the 29th ACM International Conference on Information and Knowledge Management* (2020), pp. 1893–1902
13. T. Zhang, P. Zhao, Y. Liu, V.S. Sheng, J. Xu, D. Wang, G. Liu, X. Zhou, in *International Joint Conference on Artificial Intelligence* (2019), pp. 4320–4326
14. K. Cho, B. Van Merriënboer, C. Gulcehre, D. Bahdanau, F. Bougares, H. Schwenk, Y. Bengio, arXiv preprint arXiv:1406.1078 (2014)
15. J. Devlin, M.W. Chang, K. Lee, K. Toutanova, arXiv preprint arXiv:1810.04805 (2018)
16. M. Zhang, S. Wu, X. Yu, Q. Liu, L. Wang, IEEE Trans. Knowl. Data Eng. **35**(5), 1–1 (2022). https://doi.org/10.1109/tkde.2022.3151618
17. L. Huang, Y. Ma, Y. Liu, B. Danny Du, S. Wang, D. Li, ACM Trans. Inf. Syst. **41**(1), 1 (2023)
18. D. Xu, C. Ruan, E. Korpeoglu, S. Kumar, K. Achan, arXiv preprint arXiv:2002.07962 (2020)
19. Z. Fan, Z. Liu, J. Zhang, Y. Xiong, L. Zheng, P.S. Yu, in *Proceedings of the 30th ACM International Conference on Information and Knowledge Management* (2021), pp. 433–442
20. M. Quadrana, P. Cremonesi, D. Jannach, ACM Comput. Surv. (CSUR) **51**(4), 1 (2018)
21. B. Hidasi, A. Karatzoglou, L. Baltrunas, D. Tikk, arXiv preprint arXiv:1511.06939 (2015)
22. W. Zaremba, I. Sutskever, O. Vinyals, arXiv preprint arXiv:1409.2329 (2014)
23. S.Y. Shih, F.K. Sun, H.y. Lee, Mach. Learn. **108**(8), 1421 (2019)
24. J. Tang, K. Wang, in *Proceedings of the 11th ACM International Conference on Web Search and Data Mining* (2018), pp. 565–573
25. W.C. Kang, J. McAuley, in *Proceedings of the 18th IEEE International Conference on Data Mining* (IEEE, New York, 2018), pp. 197–206
26. F. Sun, J. Liu, J. Wu, C. Pei, X. Lin, W. Ou, P. Jiang, in *Proceedings of the 28th ACM International Conference on Information and Knowledge Management* (2019), pp. 1441–1450. https://doi.org/10.1145/3357384.3357895
27. Q. Liu, Y. Zeng, R. Mokhosi, H. Zhang, in *Proceedings of the 24th ACM SIGKDD International Conference on Knowledge Discovery and Data Mining* (2018), pp. 1831–1839
28. R. Qiu, H. Yin, Z. Huang, T. Chen, in *Proceedings of the 43rd International ACM SIGIR Conference on Research and Development in Information Retrieval, SIGIR 2020, Virtual Event, China, July 25–30, 2020* (2020), pp. 669–678. https://doi.org/10.1145/3397271.3401109
29. S. Wu, Y. Tang, Y. Zhu, L. Wang, X. Xie, T. Tan, in *Proceedings of the 33rd AAAI Conference on Artificial Intelligence* (2019), pp. 346–353. https://doi.org/10.1609/aaai.v33i01.3301346
30. B. Hidasi, M. Quadrana, A. Karatzoglou, D. Tikk, in *Proceedings of the 10th ACM Conference on Recommender Systems* (2016), pp. 241–248
31. C. Liu, X. Li, G. Cai, Z. Dong, H. Zhu, L. Shang, in *Proceedings of the 35th AAAI Conference on Artificial Intelligence*, vol. 35 (2021), pp. 4249–4256
32. Y. Xie, P. Zhou, S. Kim: arXiv preprint arXiv:2204.11046 (2022)
33. X. Wang, X. He, Y. Cao, M. Liu, T.S. Chua, *Proceedings of the 25th ACM SIGKDD International Conference on Knowledge Discovery and Data Mining* (2019), pp. 950–958
34. X. He, K. Deng, X. Wang, Y. Li, Y. Zhang, M. Wang, in *Proceedings of the 43rd International ACM SIGIR Conference on Research and Development in Information Retrieval* (2020), pp. 639–648
35. G.H. Nguyen, J.B. Lee, R.A. Rossi, N.K. Ahmed, E. Koh, S. Kim, in *Companion proceedings of the Web Conference 2018* (2018), pp. 969–976
36. X. Chang, X. Liu, J. Wen, S. Li, Y. Fang, L. Song, Y. Qi, in *Proceedings of the 29th ACM International Conference on Information and Knowledge Management* (2020), pp. 145–154
37. Y. Fang, L. Si, in *Proceedings of the 2nd International Workshop on Information Heterogeneity and Fusion in Recommender Systems* (2011), pp. 65–69

38. F. Johansson, U. Shalit, D. Sontag, in *International Conference on Machine Learning* (PMLR, New York, 2016), pp. 3020–3029
39. U. Shaham, X. Cheng, O. Dror, A. Jaffe, B. Nadler, J. Chang, Y. Kluger, in *International Conference on Machine Learning* (PMLR, New York, 2016), pp. 30–39
40. T. Yang, L. Hu, C. Shi, H. Ji, X. Li, L. Nie, ACM Trans. Inf. Syst. **39**(3), 32:1 (2021). https://doi.org/10.1145/3450352
41. H. Linmei, T. Yang, C. Shi, H. Ji, X. Li, *Proceedings of the 2019 Conference on Empirical Methods in Natural Language Processing and the 9th International Joint Conference on Natural Language Processing (EMNLP-IJCNLP)* (2019), pp. 4821–4830
42. D. Xu, C. Ruan, E. Korpeoglu, S. Kumar, K. Achan, Adv. Neural Inf. Proces. Syst. **32**, 15889–15899 (2019)
43. S.M. Kazemi, R. Goel, S. Eghbali, J. Ramanan, J. Sahota, S. Thakur, S. Wu, C. Smyth, P. Poupart, M. Brubaker, arXiv preprint arXiv:1907.05321 (2019)
44. Z. Fan, Z. Liu, J. Zhang, Y. Xiong, L. Zheng, P.S. Yu, in *Proceedings of the 30th ACM International Conference on Information and Knowledge Management*, pp. 433–442 (2021). https://doi.org/10.1145/3459637.3482242
45. F.M. Harper, J.A. Konstan: ACM Trans. Interact. Intell. Syst. (TIIS) **5**(4), 1 (2015)
46. T. Cover, P. Hart, IEEE Trans. Inf. Theory **13**(1), 21 (1967)
47. J. Li, Y. Wang, J. McAuley, in *Proceedings of the 13th International Conference on Web Search and Data Mining* (2020), pp. 322–330
48. T. Zhu, L. Sun, G. Chen, IEEE Trans. Knowl. Data Eng. **35**(1), 496 (2023). https://doi.org/10.1109/TKDE.2021.3073411
49. W.X. Zhao, Y. Hou, X. Pan, C. Yang, Z. Zhang, Z. Lin, J. Zhang, S. Bian, J. Tang, W. Sun, et al., arXiv preprint arXiv:2206.07351 (2022)
50. D. Song, D.A. Meyer, D. Tao, in *Proceedings of the 21th ACM SIGKDD International Conference on Knowledge Discovery and Data Mining* (2015), pp. 1105–1114

Chapter 7
Collaborative Behavioral Protection in Anti-fraud Systems

Abstract Anti-fraud engineering for online credit loan (OCL) platforms is getting more challenging due to the developing specialization of gang fraud. State-of-the-art solutions employ graph-based methods to mine hidden associations among loan applications effectively. They perform well based on the information asymmetry that is guaranteed by the huge advantage of platforms over fraudsters in terms of data quantity and quality at their disposal. The inherent difficulty that can be foreseen is the *data isolation* caused by mistrust between multiple platforms and data control legislations for privacy preservation. To maintain the advantage owned by the platforms, we design a privacy-preserving distributed graph learning framework that ensures critical association repairs by merging parameter sharing and data sharing. Specially, we propose the *association reconstruction mechanism* (ARM) that consists of the devised exploration, processing, transmission, and utilization schemes to realize data sharing. For parameter sharing, we design a hybrid encryption technique to protect privacy during collaboratively learning graph neural network models among different financial client platforms. We conduct the experiments over real-life data from large financial platforms. The results demonstrate the effectiveness and efficiency of our proposed methods.

7.1 Introduction to Critical Association Repair

Risk control is the core problem in the financial field. As one of the major financial businesses, the credit loan business is facing more challenges in risk control, while it is booming over the Internet [1]. Large volume and quick processing of business directly poses a greater challenge to its risk control. Especially for its newly developed pattern, say *online credit loan* (OCL) business [2], behind providing high efficiency for platforms and high convenience for clients, it also provides fraudsters more opportunities to submit the forged or misused information to defraud the loan [3] since they are usually not required to be personally identified and assessed offline. Thwarting the loan frauds that cause a significant loss of assets is indeed one of the key parts of OCL risk control. This is just the task of this work.

C. Wang, H. Zhu, *Universal Behavior Computing for Security and Safety*,
https://doi.org/10.1007/978-981-97-9014-2_7

Actually, there are rich associations between different entities in financial services, which are beneficial to the fraud detection problem [4]. Graph-structured data is an efficient carrier of significant association information. The graph-based models, e.g., graph neural networks (GNNs), have been widely applied to capture and mine associations in OCL business [5, 6]. Their superior performance is significantly affected by data quantity and quality, i.e., abundant and complete associations observed by OCL platforms. To reduce the platforms' advantage in terms of information asymmetry, fraudsters distribute load applications across different platforms to hide associations. This tactic of fraudsters seems straightforward but can effectively fool the existing solutions by exploiting data isolation that is caused by the mistrust among multiple platforms and the strict constraint of privacy protection, such as CCPA [7], GPDR [8], and DSL [9]. This is especially fatal for OCL fraud prediction that is an extremely association-sensitive problem.

To combat this multi-platform fraud, a feasible solution seems to be reached: Graph-based learning methods under federated learning architecture, also named federated graph learning (FGL) [10, 11], could be utilized to counter the above tactic of fraudsters. However, this conclusion is obtained by neglecting a fact that fraud prediction of OCL business is a typically association-sensitive problem. It means that such a prediction model learning relies on data associations heavily, while some critical associations are severed by the distributed learning architecture of FGL. The performance of FGL is often greatly impaired by the absence of critical associations. As a matter of fact, the parameter sharing scheme of FGL can alleviate the performance degradation caused by decentralized graph data to some extent, but it cannot solve the loss of associations across different platforms [12]. Therefore, the key challenge can be reduced to how to energize GNNs by recovering these critical associations for each client platform without violating their requirement for privacy preservation.

To address this key challenge, we design a privacy-preserving distributed graph learning framework for cross-platform OCL fraud prediction by a fusion of parameter sharing and data sharing. The core of our framework is the *association reconstruction mechanism* (ARM), under which, without sacrificing privacy, each client platform can recover the explicit associations split by cross-platform loan applications. Then, those recovered associations can energize GNN models on client platforms to further mine hidden associations.

The main idea of ARM is to share the appropriate data with high utility for association recovery and little impact on privacy concerns. It selects the fraudulent application information at every client platform as the data to share, since they are usually critical to fraud prediction but represent a very small percentage of the overall data according to the characteristics of OCL applications [13]. Even so we must state that the fraudulent application information also presents privacy concerns, though sharing information about fraudsters (e.g., blacklists) is often adopted by the joint risk control among multiple financial platforms [14].

To mitigate privacy breaches in the shared fraud data, before data sharing, the ARM creates a four-step data operation, called *selecting-shattering-shuffling-desensitizing* (SSSD):

Step 1. Selecting the data to be shared from the heterogeneous associated graph of applicant information at every client platform according to the specific data requirements of platforms. An on-demand graph search algorithm is designed to realize the selection by customizing meta-paths for specific data requirements. It ensures that as little information as possible is shared while the useful information is shared.

Step 2. Shattering the selected pieces of applicant information records into field values by disassociating the selected application IDs and corresponding properties. It makes the shared data field-level instead of record-level.

Step 3. Shuffling the scattering field values randomly. It makes the transferred/shared field-level data out-of-order to avoid some information associations that involve privacy are inferred by sequential analysis.

Step 4. Desensitizing the selected field values to ensure transmission security. It makes the transferred/shared data not the content of field values but their identification in the information association network.

By the SSSD operation of ARM, the shared data can help client platforms learn more association information, but are of no use to attackers since they do not contain user's personal information, e.g., the attributes of applicants. From there, based on the shared data, a two-step association operation, called *exploring-repairing* (ER), is designed for the ARM to improve the utility of shared data on client platforms. It explores the useful shared data from those aggregated in the central server according to the features required by GNN models of client platforms. Based on the explored data, each client platform repairs the corresponding associations.

Under the proposed framework centered on the ARM, for all client platforms, their GNN models for heterogeneous associated graphs of applicant information are trained cooperatively by parameter sharing and data sharing via the central server. Data sharing is an iterative process. In each iteration, there are two interactions between the central server and client platforms. In the first interaction, client platforms submit their data requests to the central server that aggregates these requests and determines the set of meta-paths and corresponding search rules by analyzing the structure of the local model and then broadcasts the information of reconstruction tasks to different platforms. In the second interaction, after receiving the information of reconstruction tasks, each client platform performs the SSSD operation and returns its shared data to the central server which aggregates the shared data and returns the corresponding increment of the associated information and repair rules to client platforms. After receiving the shared data, each client platform performs the ER operation, updates data requests, and launches the next iteration. The data sharing process continues to iterate until all client platforms have no new data requirements. Parameter sharing starts after the data sharing process. The shared parameter also contains privacy information. Thus, we apply a hybrid encryption technique to preserve the privacy of both the parameter and the key used to encrypt the parameter during model training process. The central server sends current encrypted parameters to each client platform, and each client computes an update to the current global GNN model, and then the central server aggregates these updates. The parameter sharing process repeats until convergence.

The main contributions are summarized as follows:

- We propose a novel privacy-preserving distributed graph learning framework for OCL fraud prediction by merging parameter sharing and data sharing.
- Under our framework, the association reconstruction mechanism (ARM) is devised to recover the key associations for different client platforms, and the parameter protection with a hybrid encryption technique is realized in collaborative model training.
- We propose selecting-shattering-shuffling-desensitizing (SSSD) operations to alleviate privacy concerns before sharing data and exploring-repairing (ER) operations to ensure the utility after receiving the shared data.
- Extensive experiments have been conducted on real-world commercial client platforms. The results show that the average Kolmogorov-Smirnov (KS) value [15] of different client platforms is 45.6%,[1] which demonstrates that our method can efficiently detect cross-platform OCL frauds.

7.2 Related Work

We first introduce methods of anti-fraud engineering in the financial field and then describe some graph learning algorithms for anti-fraud. Finally, we introduce the recent researches about federated graph learning.

7.2.1 Financial Anti-fraud Methods

Financial crime is a growing class of criminal activity which poses a serious threat to a country's economic development and social security. The traditional anti-fraud methods include black/white lists and rule engines. These methods rely on expert knowledge heavily and have limited detection effect. The machine learning methods have gradually become the new favorite of anti-fraud [16] in recent years due to their ability to capture new types of frauds and a high degree of automation. E.T.-RNN[17] uses deep neural networks to process customers' transactional data for credit scoring applications in the banking industry. Three independent classification modules are ensembled [18] to analyze different aspects of public profiles, which can help understand the characteristic distinctions between scammer profiles and

[1] The achieved performance, i.e., the KS value of 45.6%, is indeed competitive according to the de facto standards for OCL fraud prediction [5]. Note that this performance is obtained in a decentralized pattern under the precondition that different datasets cannot be aggregated from different client platforms, which meets the stricter legal rules of data control.

those of regular users. FlowScope [19] uses a multipartite graph to model the transactions for detecting the complete flow of money from source to destination. An effective shallow model called Encoding Broad Learning System (EBLS) is designed [20] for achieving a faster training speed and ensuring the accuracy and timeliness of online fraud identification.

These methods inefficiently detect complex and conceal fraud patterns for their inability of representing and mining associations between different entities.

7.2.2 Graph Learning Algorithms for Anti-fraud

Due to the good ability of graph neural networks (GNNs) to handle and mine relationships between objects, some methods have applied them to anti-fraud management for OCL business. Some inductive GNN models are widely used to capture relationships between objects, e.g., GraphSAGE [21] and GAT [22]. The dynamic default prediction framework [23] adopts gated recursive neural networks and uses the attention mechanism to integrate both the loan behavior and network information. But most real-world graphs are heterogeneous information networks with multi-typed nodes and edges, and some methods design specialized module to aggregate embeddings of them, such as HetGNN [24] and HAN [25]. A novel data enhancement scheme [26] adopts heterogeneous relation networks for behavioral modeling by representing and mining more fine-grained attribute-level co-occurrences. The hierarchical attention mechanism [5] is applied to correlate different neighbors and different views for financial fraud detection.

These algorithms are effective in a centralized pattern where graph data can be aggregated from multiple platforms. If loan applications are submitted on different platforms separately, the performance of fraud prediction may degenerate because of insufficient local graph data.

7.2.3 Distributed Graph Learning

Recently, some researchers have made progress in federated graph learning. Several existing frameworks are proposed to address graph data learning tasks [27–30]. Some of them [27, 28] enable collaborative training for handling graph-level tasks and characterize the property of the entire input graph, which are inapplicable to applicant node classification task in the anti-fraud system. Suzumura [29] proposes an efficient solution for the collaborative training of the financial crime detection model across multiple financial institutions. This method does not consider the impact of missing critical associations on fraud detection performance and lacks a mechanism for enhancing associations of different subgraphs. Another federated learning framework called FedGNN [30] employs a pseudo interacted item sampling technique to expand local subgraph. It enriches local graph information by directly sharing coarse-grained embeddings of neighbors' nodes based on the item sampling technique during the training process, but it is unable to repair key associations accurately based on field-level data before the training process.

To protect the parameters of models, some existing methods apply homomorphic encryption [31, 32] to parameters. The homomorphic encryption needs huge storage and communication costs during model training [33], and the structure of heterogeneous GNN is complex with more parameters that increase the challenge of performing homomorphic encryption. The local differential privacy (LDP) [34] is an efficient strategy to protect parameters [35, 36], but the randomly generated noise may affect the performance of models [37–39].

Distributed anti-fraud prediction is challenging because cross-platform associations are completely missing and must be accurately reconstructed without causing privacy leakage. We focus on building a privacy-preserving distributed graph learning system through parameter sharing and accurate association repair.

7.3 Background

7.3.1 Scenario

Online credit loan (OCL) business is one of the most basic functional patterns in the irreversible trend of the integration of digital technology and finance. It has provided convenience for customers and made great contributions to the improvement of the diversified financial market. However, the rapid development of the OCL business has been accompanied by an increasing number of frauds. The main cause that leads to the decline of profits in OCL business is bad debts, which are also called *bad loans*. It mainly consists of two types: (1) the loan whose borrower is unable to repay the due loan for certain reasons and (2) the loan whose beneficiary deliberately denies it. The latter is the *fraudulent loan*, i.e., the so-called *fraud*, where its beneficiary, i.e., the *fraudster*, uses forged or misused information to defraud the loan. The latter type results in a huge loss of profits for commercial banks because the fraudsters do not pay the money back subjectivity. The fraudsters usually use misused application information or even use illegal means to obtain other people's information and then impersonate others' identities to apply for loans. We give a representative example to illustrate how fraudsters steal information.

As shown in Fig. 7.1, the fraud group pretends to be a legitimate agency under the guise of recruiting good-pay click farming workers (Fig. 7.1, ①). The fraud group gives each worker a mobile phone card and asks them to apply for a payroll card in the bank (Fig. 7.1, ②). Under the pretext of registration, the fraudsters use bank cards and mobile phone numbers to obtain the worker's ID card, home address, and other information (Fig. 7.1, ③). Finally, the fraudsters use the identity and information of workers to apply for a number of loan applications on the online loan platform by binding the payroll card (Fig. 7.1, ④).

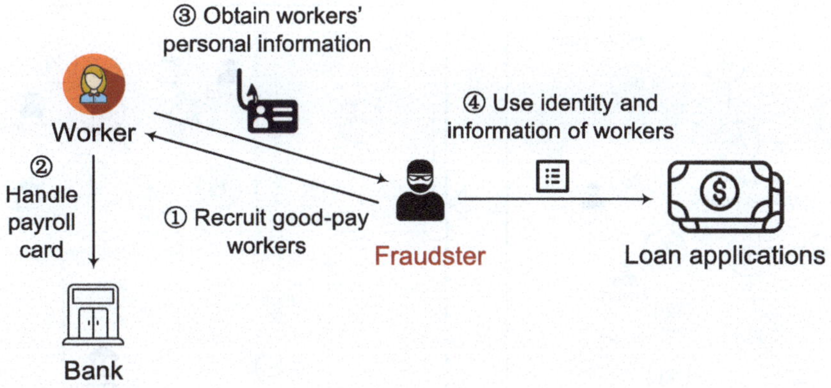

Fig. 7.1 A representative example of how fraudsters collect information from victims

To detect frauds, the financial client platforms spend substantial resources to develop more effective anti-fraud systems. The main goal of commercial bank anti-fraud management can be summarized as the process of maximizing the profit of credit loan business within the tolerable risk fluctuation range [40, 41]. For OCL business, the fraction of the predicted fraudulent applications which are actually fraudulent among all actually fraudulent loan applications is measured by recall [42]. Due to the complexity and concealment of the fraud, anti-fraud systems may generate a large number of false positive alerts during the process, and the fraction of predicted fraudulent applications which are actually normal among all actually normal loan applications is measured by Disturbance [43]. The purpose of anti-fraud engineering for OCL business is to detect more frauds (high recall) while reducing the effects on normal loan applications (low disturbance) which is measured by KS value [44].

7.3.2 *Problem Analysis*

For loan fraudsters, the unidentifiable forged or misused application information is the rare and precious resource with a cost. Therefore, it is likely that a piece of applicant information is partially even integrally used multiple times, which inevitably generates associations among application information of some fraudulent loans. Then, these associations actually become the telltale signs of frauds, which is really good for the platform side. In fact, this is the reason why the OCL anti-fraud problem is generally considered to be association-sensitive. Fraudsters have been making efforts to hide these associations spontaneously by exploiting data isolation. The main approach they take is as follows.

The first is to submit the derived similar information instead of the same one. For example, an address derived from a real address can be adopted to break the direct

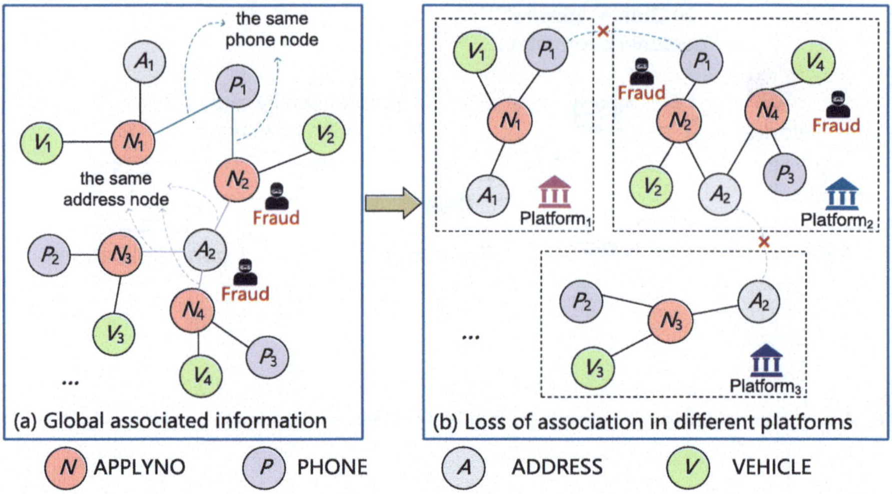

Fig. 7.2 An example of severed associations among cross-platform loan applications

associations between the same addresses of different applications. To mine these hidden associations, graph-based learning methods, e.g., graph embeddings (GEs) [45] and graph neural networks (GNNs) [46], are almost certainly the preferred methods of choice. GEs, as a special case of *transductive* learning, are difficult to satisfy the low latency demand for OCL business [47], since the embeddings need to be relearned whenever the graph is updated. Different from GEs, GNNs methods offer two types of models, viz., inductive models and transductive models for various targets [48]. Inductive GNN methods do not need to be retrained every time when new data come, and therefore they are more suited to fraud prediction by mining anomalous associations in real time for OCL business.

The second is to distribute their load applications across different platforms. As illustrated in Fig. 7.2, if four loan applications consisting of the unique identifier and three different attributes are submitted on the same platform, when some of them are judged to be fraud, the risk of associated applications can be analyzed deeper and more comprehensively. N represent applicant nodes, which are identifiers of loan applications. P, A, and C represent different attribute nodes. Figure 7.2a describes that four different applications are submitted on the same client platform. If some of the loan applications are fraud, the risk of related applications can be analyzed more accurately according to associations between the applications. Figure 7.2b shows that four loan applications are submitted on different client platforms. If some of the loan applications are fraud, other client platforms cannot evaluate the risk of loan applications adequately due to losing associations. It may result in a high probability that these fraudulent loan applications are approved as normal loan applications. To address this problem, a straightforward solution is to make different platforms collaborate to share data. Unfortunately, this is too idealistic to be realistic due to the lack of mutual trust. For the sake of data security and commercial confidentiality, financial institutions tend to separate data from each other and are reluctant to share data.

7.3.3 *Challenges*

The Limit of Local Graph Because of tighter data constraints and more privacy concerns brought by CCPA [7], GPDR [8], and DSL [9], financial data from different client platforms are hard to be transmitted and stored with the traditional centralized storage method. Different client platforms store the graph data locally and train the anti-fraud model separately based on local data. However, it is very hard for different client platforms to train an accurate GNN model based on insufficient local graph data. Besides, these local GNN models cannot capture global graph distribution.

The Loss of Associations The essence of data-driven anti-fraud engineering on OCL platforms is that the platform side has the advantage of information asymmetry when compared with the fraudster side. This leads to a heavy dependency on data. As a matter of fact, the advantages of the platform are somewhat weakened due to the loss of associations. The submitted cross-platform loan applications [49] with similar applicant information are dissociated due to the isolation of different platforms.

7.4 Association Reconstruction Mechanism

In this part, we first provide preliminary knowledge. Then we introduce the customized local models of the proposed privacy-preserving distributed graph learning framework. After that, we illustrate the association reconstruction mechanism (ARM) for repairing critical associations in cross-platform online credit loan (OCL) fraud prediction and describe collaborative training of heterogeneous graph neural networks from multiple local graph data. Furthermore, we give the analysis on privacy protection. The framework is shown in Fig. 7.3. It helps different client platforms to recover the critical associations and realizes collaborative graph learning for cross-platform OCL fraud prediction by a fusion of data sharing and parameter sharing.

7.4.1 *Preliminary*

Heterogeneous Graph [50] A heterogeneous graph is defined as $G = \{V, E\}$, where V denotes the node set and E denotes the edge set, respectively. Each node

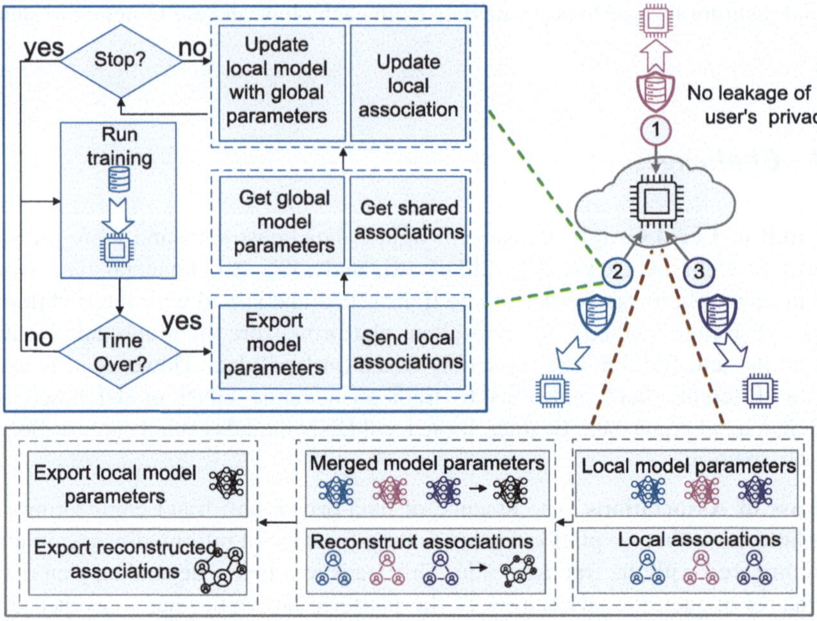

Fig. 7.3 The proposed privacy-preserving distributed graph learning framework

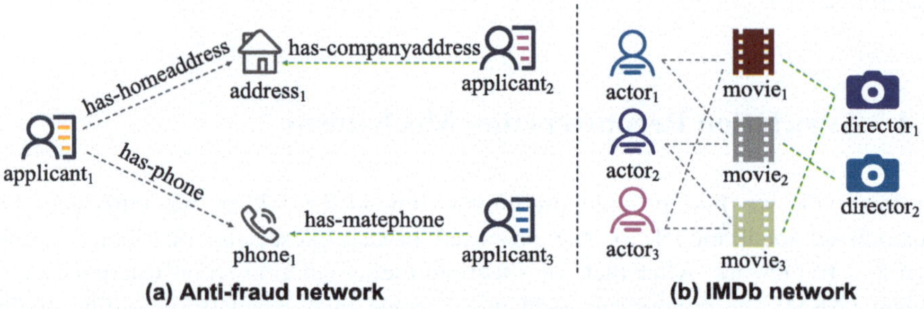

Fig. 7.4 The difference between anti-fraud network and IMDb network

$v \in V$ and each edge $e \in E$ are associated with a node type mapping function $\psi_v : V \rightarrow \mathscr{A}$ and an edge-type mapping function $\psi_e : E \rightarrow \mathscr{R}$. $\mathscr{A}$ and $\mathscr{R}$ denote the sets of predefined node types and edge types, respectively, where $|\mathscr{A}|+|\mathscr{R}| > 2$.

In the constructed anti-fraud network, it consists of multiple types of nodes (e.g., *applicant*, *phone*, and *address*) and relations (e.g., *has-phone*, *has-matephone*, *has-homeaddress*, and *has-companyaddress*) which are shown in Fig. 7.4a. There are more types of edges in anti-fraud graph, not determined by the node pairs. For the same node pair, the edges can also be a more fine-grained type. The node set V consists of two parts, i.e., the sets of applicant nodes and attribute nodes, denoted by U and T, respectively. Here, attribute nodes can be further divided into different types. At the same time, different types of nodes also have their own attributes. Different applicant nodes are connected to other attribute nodes through different edges.

Meta-path [25] A meta-path Φ is defined as a path in the form of $A_1 \xrightarrow{R_1} A_2 \xrightarrow{R_2} \ldots \xrightarrow{R_l} A_{l+1}$, abbreviated as $A_1 A_2 \ldots A_{l+1}$. This meta-path describes a composite relation $R = R_1 \circ R_2 \circ \cdots \circ R_l$ between objects A_1 and A_{l+1}, where $\circ$ denotes the composition operator on relations.

As shown in Fig. 7.4a, the applicants are connected with multiple meta-paths, e.g., *applyno-address-applyno* and *applyno-phone-applyno*. These meta-paths usually have different semantics in the constructed anti-fraud network.

Repair Rule A repair rule $r^{j}_{\Phi_v}$ represents utilizing a (set of) hashed value(s) based on a specific meta-path Φ_v (e.g., *applyno-address-applyno*). According to rule j (e.g., an attribute of the node's first-order neighbors matches a shared hash value), the local platform updates (e.g., increments) the node's corresponding associated feature (e.g., ApplOne) under the precondition that such a meta-path Φ_v exists within the local heterogeneous graph G.

Association Information Exploitation The central server utilizes the aggregated associated information from different platforms and then calculates the incremental hashed values for local platform i by $S^i_g = \bigcup_{j=1}^{n} S_j - S_i$, where S_j denotes a set of hashed values from local platform j. The central server determines meta-paths that are involved with incremental information and establishes the corresponding repair rules. The incremental hashed values and repair rules collectively constitute $\Delta profile_i$.

7.4.2 *Customized Local Model*

For each client platform, we design a customized method to construct an applicant information associated heterogeneous graph. The overview of the local model is shown in Fig. 7.5: (1) It initializes the embedding of the applicant node (node u in this case) by the aggregation of its neighbor attribute nodes and deep features. (2) It aggregates the embedding of neighbor nodes through the attention mechanism on every specific meta-path, and finally aggregates the embeddings on different meta-paths.

7.4.2.1 Feature Engineering

Traditional machine learning methods often rely heavily on manual features constructed by experts in the field of OCL anti-fraud engineering. Therefore, we

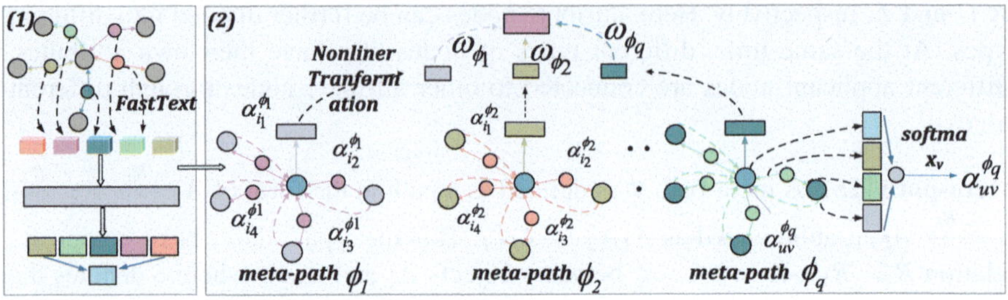

Fig. 7.5 The scheme of local models

Table 7.1 The extracted edges in loan transactions

Features	Description
Degree	The number of attribute nodes of an applicant node
ApplOne	The number of first-order neighbors of an applicant node
ApplTwo	The number of second-order neighbors of an applicant node
FraudOne	The number of fraud neighbors in an applicant node's first-order neighbors
FraudTwo	The number of fraud neighbors in an applicant node's second-order neighbors

adopt a combination of a small number of expert features and automated feature engineering, e.g., Deep Feature Synthesis (DFS)[51], to reduce the time consumed by manpower and improve the utility of features at the same time. It should be pointed out that the so-called manual features are mainly some of graphical features in the graph structure. Table 7.1 shows the main graphical features we extracted. The next job is to enable automated feature engineering and use deep synthesis algorithms to automatically construct other risk features. The DFS is an algorithm for automatic feature construction. Its input has two main parts. One of the inputs is a collection of related entities, and the other input is a basic element function that can obtain the underlying structure of construction feature. There are many calculation conversion functions required to construct features. The basic aggregate function takes a column features of related instances in one or more entity tables as input and returns a value at output. The basic conversion function takes a column or certain columns of features of all instances in an entity table as input and returns a new column of features at output. We use the Pearson correlation coefficient (PCC) to measure the linear correlation between the generated features, and let pcc denote the threshold of density clustering to select features. We denote the feature vector obtained in this step as h_{DFS}.

7.4.2.2 Encoding Attribute Nodes

Although automated feature engineering has obtained some useful features, in order to make full use of the attribute information contained in the nodes, we encode these attribute nodes. The attribute nodes contain text information. For example, the *address* node contains the family, household, or company address, and the *company* node contains the company name and a brief introduction. For each attribute node $t \in T$, we utilize FastText[52] to pre-train the text information and denote the embedding of t as $h_t \in \mathbb{R}^{d_h \times 1}$. After we have the embedding of attribute nodes, there are several choices in how to represent the applicant nodes according to the attribute embedding. For the associations between attribute nodes, directly using simple concatenation or averaging for attribute embedding might lose these correlations. Inspired by HetGNN, we utilize bidirectional LSTM (bi-LSTM)[53] to capture the deep interaction information between attribute nodes. Given an applicant node $u \in U$ and its neighbor attribute nodes T_u, its initial embedding $z_u \in \mathbb{R}^{d \times 1}$ is calculated as follows:

$$
\begin{aligned}
h_u &= \frac{1}{|T_u|} \cdot \sum\nolimits_{t \in T_u} tanh(\overrightarrow{LSTM}(h_t) \oplus \overleftarrow{LSTM}(h_t)), \\
z_u &= h_u \oplus h_{DFS},
\end{aligned}
$$

where $\oplus$ denotes the concatenation. The form of LSTM is described as:

$$
\begin{aligned}
z_j &= \sigma(U_z h_t + W_z q_{j-1} + b_z), \\
f_j &= \sigma(U_f h_t + W_f q_{j-1} + b_f), \\
o_j &= \sigma(U_o h_t + W_o q_{j-1} + b_o), \\
\widehat{c_i} &= tanh(U_c q_j + W_c q_{j-1} + b_c), \\
c_j &= f_i \otimes c_{j-1} + z_j \otimes \widehat{c_j}, \\
q_j &= tanh(c_j) \otimes o_j,
\end{aligned}
$$

where U_j, W_j, b_j $(j \in \{z, f, o, c\})$ are learnable parameters, $q_j \in \mathbb{R}^{d \times 1}$ denotes the output hidden state of t_j, and σ is activation function and $\otimes$ is Hadamard product. After the deep interaction of bi-LSTM and nonlinear transformation, the correlations are captured deeply.

7.4.2.3 Aggregating Heterogeneous Neighbors

The previous GNNs usually study graphs with a single edge type between nodes, and the influence of multiplex edge types is rarely considered. As shown in Fig. 7.4, Internet Movie Database (IMDb) is an online database of information related to films, actors, etc. In IMDb network, it consists of three types of nodes including actor, movie, and director, and there is only one edge type between *actors* and *movies*. In our applicant information associated network, the type of edges between nodes is not only determined by the types of node pairs. With the meta-path *applyno-address-applyno*, although two applicant nodes are connected to the same attribute node *address*, the relationship categories of the two edges are *has-homeaddress* and *has-companyaddress*. So it is necessary to add edge types to consider the cross-correlation between applicant nodes. Given an applicant node $u \in U$ and a meta-path Φ, we denote N_u^{Φ} as the meta-path-based neighbors of node u (including itself). Since the type number of edges is fixed, we use one-hot to encode the edges between the applicant nodes and attribute nodes. For example, given a pair of applicant nodes u and v, there are two edges e_u and e_v connecting from node u and node v to the same attribute node, and we denote h_e^u, h_e^v as the one-hot embedding of e_u and e_v, respectively.

Then the attention score α_{uv}^{Φ} between nodes u and v is formulated as follows:

$$e_{uv}^{\Phi} = att(z_u, z_v, h_e^u, h_e^v), \qquad \alpha_{uv}^{\Phi} = \frac{exp(e_{uv}^{\Phi})}{\sum_{k \in N_u^{\Phi}} exp(e_{uk}^{\Phi})}, \tag{7.1}$$

where the attention mechanism att is calculated as:

$$att(z_u, z_v, h_e^u, h_e^v) = LeakyReLU(a_{\Phi}^T[z_u \oplus z_v \oplus h_e^u \oplus h_e^v]),$$

where a_{Φ}^T is a learnable vector. Now we can get the edge-type-based embedding of node u which is denoted as z_u^{Φ}:

$$z_u^{\Phi} = \sigma\left(\sum_{v \in N_u^{\Phi}} \alpha_{uv}^{\Phi} \cdot z_v\right). \tag{7.2}$$

To capture more information from different representation subspaces, we use multi-head attention, i.e., K independent edge-type-based attentions. Then we concatenate the K embeddings as edge-type-based embedding:

$$z_u^{\Phi} = \bigoplus_{k=1}^{K} \sigma\left(\sum_{v \in N_u^{\Phi}} \alpha_{uv}^{\Phi} \cdot z_v\right). \tag{7.3}$$

For a given set of meta-paths $\{\Phi_1, \Phi_2, \cdots, \Phi_q\}$, we have a corresponding set of embeddings $\{z_u^{\Phi_1}, z_u^{\Phi_2}, \cdots, z_u^{\Phi_q}\}$. The final applicant node embedding is represented by this set of edge-type-based embeddings. Then the final node embedding is calculated as:

$$z_u = \frac{1}{q} \sum_{i=1}^{q} d^{\mathrm{T}} \cdot \sigma(W \cdot z_u^{\Phi_i} + b), \tag{7.4}$$

where W is the weight matrix, b is the bias vector, and d is the semantic-level attention vector.

7.4.2.4 Loss Function

In the traditional graph node classification tasks, the most commonly used loss function is cross-entropy. But like most anti-fraud engineering, our data distribution is extremely uneven. Fraud data accounts for only a small part of all data. Therefore, the original cross-entropy function is no longer suitable. For all labeled nodes, we use the *Focal Loss* [54] to solve the unbalanced problem. That is,

$$a_t = \begin{cases} a & if\ y = 1 \\ 1 - a & \text{otherwise,} \end{cases} \tag{7.5}$$

$$p_t = \begin{cases} p & if\ y = 1 \\ 1 - p & \text{otherewise,} \end{cases} \tag{7.6}$$

$$p = \sigma(\Theta \cdot z_u), \tag{7.7}$$

$$Loss = -a_t(1 - p_t)^{\gamma} log(p_t), \tag{7.8}$$

where Θ is a learnable parameter, $a \in [0, 1]$ is a weighting factor, $p \in [0, 1]$ is the model's estimated probability for the class with label $y = 1$, and γ is a focusing parameter.

7.4.3 Association Reconstruction Mechanism

Excavating the associations is an important means of anti-fraud engineering. The client platform summarizes the characteristics of these confirmed fraud gangs and learns from these experiences to discover new frauds. Locating potential fraud groups is based on graph data mining technology, but the effectiveness is weakened due to data isolation.

The association reconstruction mechanism (ARM) is to resolve the problem of losing key associations in OCL fraud prediction. The diagram of this process is shown in Fig. 7.6. (1) The central server determines the set of meta-paths and corresponding search rules after aggregating data requests. (2) Client platforms perform the SSSD operation and return their shared data to the central server. (3) The central server aggregates the shared data and returns the corresponding increment and repair rules to client platforms. (4) Client platforms perform the ER operation and update the local association information. Plf_i is an acronym for platform_i in this figure.

At the beginning of each iteration for reconstructing association information, first of all, the central server aggregates the data requests and the meta-path set $\Phi = \{\Phi_1, \Phi_2, \ldots, \Phi_q\}$ after different client platforms have submitted data requests and analyzed the structure of local models. Then the central server determines the meta-path set $\Phi_a = \{\Phi_1, \Phi_2, \ldots, \Phi_m\}$ and defines the set of search rules

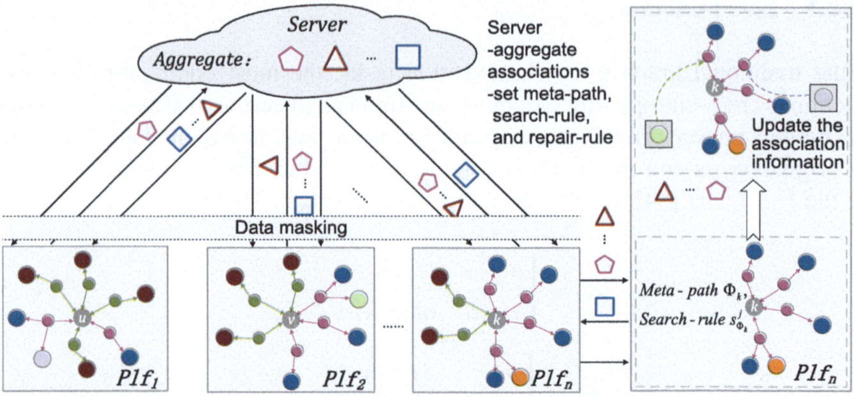

Fig. 7.6 Diagram of our proposed association reconstruction mechanism

$s_{\Phi_v} = \{s^1_{\Phi_v}, s^2_{\Phi_v}, \ldots, s^p_{\Phi_v}\}$ according to all requirements of repairing associations, where $\Phi_a \subseteq \Phi$. After that, the central server distributes semaphore of association reconstruction sem and the search rule $s^j_{\Phi_v}$ by a certain meta-path Φ_v to different platforms to ensure consistency of information collection across platforms, where $s^j_{\Phi_v} \in s_{\Phi_v}$.

For each client platform, after receiving the semaphore sem, the search rule $s^j_{\Phi_v}$ and a certain meta-path Φ_v, if $sem = 1$, each platform performs association mining parallelly according to the same search rule $s^j_{\Phi_v}$ from a particular meta-path Φ_v to select association information. Then, the client platforms shatter the selected pieces of applicant information records. After that, the client platforms shuffle the scattering field values randomly to ensure the shared field-level data out-of-order. To further protect the data, the client platforms desensitize the shared data by applying one hash function before transmitting it to the central server finally. As a result of these steps, the final product of this progress is the local shared information.

Once the central server has received all local shared information from different platforms, it aggregates them for forming the global associated information and defines repair rules $r_{\Phi_v} = \{r^1_{\Phi_v}, r^2_{\Phi_v}, \ldots, r^t_{\Phi_v}\}$. After that, the central server exploits the global associated information and compares it with the local shared information from different platforms in order to calculate the increment of the associated information for each platform. After the calculation is completed, the central server combines the corresponding increment of hashed values and $r^h_{\Phi_v}$ to form $\Delta profile_i$ and then transmits it to the corresponding platform.

After the local client platform has received the corresponding $\Delta profile_i$, it analyzes required features after exploring the useful shared data from $\Delta profile_i$ and repairs them of the local heterogeneous graph structure by updating the associated features based on the repair rules. After this process is completed, different client platforms update their data requests. The data sharing process iterates continuously until all client platforms send no new data requirements to the central server.

We give an example to demonstrate the process of reconstructing important fraud associations from different platforms specifically. Suppose there are four loan applications N_1, N_2, N_3 and N_4, submitted through three different client platforms. Each application node is associated with different types of attribute nodes including phone number, address, and chassis number. N_1 is submitted on platform$_1$. N_2 and N_4 are submitted on platform$_2$. N_3 is submitted on platform$_3$. N_1 and N_2 are associated with the same phone number node. N_2 and N_3 are associated with the same address node. N_4 and N_3 are associated with the same address node. These four loan applications are applied on three client platforms, and client platform$_2$ detects that N_2 and N_4 are fraudulent applications.

For reconstructing these important associations about fraud loan applications, the central server determines all meta-paths Φ_a including *applyno-address-applyno*, *applyno-phone-applyno*, and *applyno-vehicle-applyno*. For meta-path *applyno-address-applyno*, we define one search rule $s^1_{\Phi_v}$ which means that each platform searches for information about fraud loan application nodes and the degree of these nodes must be greater than 1 (when we calculate the degree of application nodes, only attribute nodes are counted). After each client platform has received Φ_v *applyno-address-applyno* and $s^1_{\Phi_v}$, they perform association mining parallelly and search for information about fraud loan application nodes satisfying the degree of these nodes are greater than 1. In this example, $profile_1$ and $profile_3$ are empty because there are no fraud loan applications on platform$_1$ and platform$_3$ by Φ_v. The first step of platform$_2$ is to select address records of N_2 and N_4 by Φ_v. Then platform$_2$ shatters selected records into IDs and corresponding properties. After that, platform$_2$ shuffles the scattering field values. Finally, platform$_2$ uses a hash function $H(\cdot)$ to desensitize the shared information and forms $profile_2$.

After the central server has collected various local shared information from these client platforms, it further calculates the increment of the associated information for each client platform. In this example, $\Delta profile_2$ is empty; $\Delta profile_1$ and $\Delta profile_3$ contain information about fraudulent loan applications from platform$_2$. The repair rule $r^1_{\Phi_v}$ means repairing fraud neighbors in the node's first-order neighbor in meta-path *applyno-address-applyno*. The platform$_1$, platform$_2$, and platform$_3$ explore the useful shared data and repair the local graph associations by corresponding $\Delta profile_i$, so the number of fraud first-order neighbors of N_3 is updated.

7.4.4 Collaborative Graph Learning

Each client platform uses the same local model based on the local heterogeneous graph $G(V, E)$, takes the same initialization parameters (learning rate, loss function, hidden_units, drop_out, activation function, etc.), and updates the parameters through the back propagation.

Specifically, S_t is the set of client platforms, n_k is data owned by client platform k, and ∇g_k represents gradient of client platform k. If the learning rate of the platform is l when the global parameters are updated in the t round, each client platform performs parameter updates locally:

$$w_{t+1}{}^k = w_t{}^g - l\nabla g_k. \tag{7.9}$$

The transmission of parameters of GNN model may also lead to privacy leakage during training process [55]. To tackle this challenge, we employ a hybrid encryption technique to preserve the privacy of both the parameters and the key used to encrypt the parameters.[2] Then the platform sends the encrypted parameters $w_{t+1}{}^k$ along with the encrypted key to the central server. After the central server has received the encrypted parameters and encrypted key of symmetric-key algorithm, the central server uses the secret-key to decrypt the key of symmetric algorithm and gets key of symmetric-key algorithm and then uses it to decrypt the parameters. After that, the central server computes a weighted average of the parameters as follows:

$$w_{t+1}{}^g = \sum\nolimits_{k=1}^{K} \frac{n_k}{\sum_{k=1}^{K} n_k} \cdot w_{t+1}{}^k, \tag{7.10}$$

and then returns the encrypted $w_{t+1}{}^g$ to each platform. The platform decrypts the parameters from the central server and then updates the local models with new parameters for the next training round. We take $l(x_i, y_i; w)$ as the loss of prediction on $(x_i; y_i)$ with model parameters w. Finally, the algorithm defines the following objective function:

$$\min L(x_i, y_i; w) = \frac{1}{n_k} \sum\nolimits_{i=1}^{n} l\left(x_i^k, y_i^k; w\right). \tag{7.11}$$

Combining this part with association reconstruction mechanism, the training process is described in Algorithm 6.

7.4.5 Analysis on Privacy Protection

User privacy is protected by four aspects in our framework. First, only parameters of local GNN models are sent to the central server during the training process. The parameters contain less privacy information according to data processing inequality

[2] We use Blowfish encryption algorithm [56] during training process, and the key is generated randomly each training round to achieve better privacy protection. RSA encryption algorithm [57] is applied to preserve the key used to encrypt the parameters. This combination gives the good throughput for both encryption and decryption [58].

Algorithm 6: Association reconstruction mechanism

Input: The private data set of different platforms
Output: The model of each platform
Server executes:
Initialize the local model and its parameters ω
for *each round* $t = 1, 2, \ldots, N$ **do**
  for *each platform* $k \in S_t$ ***in parallel*** **do**
    ${w_{t+1}}^k \leftarrow \text{PlatformUpdate}(k, w_t, sem)$
    if sem **then**
      `// set meta-path, search rule, repair rule of ARM`
      for $\Phi_v \in \Phi_a, s^j_{\Phi_v} \in s_{\Phi_v}, r^h_{\Phi_v} \in r_{\Phi_v}$ **do**
        $\text{AggregateAssociation}(\Phi_v, s^j_{\Phi_v}, r^h_{\Phi_v})$
      end
    end
  end
  `// update the global model`
  ${w_{t+1}}^g = \sum_{k=1}^{K} \frac{n_k}{\sum_{k=1}^{K} n_k} {w_{t+1}}^k$
end

PlatformUpdate(k, w, sem):

if sem **then**
  `// get meta-path, search rule, repair rule of ARM`
  $\text{SearchAssociation}(\Phi_v, s^j_{\Phi_v})$
  $\text{RepairAssociation}(\Phi_v, r^h_{\Phi_v})$
end
Compute model gradients ∇g_k
${w_{t+1}}^k = {w_t}^g - l\nabla g_k$
return ${w_{t+1}}^k$

[59] compared with the raw data of different platforms. Second, during the process of repairing associations, the central server cannot know the values of the shared data that have been desensitized before sending to the central server since it does not have the raw data. Third, we employ a hybrid encryption technique to preserve the privacy of both the parameters and the key used to encrypt the parameters for strong privacy preservation during transmission of model parameters. Fourth, different from many works that assume that the client platforms are honest participants, we consider the extreme situation that the platforms may disclose privacy after receiving the shared data from other platforms. Fortunately, we apply the SSSD operation before sharing the information about fraud loan applications, and the shattering and shuffling steps of SSSD make it hard to infer correct information from the observation. Furthermore, we discuss the solution when there are concerns about data trust between client platforms.

We quantify the impact on the privacy of shared data of platform i to platform j by measure $\mathscr{L}(S_i \rightarrow O_j)$. It reflects the probability of the shared fraud loan records S_i in platform i can be inferred correctly from the observation O_j in platform j in a single attack. For associations in each fraud record, we denote it as $p_z = (a_1, a_2, \ldots, a_{k_z})$, where k_z denotes the dimension of associations of the shared record z; a_1, a_2, and a_{k_z} are the detailed values of associations in this shared fraud loan record. For each field data, we use $u(a_t|H(a_t))$ to reflect whether the value of one field-level data can be inferred after one platform has received the shared data, where $H(a_t)$ is the result of hash mapping. If this client platform has the association related to the data, it can know the identity of the shared data before the hash function, and as a result, it will know the attributes of this node, i.e., $u(a_t|H(a_t)) = 1$; otherwise, if this platform has no association related to the data, it cannot know the identity of the shared data before the hash function, and as a result, it cannot know the attributes of this node, i.e., $u(a_t|H(a_t)) = 0$.

We define $I_{p_z} = \sum_{t=1}^{k_z} u(a_t|H(a_t))$ as the field values of one fraud loan record can be inferred in platform j, and the complete information of one fraud loan record can be inferred if $|I_{p_z} = k_z|$. We measure the theoretical maximum reducibility for the shared data from platform i in platform j by $W(S_i \rightarrow \mathcal{S}_j)$, where S_i denotes the shared data in platform i and $\mathcal{S}_j$ denotes the shared data that has been inferred in platform j. As a consequence, the value of $W(S_i \rightarrow \mathcal{S}_j)$ is calculated as follows:

$$W(S_i \rightarrow \mathcal{S}_j) = \sum_{z=1}^{n} I_{p_z}. \tag{7.12}$$

Further, we can calculate $\mathcal{L}(S_i \rightarrow O_j)$ by:

$$\mathcal{L}(S_i \rightarrow O_j) = \frac{W(S_i \rightarrow \mathcal{S}_j)}{O_j}, \tag{7.13}$$

where the value of O_j is calculated differently according to different sharing strategies.

If different platforms share data directly without the shattering and shuffling steps of SSSD, it means each platform knows the record-level shared data. So the value of O_j is calculated as follows: $O_j = \sum_{z=1}^{n} k_z$, and we can get

$$\mathcal{L}_a(S_i \rightarrow O_j) = \frac{\sum_{z=1}^{n} \sum_{t=1}^{k_z} u(a_t|H(a_t))}{\sum_{z=1}^{n} k_z}. \tag{7.14}$$

After applying the shattering and shuffling steps of SSSD, each platform only knows the field-level data that are out-of-order, and the observation of O_j is calculated as follows: $O_j = \prod_{t=1}^{k_p} cum(H(a_t))$, where $cum(H(a_t))$ means the number of different values of a_t and k_p is the number of all types of associations. So we can get

$$\mathcal{L}_b(S_i \rightarrow O_j) = \frac{\sum_{z=1}^{n} \sum_{t=1}^{k_z} u(a_t|H(a_t))}{\prod_{t=1}^{k_p} cum(H(a_t))}. \tag{7.15}$$

Table 7.2 The extracted edges in loan transactions

Edges	Description
R_ADDR	The loan applicant's address
R_DLPN	The loan applicant's driving license plate number
R_CO	The mobile number of the applicant's colleagues
R_CN	The chassis number of loan applicant's car
R_CO_ADDR	The address of the applicant's company
R_EN	The engine number of loan applicant's car
R_CO_NAME	The loan applicant's company name
R_LRMP	The mobile phone number of the applicant's immediate family members
R_CO_TEL	The telephone number of the applicant's company
R_MAIDNO	The identity card number of the applicant's spouse
R_CPADDR	The loan applicant's residential address
R_MP	The loan applicant's mobile phone number
R_CRMP	The mobile phone number of the applicant's general relatives
R_OCMP	The mobile phone number of the applicant's other contacts
R_FCMP	The mobile phone number of the applicant's friends
R_TEL	The loan applicant's telephone number
R_IDNO	The loan applicant's identity card number
R_REGADDR	The loan applicant's native place

Table 7.3 The division of the datasets

Platform	Legitimate	Fraudulent	Unlabeled	Total
Plf_1	13,485	1546	93,054	108,085
Plf_2	13,673	1558	93,081	108,312
Plf_3	13,632	1543	93,078	108,253

We assume that different associations have the same dimension which is denoted by k_O, and the values of the shared data are different. Thus, we can get $O_j = nk_O$ if different client platforms share data directly without the shattering and shuffling steps of SSSD. We can get $O_j = \prod_{t=1}^{k_p} cum(H(a_t)) \geq n^{k_O}$ if different client platforms share data with the shattering and shuffling steps of SSSD. In our case, $k_O \leq 18$ (according to Table 7.2) and obviously $n \geq 5$ (according to Table 7.3). Inserting them into Eqs. (7.14) and (7.15), then $\mathscr{L}_b(S_i \rightarrow O_j) < \mathscr{L}_a(S_i \rightarrow O_j)$.

When there are concerns about data trust between different client platforms, our framework can address this issue by introducing blockchain technology [60]. It is orthogonal to our method and thus can be incorporated into our framework for sharing critical associations. As shown in Fig. 7.7, the blockchain is established among different client platforms. Each platform shares the critical associations in the blockchain. Using the characteristics of the shared ledger of the blockchain, the critical associations on the chain are synchronized to the blockchain node of each client platform. Of course, due to the requirements of privacy protection and laws and regulations, it will not directly upload the chain in plaintext. It can utilize some

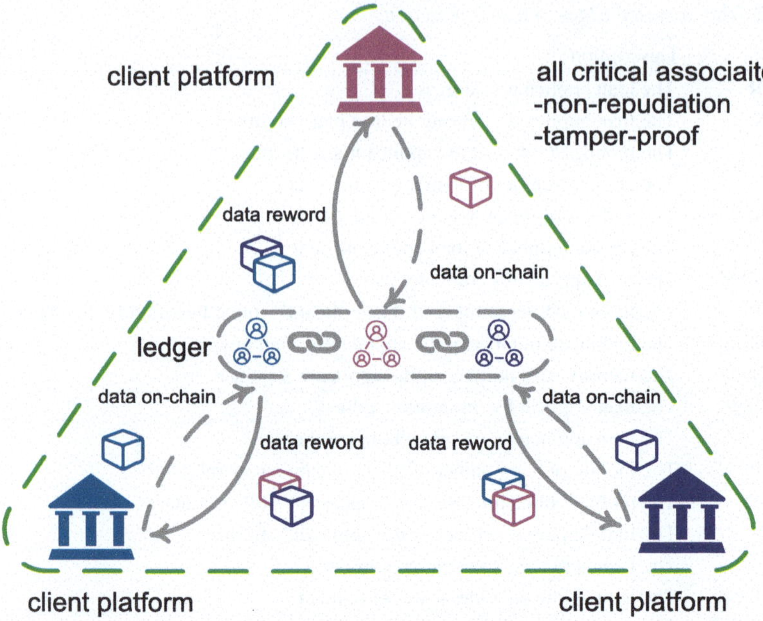

Fig. 7.7 The block-chained process for sharing critical associations

cryptographic methods to share data among various client platforms, ensuring that private information is not leaked.

7.5 Experiments

We perform experiments to validate the effectiveness of our proposed framework for fraud prediction of online credit loan (OCL) business by investigating the research questions as follows:

- (**RQ1**) How does our proposed framework perform for OCL fraud prediction?
- (**RQ2**) How do key components in our framework affect the model performance?
- (**RQ3**) Does our method lead to unacceptable risk of privacy leakage?
- (**RQ4**) How much traffic and time does our framework consume? How do different hyperparameter settings affect the performance?

7.5.1 Experimental Setup

7.5.1.1 Datasets

Our dataset comes from large-scale Internet financial lending platforms. The dataset consists of the loan transactions from June 1, 2017, to July 31, 2018. After excluding isolated samples, there are a total of 324,650 application records. To avoid the problem of time crossing, we select the records from June 1, 2017, to June 30, 2018, as the training set, and the data from July 1, 2018, to July 31, 2018, as the test set. We choose graph database Neo4j to store structured data in the graph. The stored heterogeneous graph includes nine different types of nodes (as shown in Table 7.4) and 18 different types of edges (as shown in Table 7.2), and the nine different types of nodes consist of one type of applicant nodes and eight types of attribute nodes.

We simulate the datasets owned by different platforms in real distributed situations by evenly sampling from the original datasets. The division of datasets is shown in Table 7.3.

7.5.1.2 Metrics

To evaluate the performance of our system, we use three widely used metrics in anti-fraud for OCL business including recall [61] (true positive rate), disturbance [62] (false positive rate), and KS value. The higher recall reflects the better ability of the system to detect fraud. The lower the disturbance is, the less impact it will have on normal applications. The definition of prediction is not black or white (discrete type) in the field of anti-fraud, so it is more reasonable to measure it with probability distribution (continuous type). KS value balances the models' ability to catch more

Table 7.4 The selected nodes in application transactions

Nodes	Attribute	Description
APPL	Name, time	The identifier of a transaction. Extract the applicant's name and the time of the transaction as attributes
ADDR	Province, city, district	The detailed address. Extract three levels of administrative areas as attribute description
MPN	Name	The mobile phone number. Use holder's name as an attribute
TEL	Name	The telephone number. Use holder's name as an attribute
CONAME	None	The name of the company where the online loan applicant works
IDNO	Name	The identity card number. Use holder's name as an attribute
DLPN	None	The driving license plate number of the applicant
CN	None	The chassis number of the loan applicant
EN	None	The engine number of the loan applicant

frauds (high recall) while reducing the effects on normal loan applications (low disturbance), which is the de facto standard in the field of loan anti-fraud [5, 44].

7.5.1.3 Implementation

The embedding dimension of FastText and edge-type-based attention is set to 256 and 128, respectively. We set the learning rate to 0.01, the regularization parameter to 0.3, the dimension d of the semantic level attention vector to 128, the number of the attention head K to 4, and the dropout parameter of the attention layer to 0.3. The length of bi-LSTM is variable because the number of neighbor attribute nodes of the applicant node is not fixed. The training round number is 300. For the association reconstruction mechanism, we determine the set of meta-paths with eight different meta-paths and the corresponding search rules about fraud applicant nodes and define two repair rules for repairing the fraud neighbors' associations in the node's first-order neighbors and the fraud neighbors' associations in the node's second-order neighbors. The hash function that we adopt is MD5. There are three different client platforms. Plf_1, Plf_2, and Plf_3 represent platform_1, platform_2, and platform_3, respectively; Plf_g represents the model that is trained with the original graph data.

We compare **FedARM** with other methods including the following: (1) **Dis**: Different client platforms use the same local models, and these local models are trained separately with their distributed local graph data. (2) **Fed** [29]: Different client platforms with their local graph data are federated to train GNN models by sharing model parameters and the central server aggregates these parameters[3] during communication rounds, and this method does not apply the association reconstruction mechanism (ARM). (3) **FedGNN** [30]: It exploits decentralized local graph data to collectively train GNN models and expands the local graph information by sharing the embeddings of neighbors' nodes. (4) **Glo**: It uses original global graph for training without losing associations (as the upper bound of our framework).

FedARM means the method where different client platforms apply ARM before communication rounds, and these client platforms are federated to train the model by sharing model parameters and the central server aggregate parameters in communication rounds.

[3] During the training process, we apply Federated Averaging Algorithm [63] to deal with the shared parameters.

7.5.2 Contrast Experiments

In this part, we conduct experiments to validate the effectiveness of FedARM compared with other methods. We use the customized HAN [25] as the local models of different client platforms. The performance is shown in Table 7.5.

In the second row of the table, Plf_1, Plf_2, and Plf_3 represent $platform_1$, $platform_2$, and $platform_3$ with local subgraph data, respectively; Plf_g represents the model is trained with the original graph data. As observed, FedARM not only outperforms the Dis by an average of 27.8% in recall and by an average of 6.77% in KS value but also alleviates the impact brought by the loss of associations since it reduces the recall drop from 37.6% to 9.8% and the KS value drops from 8.5% to 1.7% compared with Glo. FedARM also has a better performance than FedGNN, and it demonstrates that the association reconstruction mechanism (ARM) of FedARM is more effective for enhancing local graph information, which is beneficial for anti-fraud prediction. The sampling strategy adopted by FedGNN to enrich local graph information is not very satisfactory in the fraud prediction scenario. It is mainly because the fraudster's information has already been compressed when exchanging node embeddings between different client platforms, which makes it difficult to accurately distinguish and repair key associations in heterogeneous graphs.

In Fig. 7.8, the colored curve (which is denoted by KS_g) represents the KS curve of the original global graph. We also mark the KS value for each platform which tells how to set the threshold for them to get the best model discrimination ability. We can know that the KS value of each client platform is about an average of 38.8% with Dis, and the models have a mediocre ability to distinguish between positive and negative samples. The KS value has a certain improvement of about an average of 1.4% through federated training after comparing Fed and Dis, and it proves that federated training by sharing parameters can alleviate the performance degradation caused by the decentralized graph data in some ways. There is a further improvement in KS value about an average of 5.4% by applying ARM before the training rounds, but the FedGNN only brings an average of 2.1% in KS value by sharing the embeddings of nodes during training process. The significant gap between FedARM and Fed corroborates the degeneration of performance caused by missing the key associations and the effectiveness of the ARM for association-sensitive fraud prediction. The KS value of Glo is greater than others, and its performance is the upper bound of our proposed framework.

Table 7.5 Performance of different methods in terms of KS, recall, and disturbance (%)

Methods	Dis			Fed			FedGNN			FedARM			Glo
Platform	Plf_1	Plf_2	Plf_3	Plf_1	Plf_2	Plf_3	Plf_1	Plf_2	Plf_3	Plf_1	Plf_2	Plf_3	Plf_g
KS	39.7	38.2	38.5	40.8	39.5	40.2	42.9	42.3	41.7	**46.3**	**45.4**	**45.0**	47.3
Recall	46.1	42.4	43.8	54.7	51.0	55.2	65.3	63.0	63.3	**74.4**	**71.7**	**69.6**	81.7
Disturbance	6.4	4.2	5.3	13.9	11.5	15.0	22.4	20.7	21.6	**28.1**	**26.3**	**24.6**	34.4

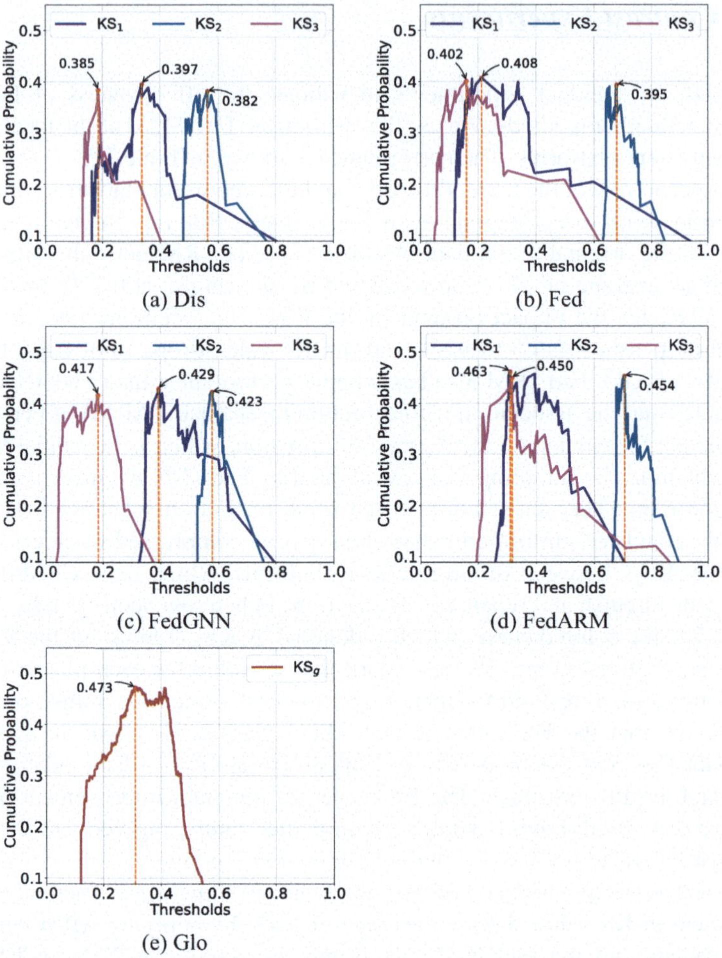

Fig. 7.8 The KS curves of different platforms

7.5.3 Ablation Study

We conduct a comprehensive evaluation from two aspects, i.e., the change of local model and the change of repair rule.

7.5.3.1 The Impact of Local Model

To verify the compatibility of our proposed distributed graph learning framework under other typical GNN models, we further compare the results of our method with different training methods under some typical GNN models such as GAT,

GraphSAGE, and SemiGNN. The spatial domain-based method expresses the graph convolution as the aggregation of feature information from the neighbors. It does not rely on the graph convolution theory and directly defines the convolution operation in space, and the ARM can reconstruct associations for these models. We replace the local model with other models as follows:

- GAT [22]: It is a typical model of graph neural networks in the airspace which uses the attention mechanism. We test all the meta-paths for GAT and report the best performance.
- GraphSAGE [21]: It is a general inductive learning model that leverages node feature information by using an aggregator instead of learning the representation for each node to improve the flexibility and generalization of the model.
- SemiGNN [5]: It uses a two-level attention mechanism that can deal with various types of nodes and various views of the heterogeneous graphs.

We present the results with different local models in Table 7.6, where we change the local model of different methods to other typical GNN models, including GAT, GraphSAGE, and SemiGNN, and the subscript denotes a specific GNN model. We further summarize several observations as follows: (1) We find that our proposed ARM also improves the performance of all other local models in some ways. (2) Although we have tested all the meta-paths and reported the best performance for Dis_{GAT} and $\text{Dis}_{\text{GraphSAGE}}$, the performance of them is still not good because the related information is lost after the heterogeneous graph converted into homogeneous graph. (3) The performance of $\text{FedARM}_{\text{GAT}}$ and $\text{FedARM}_{\text{GraphSAGE}}$ is lower than $\text{FedARM}_{\text{SemiGNN}}$ because they can only use a specific type of edges which limits the number of associations that can be repaired by ARM. (4) $\text{FedARM}_{\text{SemiGNN}}$ gets the highest recall because it can deal with heterogeneous graphs and repair associations adequately, but it also brings a higher disturbance, which may be due to the similarity between the embeddings of normal nodes which are connected to fraud nodes and the embeddings of fraud nodes. (5) Our FedARM method also consistently outperforms FedGNN with different local models, and its advantages are more significant on heterogeneous graph models. This is because the mechanism of our method, i.e., ARM, can repair key associations more accurately through different types of meta-paths than the mechanism of FedGNN, which shares the sampled embeddings of neighbors' nodes from local graph data.

As a consequence, using the customized HAN as local model gets the highest KS value compared with other local models that demonstrate the superiority of the customized HAN. The models that can deal with the heterogeneous graph have generally better performance which proves that the multiple edge types between different entities of applications are also very important for fraud detection in OCL business. Besides, we can know that the proposed ARM depends on the meta-paths to repair associations, and it can repair associations more efficiently for the GNN models that use multiple types of edges.

Table 7.6 The performance (%) with different local models

Methods	Dis_{GAT}			Fed_{GAT}			$FedGNN_{GAT}$			$FedARM_{GAT}$			Glo_{GAT}
Platform	Plf_1	Plf_2	Plf_3	Plf_1	Plf_2	Plf_3	Plf_1	Plf_2	Plf_3	Plf_1	Plf_2	Plf_3	Plf_g
KS	33.0	31.2	32.1	33.7	32.1	32.9	34.6	32.4	33.2	**35.1**	**33.4**	**34.9**	36.6
Recall	43.3	41.1	41.3	47.6	44.9	42.7	48.1	47.5	45.3	**48.5**	**51.3**	**54.2**	55.4
Disturbance	10.3	9.9	9.2	13.9	12.8	9.8	13.5	15.1	12.1	**13.4**	**17.9**	**19.3**	18.8
Methods	$Dis_{GraphSAGE}$			$Fed_{GraphSAGE}$			$FedGNN_{GraphSAGE}$			$FedARM_{GraphSAGE}$			$Glo_{GraphSAGE}$
Platform	Plf_1	Plf_2	Plf_3	Plf_1	Plf_2	Plf_3	Plf_1	Plf_2	Plf_3	Plf_1	Plf_2	Plf_3	Plf_g
KS	32.6	31.8	31.3	33.1	32.6	32.4	33.4	32.7	32.6	**33.7**	**33.1**	**32.8**	36.1
Recall	42.1	41.7	39.5	45.6	45.3	44.5	45.8	45.6	44.8	**49.4**	**48.7**	**47.2**	53.6
Disturbance	9.5	9.9	8.2	12.5	12.7	12.1	12.4	12.9	12.2	**15.7**	**15.6**	**14.4**	17.5
Methods	$Dis_{SemiGNN}$			$Fed_{SemiGNN}$			$FedGNN_{SemiGNN}$			$FedARM_{SemiGNN}$			$Glo_{SemiGNN}$
Platform	Plf_1	Plf_2	Plf_3	Plf_1	Plf_2	Plf_3	Plf_1	Plf_2	Plf_3	Plf_1	Plf_2	Plf_3	Plf_g
KS	38.7	38.2	38.3	39.3	38.9	38.8	41.8	42.2	41.9	**44.9**	**44.1**	**44.5**	46.3
Recall	49.6	48.4	47.9	55.7	54.3	55.3	64.5	66.3	65.7	**73.8**	**73.2**	**71.8**	82.2
Disturbance	10.9	10.2	9.6	16.4	15.4	16.5	22.7	24.1	23.8	**28.9**	**29.1**	**27.3**	35.9

Table 7.7 The performance (%) with different repair rules

Rule	R_1			R_2			R_1+R_2		
Platform	Plf_1	Plf_2	Plf_3	Plf_1	Plf_2	Plf_3	Plf_1	Plf_2	Plf_3
KS	42.1	41.6	42.3	43.2	42.8	43.5	46.3	45.4	45.0
Recall	55.2	51.9	56.8	58.9	57.7	60.2	74.4	71.7	69.6
Disturbance	13.1	10.3	14.5	15.7	14.9	16.7	28.1	26.3	24.6

7.5.3.2 The Impact of Repair Rules

To explore the effect of different repair rules for ARM, we define two repair rules for repairing the fraud neighbors' associations. The repair rule R_1 refers to repairing the fraud neighbors' associations in the node's first-order neighbors, and the repair rule R_2 refers to repairing the fraud neighbors' associations in the node's second-order neighbors.

We use the repair rules R_1 and R_2 in ARM, and the results are shown in Table 7.7. R_1 refers to repairing the fraud neighbors' associations in the node's first-order neighbors, and R_2 refers to repairing the fraud neighbors' associations in the node's second-order neighbors. The performance of FedARM (R_1) and FedARM (R_2) is better than Fed but lower than FedARM (R_1+R_2), which validates that both the fraud neighbors' associations in the node's first-order neighbors and those in the node's second-order neighbors are critical for fraud predicting, and using a single repair rule in ARM cannot improve the performance of cross-platform fraud prediction very well.

7.5.4 Privacy Analysis

In this work, we measure the impact on privacy from two aspects. First, we quantify the information that needs to be shared by calculating the proportion of shared data compared with the local graph data in terms of the numbers of application IDs and attribute values, respectively. Second, we quantify the impact on privacy of the shared data of platform i to platform j according to Eqs. (7.14) and (7.15) to evaluate the effect of the shattering and shuffling steps in the ARM.

7.5.4.1 The Proportion of Shared Information

For each platform, the number of fraud applications that need to be shared (N_1) accounts for an average of 0.3% of the total number of applications, which is shown in Fig. 7.9a. Furthermore, we analyze the shared important desensitized associations involving frauds. On average, the total amount of associations that need to be shared

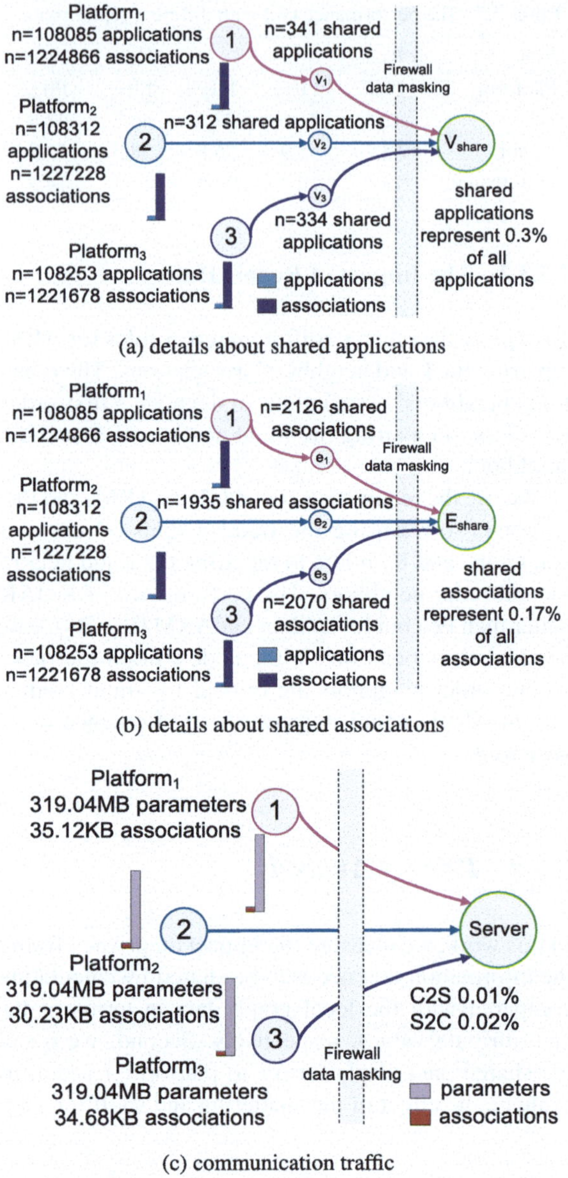

Fig. 7.9 Details about the shared applications and shared associations

(N_2) accounts for an average of 0.17% of the total number of attribute values, which is shown in Fig. 7.9b. The details about the information shared by each client platform are shown in Table 7.8. N_1 and N_2 mean the number of shared applications and shared associations, respectively. P_1 and P_2 mean the corresponding percentage.

Table 7.8 The shared information in different platforms

Platform	N_1	N_2	P_1	P_2
Plf_1	341	0.32%	2126	0.17%
Plf_2	312	0.29%	1935	0.16%
Plf_3	334	0.31%	2070	0.17%

Table 7.9 The impact on privacy with different strategies before sharing data

	ARM-NO		ARM-SSSD	
Plf_1	$\mathscr{L}_a(S_1 \to O_2)$	$\mathscr{L}_a(S_1 \to O_3)$	$\mathscr{L}_b(S_1 \to O_2)$	$\mathscr{L}_b(S_1 \to O_3)$
	0.1924	0.1591	1.07×10^{-13}	8.81×10^{-14}
Plf_2	$\mathscr{L}_a(S_2 \to O_1)$	$\mathscr{L}_a(S_2 \to O_3)$	$\mathscr{L}_b(S_2 \to O_1)$	$\mathscr{L}_b(S_2 \to O_3)$
	0.3265	0.1831	1.63×10^{-12}	9.14×10^{-13}
Plf_3	$\mathscr{L}_a(S_3 \to O_1)$	$\mathscr{L}_a(S_3 \to O_2)$	$\mathscr{L}_b(S_3 \to O_1)$	$\mathscr{L}_b(S_3 \to O_2)$
	0.2897	0.1965	4.71×10^{-13}	3.20×10^{-13}

7.5.4.2 The Improvement of Privacy Protection by SSSD

Although the associations involving frauds are only 0.17% of the total number of attribute values, different from many other works that assume the client platforms are honest participants, we consider the extreme situation that the platforms may disclose privacy after receiving the shared data from other platforms.

The shattering and shuffling steps of SSSD can protect privacy during data sharing process and reduce the probability of inferring correctly about fraud applications in a single attack if some platforms disclose the shared data. We calculate the $\mathscr{L}_a(S_i \to O_j)$ and $\mathscr{L}_b(S_i \to O_j)$ for different platforms according to Eqs. (7.14) and (7.15), respectively. Here $\mathscr{L}_a(S_i \to O_j)$ reflects the probability of the shared fraud loan records in platform i can be inferred correctly from the observation in platform j in a single attack with the strategy of sharing data directly, and $\mathscr{L}_b(S_i \to O_j)$ reflects the probability of the shared fraud loan records in platform i can be inferred correctly from the observation in platform j in a single attack after applying the shattering and shuffling steps of SSSD. The results are shown in Table 7.9.

ARM-NO is the version without the shattering and shuffling steps of SSSD, and ARM-SSSD is the version with the four-step data operation. We can find that it increases the risk of privacy leakage with the strategy of sharing data directly because the minimum value of $L_a(S_i \to O_j)$ of different client platforms is still greater than 0.15. Fortunately, $\mathscr{L}_b(S_i \to O_j)$ is far less than $\mathscr{L}_a(S_i \to O_j)$ in all client platforms which proves that the shattering and shuffling steps in the ARM alleviate privacy concerns significantly.

7.5.5 Simulation Evaluation

7.5.5.1 Communication Traffic

We examine the total amount of communication traffic including associations and parameters that need to be transmitted from client-to-server (C2S) and server-to-client (S2C), respectively.

As shown in Fig. 7.9c, we observe that the proposed ARM only increases the communication traffic by 35.12KB, 30.23KB, 34.68KB in different platforms. It is significantly communication-efficient because it does not bring a huge communication burden, and the total amount of communication traffic about associations only accounts for 0.01% (C2S) and 0.02% (S2C) of the total amount of communication traffic about parameters.

7.5.5.2 Analysis of Time Cost

We consider the runtime of different methods in this part. Fed, FedGNN, and FedARM have applied a hybrid encryption technique to gradients for privacy protection during training rounds. To evaluate the time cost of encryption strategy, we remove the encryption strategy and denote the resulting methods as Fed*, FedGNN*, and FedARM*, respectively.

The time cost of all these methods is shown in Table 7.10. Fed*, FedGNN*, and FedARM* are the methods without the hybrid encryption strategy. Due to the additional communications and computations for repairing associations, FedARM consumes more time compared with Fed. Comparing the total time of FedARM and FedARM*, there is a slight increase in time cost because of the encryption of local parameters, and we can also draw this conclusion by comparing the total time of Fed and Fed*. FedGNN consumes more time to collectively train the models than our method even when using the sampling technique, and the increase in time cost is mainly because of the frequent and time-consuming node matching operations during the training process. Our method is more suitable for training the distributed anti-fraud prediction model with lower time cost.

7.5.5.3 Parameter Sensitivity

We conduct the experiments on the hyperparameters. The results are shown in Fig. 7.10. We select the best hyperparameters by the average performance of all

Table 7.10 The time cost of different methods

Method	Dis	Fed*	Fed	FedGNN*	FedGNN	FedARM*	FedARM	Glo
Time (s)	1578	1752	1876	3689	3815	2075	2197	4689

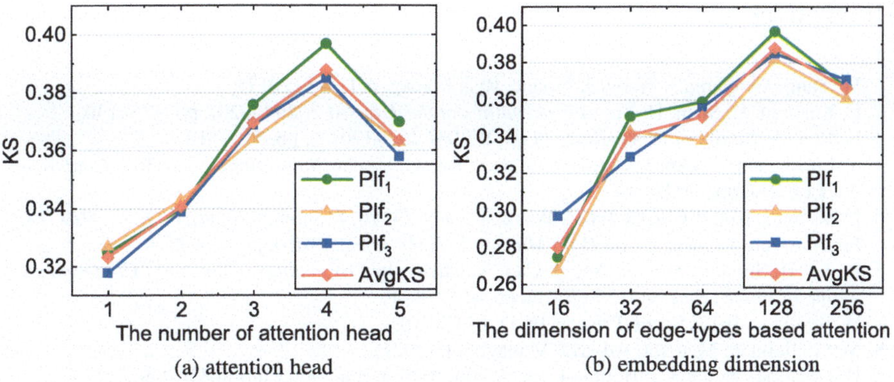

Fig. 7.10 Parameter sensitivity including the number of attention head and the dimension of edge-type-based embedding

client platforms. By analyzing the experimental results, we can get the following conclusions: (1) The best number of attention heads is 4. This number ensures the stability of the model. The model performance begins to decay as this number increases further. (2) The embedding dimension also has a great impact on experimental performance. The best performance is achieved when the dimension is set to be 128. As the dimension is further increased, it causes overfitting.

7.6 Conclusions

We have designed a privacy-preserving distributed graph learning framework for multi-platform online credit loan fraud prediction by merging parameter sharing and data sharing methods. The *association reconstruction mechanism* (ARM) is proposed to realize the novel idea that some key associations can be repaired by sharing a bit of data without compromising privacy beyond limits for identifying frauds. Our framework alleviates privacy concerns about the shared data and achieves parameter protection simultaneously. The experimental results validate the effectiveness of our framework.

There are several issues worthy of investigation. For example, the designed framework uses hash functions to protect data privacy during the sharing process. When facing attacks with sufficient storage and computing resources (e.g., rainbow table attacks), incorporating other techniques into the designed framework may further enhance privacy protection. Moreover, the selection scheme of shareable data may be further improved to meet specific privacy-utility requirements.

References

1. W. Yang, W. Wang, J. Wang, J. Financ. Risk Manag. **6**(1), 48 (2017)
2. E. Kurshan, H. Shen, H. Yu, in *Proceedings of the TransAI 2020* (2020), pp. 125–130
3. K. Ren, A. Malik, in *Proceedings of the WSDM 2019* (2019), pp. 510–518
4. Y. Liu, X. Ao, Z. Qin, J. Chi, J. Feng, H. Yang, Q. He, in *Proceedings of the Web Conference 2021* (2021), pp. 3168–3177
5. D. Wang, J. Lin, P. Cui, Q. Jia, Z. Wang, Y. Fang, Q. Yu, J. Zhou, S. Yang, Y. Qi, in *2019 IEEE International Conference on Data Mining (ICDM)* (IEEE, 2019), pp. 598–607
6. B. Xu, H. Shen, B. Sun, R. An, Q. Cao, X. Cheng, in *Proceedings of the AAAI Conference on Artificial Intelligence*, vol. 35 (2021), pp. 4537–4545
7. W. Stallings, IEEE Secur. Privacy **18**(1), 61 (2020)
8. M. Godinho de Matos, I. Adjerid, Manag. Sci. (2021)
9. H. Daoli, H. Wenhua, J. Beijing Univ. Aeron. Astron. Soc. Sci. Edition **33**(6), 9
10. D. Caldarola, M. Mancini, F. Galasso, M. Ciccone, E. Rodolà, B. Caputo, in *Proc. CVPR 2021*. pp. 2749–2758
11. Z. Xie, Y. Huang, D. Yu, R.M. Parizi, Y. Zheng, J. Pang, IEEE Trans. Ind. Inform. (2022)
12. T. Suzumura, Y. Zhou, N. Baracaldo, G. Ye, K. Houck, R. Kawahara, A. Anwar, L.L. Stavarache, Y. Watanabe, P. Loyola, et al., arXiv preprint arXiv:1909.12946 (2019)
13. C. Bailey, R. Brody, M. Sokolowski, J. Financ. Crime (2021)
14. J. Jurgovsky, Context-aware credit card fraud detection. Ph.D. Thesis, Université de Lyon; Universität Passau (Deutscheland) (2019)
15. W. Feller, in *Selected Papers I* (2015), pp. 735–749
16. L. Tong, B. Li, C. Hajaj, C. Xiao, N. Zhang, Y. Vorobeychik, in *Proceedings of the USENIX 2019*. pp. 285–302
17. D. Babaev, M. Savchenko, A. Tuzhilin, D. Umerenkov, in *Proceedings of the SIGKDD 2019*. pp. 2183–2190
18. G. Suarez-Tangil, M. Edwards, C. Peersman, G. Stringhini, A. Rashid, M. Whitty, IEEE Trans. Inform. Forensics Secur. **15**, 1128 (2019)
19. X. Li, S. Liu, Z. Li, X. Han, C. Shi, B. Hooi, H. Huang, X. Cheng, in *Proceedings of the AAAI 2020*. pp. 4731–4738
20. R. Zhong, Z. Zhang, R. Lin, H. Zou, in *Proceedings of the Big Data 2020*. pp. 5496–5504
21. W. Hamilton, Z. Ying, J. Leskovec, in *Proceedings of the NIPS 2017*. pp. 1024–1034
22. P. Velickovic, G. Cucurull, A. Casanova, A. Romero, P. Liò, Y. Bengio, in *Proceedings of the ICLR 2018*
23. D. Cheng, Y. Zhang, F. Yang, Y. Tu, Z. Niu, L. Zhang, in *Proceedings of the CIKM 2019*. pp. 2547–2555
24. C. Zhang, D. Song, C. Huang, A. Swami, N.V. Chawla, in *Proceedings of the KDD 2019*. pp. 793–803
25. X. Wang, H. Ji, C. Shi, B. Wang, Y. Ye, P. Cui, P.S. Yu, in *Proceedings of the WWW 2019*. pp. 2022–2032
26. C. Wang, H. Zhu, IEEE Trans. Depend. Secure Comput. (2020). https://doi.org/10.1109/TDSC.2020.2991872
27. C. He, K. Balasubramanian, E. Ceyani, C. Yang, H. Xie, L. Sun, L. He, L. Yang, P.S. Yu, Y. Rong, et al., arXiv preprint arXiv:2104.07145 (2021)
28. H. Xie, J. Ma, L. Xiong, C. Yang, Adv. Neural Inform. Process. Syst. **34**, 18839 (2021)
29. T. Suzumura, Y. Zhou, R. Kawahara, N. Baracaldo, H. Ludwig, in *Federated Learning: A Comprehensive Overview of Methods and Applications* (Springer, 2022), pp. 455–466
30. C. Wu, F. Wu, Y. Cao, Y. Huang, X. Xie, arXiv preprint arXiv:2102.04925 (2021)
31. D. Chai, L. Wang, K. Chen, Q. Yang, IEEE Intell. Syst. **36**(5), 11 (2020)
32. S. Meftah, B.H.M. Tan, C.F. Mun, K.M.M. Aung, B. Veeravalli, V. Chandrasekhar, IEEE Trans. Inform. Forensics Secur. **16**, 3740 (2021)

33. N.J.H. Marcano, M. Moller, S. Hansen, R.H. Jacobsen, in *2019 IEEE Globecom Workshops* (IEEE, 2019), pp. 1–6
34. G. Cormode, S. Jha, T. Kulkarni, N. Li, D. Srivastava, T. Wang, in *Proceedings of the 2018 International Conference on Management of Data* (2018), pp. 1655–1658
35. Y. Zhao, J. Zhao, M. Yang, T. Wang, N. Wang, L. Lyu, D. Niyato, K.Y. Lam, IEEE Internet Things J. **8**(11), 8836 (2020)
36. P.C.M. Arachchige, P. Bertok, I. Khalil, D. Liu, S. Camtepe, M. Atiquzzaman, IEEE Internet Things J. **7**(7), 5827 (2019)
37. E. Bagdasaryan, O. Poursaeed, V. Shmatikov, in *Advances in Neural Information Processing Systems*, vol. 32 (2019)
38. B. Jayaraman, D. Evans, in *28th USENIX Security Symposium (USENIX Security 19)* (2019), pp. 1895–1912
39. J. Abowd, R. Ashmead, R. Cumings-Menon, S. Garfinkel, D. Kifer, P. Leclerc, W. Sexton, A. Simpson, C. Task, P. Zhuravlev, Adv. Neural Inform. Process. Syst. **34**, 11883 (2021)
40. J. Aduda, J. Gitonga, J. Mod. Account. Audit. **7**(9), 934 (2011)
41. Z. Li, S. Yang, Z. Li, Int. J. Secur. Appl. **10**(3), 245 (2016)
42. Y. Wang, Y. Zhang, Y. Lu, X. Yu, Proc. Comput. Sci. **174**, 141 (2020)
43. J. Wang, J. Shi, X. Wen, L. Xu, K. Zhao, F. Tao, W. Zhao, X. Qian, Comput. Secur. **121**, 102839 (2022)
44. B. Hu, Z. Zhang, J. Zhou, J. Fang, Q. Jia, Y. Fang, Q. Yu, Y. Qi, in *Proceedings of the 29th ACM International Conference on Information & Knowledge Management* (2020), pp. 2525–2532
45. P. Cui, X. Wang, J. Pei, W. Zhu, IEEE Trans. Knowl. Data Eng. **31**(5), 833 (2018)
46. Z. Wu, S. Pan, F. Chen, G. Long, C. Zhang, S.Y. Philip, IEEE Trans. Neural Netw. Learn. Syst. **32**(1), 4 (2020)
47. D. Cheng, Z. Niu, Y. Tu, L. Zhang, in *Proceedings of the ICPR 2018*. pp. 361–366
48. Z. Wen, Y. Fang, Z. Liu, in *Proceedings of the 44th International ACM SIGIR Conference on Research and Development in Information Retrieval* (2021), pp. 1219–1228
49. X. Yu, arXiv preprint arXiv:1707.04831 (2017)
50. Y. Sun, J. Han, ACM SIGKDD Explor. Newslett. **14**(2), 20 (2013)
51. J.M. Kanter, K. Veeramachaneni, in *Proceedings of the DSAA 2015*. pp. 1–10
52. P. Bojanowski, E. Grave, A. Joulin, T. Mikolov, Trans. Assoc. Comput. Linguist. **5**, 135 (2017)
53. S. Hochreiter, J. Schmidhuber, Neural Comput. **9**(8), 1735 (1997)
54. T.Y. Lin, P. Goyal, R. Girshick, K. He, P. Dollár, in *Proceedings of the ICCV 2017*. pp. 2980–2988
55. L. Zhu, Z. Liu, S. Han, in *Advances in Neural Information Processing Systems*, vol. 32 (2019)
56. J. Thakur, N. Kumar, Int. J. Emer. Technol. Adv. Eng. **1**(2), 6 (2011)
57. X. Zhou, X. Tang, in *Proceedings of the 2011 6th international forum on strategic technology*, vol. 2 (IEEE, 2011), pp. 1118–1121
58. S. Routray, D. Raipure, S.R. Nayak K, N.P. Shetty, in *2021 Fourth International Conference on Electrical, Computer and Communication Technologies (ICECCT)* (2021), pp. 1–7. https://doi.org/10.1109/ICECCT52121.2021.9616671
59. R. Zamir, IEEE Trans. Inform. Theory **44**(3), 1246 (1998)
60. Z. Zheng, S. Xie, H.N. Dai, X. Chen, H. Wang, Int. J. Web Grid Services **14**(4), 352 (2018)
61. M.A. Sheikh, A.K. Goel, T. Kumar, in *2020 International Conference on Electronics and Sustainable Communication Systems (ICESC)* (IEEE, 2020), pp. 490–494
62. Y.Y. Lee, M.H.D.H. Yahya, M.S. Habibullah, Z.M. Ashhari, J. Financ. Econ. Policy (2019)
63. B. McMahan, E. Moore, D. Ramage, S. Hampson, B.A. y Arcas, in *Proceedings of the AISTATS 2017*. pp. 1273–1282

Chapter 8
Behavioral Authentication for Security and Safety

Abstract The issues of both system security and safety can be dissected integrally from the perspective of behavioral *appropriateness*. That is, a system is secure or safe can be judged by whether the behavior of certain agent(s) is *appropriate* or not. Specifically, a so-called appropriate behavior involves the right agent performing the right actions at the right time under certain conditions. Then, according to different levels of appropriateness and degrees of custodies, behavioral authentication can be graded into three levels, i.e., the authentication of behavioral *identity*, *conformity*, and *benignity*. In a broad sense, for the security and safety issue, behavioral authentication is not only an innovative and promising method due to its inherent advantages but also a critical and fundamental problem due to the ubiquity of behavior generation and the necessity of behavior regulation in any system. By this classification, this review provides a comprehensive examination of the background and preliminaries of behavioral authentication. It further summarizes existing research based on their respective focus areas and characteristics. The challenges confronted by current behavioral authentication methods are analyzed, and potential research directions are discussed to promote the diversified and integrated development of behavioral authentication.

8.1 Introduction to Behavioral Authentication

The advantages of frictionlessness, continuousness, synthesizability, and inherence in behavioral authentication have attracted significant attention from security research community [1–9]. Narrowly defined, behavioral authentication usually refers to *behavioral identity authentication* [10, 11]. It is a technology that has emerged as a result of the continuous development of artificial intelligence industry and the improvement of computing and storage capabilities in hardware, driven by the increasing dependence of individuals on the convenience provided by devices. Behavioral authentication collects behavioral data implicitly in the background and distinguishes them from others by analyzing unique behavior patterns, as behavior patterns are the intrinsic properties of agents and naturally possess distinct characteristics compared to others. Initially, research on behavioral identity

C. Wang, H. Zhu, *Universal Behavior Computing for Security and Safety*,
https://doi.org/10.1007/978-981-97-9014-2_8

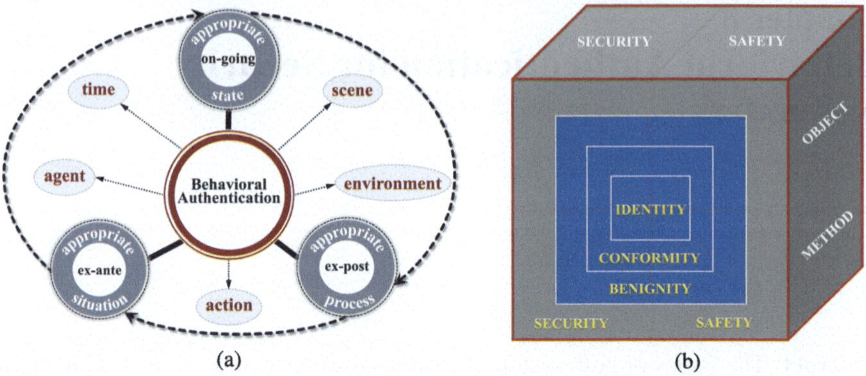

Fig. 8.1 The illustration of security and safety (**a**) and the conceptual framework of behavioral authentication (**b**)

authentication primarily focused on utilizing collected behavioral data to address specific authentication scenarios and meet the essential usability requirements [12–14]. As researchers gain a deeper understanding of behavior, they realize the close correlation between security and behavior [15, 16]. The connotation of security can be described from *a behavioral perspective*, where it involves the *right agent* performing the *right actions* at the *right time* under certain conditions (e.g., *scene* and *environment*), as illustrated in Fig. 8.1a. Therefore, the study on security can be essentially implemented by the research on behavior, and then the behavior is also regarded as the object of security research [17, 18]. Behavioral identity authentication ensures protection against external malicious attackers and safeguards security. As the scope of behavioral identity authentication applications gradually expands, the importance of nonfunctional requirements is also increasing. Researchers are paying more attention to the robustness, reliability, and timeliness aspects of behavioral identity authentication [2, 18, 19], which is of great significance to the safety.

In real-life business, after verifying the legitimacy of identity, there are still some behaviors that require further behavioral authentication to evaluate their compliance. For example, in the scenario of credit loans [20], there are some users who use their genuine identities, but after being approved, they engage in malicious cash-out activities. In the database system, users access the system through their own account, so the user's identity is normal, but when the user steals database information through this account, the behavior is illegal for the system. We call this type of problems as *behavioral conformity authentication.* Behavioral conformity authentication refers to identifying whether behavior patterns conform to the rules within the system during the process of using various intelligent information services under the precondition of user legitimate identity. In fact, some studies have already utilized behavioral data as the resource to solve different cybersecurity issues, e.g., online payment anti-fraud [20–22], network system intrusion detection [23], and social network compromised account detection [10, 24, 25].

Furthermore, security and safety issues may arise due to unknown risks and vulnerabilities associated with an agent's behavior, even if its identity is legitimate and its actions comply with regulations. For instance, in the scenario of credit loans, there is the issue of multiple loans, where some users borrow from different platforms using their genuine identities. Their identities are legitimate, and they repay the loans on time initially. However, due to the limited repayment capacity of some borrowers, there is a higher risk associated with borrowing from multiple platforms. In the short term, it may seem like there are no issues, but when the last platform is unable to borrow, it leads to overdue payments and a mounting debt burden, resulting in bankruptcy caused by a cycle of borrowing to repay existing debts. In an industrial Internet scenario, the operational state of a device manifests as an uninterrupted flow of temporal data and occasionally accompanied by abnormal signals do not trigger compliance alerts. These signals do not originate from external intrusions upon the device, but the accumulation of occasional abnormal signals over the long term can lead to systemic risks of system [26–28]. These concerns are called *behavioral benignity authentication*. It refers to detecting any potential risks in user behavior during the process of using various intelligent information services under the precondition of user identity legitimacy and behavioral compliance.

Considering the aforementioned aspects, behavioral authentication encompasses three main components: behavioral identity authentication, behavioral conformity authentication, and behavioral benignity authentication. We propose the conceptual framework of behavioral authentication, which is shown in Fig. 8.1b. In this work, we first provide an overview of the background and preliminaries of behavioral authentication. Next, we conduct a comprehensive review of existing research and categorize the studies based on their respective focus areas and distinctive characteristics within each category. Moreover, we highlight and discuss the challenges of the current behavioral authentication methods and propose future research directions to enhance the effectiveness and efficiency of behavioral authentication.

8.2 Background

According to the *52th Statistical Report on the Development of the Internet in China* [29] released by the China Internet Network Information Center in 2023, as of June 2023, China's Internet user base has reached 1.079 billion people, with an Internet penetration rate of 76.4%. The user base for instant messaging, online video, and short video stands at 1.047 billion, 1.044 billion, and 1.026 billion, respectively, with user adoption rates of 97.1%, 96.8%, and 95.2%. These new digital services have been closely related to people's lives. However, at the same time, digital transformation is also facing severe and complex security and safety challenges, posing real threats to critical infrastructure, systems, and citizen privacy [30, 31]. From the perspective of attack targets, the focus has shifted from networks and systems to business and data, with an increasing number of ransomware and application-based attacks. In terms of attack methods, the trends of automation, intelligence, and concealment in attack tools have become more prominent. Additionally, cybercriminals further enhance the success rate of attacks by illegally acquiring personal data, such as email and physical addresses, phone numbers, and other personally identifiable information. All these factors pose significant challenges to traditional authentication methods. Driven by the security and safety demands of the real world, there is a current need to adopt new technologies to address these challenges. Behavioral authentication is one of the most promising methods, and this innovative approach is moving toward a safer and more secure cyberspace.

The technique of behavior modeling is closely related to behavioral authentication. In fact, there is a rich history of research focused on predicting user characteristics, including attributes, personality, and behavior, through the utilization of behavior modeling techniques. In the years 2013 and 2015, Kosinski et al. [32, 33] explored the feasibility of using user behavioral data to infer, predict, and model user attributes, interests, and personality. In terms of user behavior prediction, as early as 2010, Song et al. [34] conducted a three-month study and analysis of the travel records of 50,000 anonymous mobile phone users in their publication in *Science* and found that users' historical travel behaviors followed specific patterns, with an accuracy of up to 93% in predicting users' potential travel behavior. There has been extensive research in the specific application of behavior modeling requirements for Internet services. For example, personalized modeling has been implemented by mining users' interests and hobbies based on their behavioral characteristics [35–38]. Behavior trend analysis has been conducted by observing changes in user behavior patterns [39–42]. Malicious accounts have been detected by analyzing account behavioral characteristics [43, 44]. Risk assessment of default has been performed by analyzing user transaction behavior records [45]. Social identity association across different platforms has been addressed by matching user behavior patterns [46–48].

In the field of identity authentication, the majority of current online methods resemble access control methods. Common methods include setting up alphanumeric or graphical passwords for accounts [49–54], utilizing security tokens [53, 55–57], and employing biometric features such as facial and iris recognition [58–62]. Setting passwords for accounts is the simplest and most widely used method of identity authentication, while linking mobile phones and emails serves as an effective measure for account protection and recovery. Security tokens are physical devices employed as a means of identity authentication and are currently extensively utilized. Biometric authentication refers to the utilization of unique biological features of individuals to verify their identities. Biometric authentication is considered a relatively reliable method of identity verification. The abovementioned methods primarily operate during the login authentication phase of an account. Due to the fact that these methods typically require additional user operations, they are regarded as intrusive authentication methods. These intrusive methods are hard to meet the high requirements of users for authentication convenience and service experience. When confronted with the issue of managing multiple account passwords, many users tend to adopt complete or partial password reuse strategies, underestimating the threat of password reuse to account security [9, 63, 64]. Furthermore, certain risks exist with certain biometric-based authentication technologies. For instance, in facial recognition, the replicability of facial data (obtainable and replicable in public environments), the instability of facial data (affected by makeup, allergies, injuries, and cosmetic surgeries leading to changes in features), and the security of backend data should be taken seriously (a breach of backend data would have devastating consequences for industries and society).

In the field of conformity authentication, the existing practices mainly rely on establishing conformity authentication systems based on rules [65–69]. The process of setting up rule libraries is time-consuming and incurs high manual costs [70]. Once authentication is granted through rule-based detection systems, these systems tend to persist in granting access for similar requests in the future. However, due to the absence of timely updates to the rule library, identifying new instances of nonconformity becomes challenging. In practical business scenarios, the limitations of rule-based construction have become increasingly apparent. Conformity authentication models sometimes require the input of hundreds of expert features. The manual construction of rules is also challenging to transfer and switch across different application scenarios, resulting in significant manual and time costs and impacting the efficiency of model development and operation. This method is constrained by human expertise and may overlook potential nonconformities, preventing the model from achieving excellent detection performance [71]. Additionally, this time-consuming and labor-intensive approach to constructing conformity models, which may overlook complex risks, no longer meets the requirements of secure, dynamically changing conformity authentication systems. It is of great significance for the development of conformity authentication models to effectively utilize and reuse knowledge, reduce manual and time costs, and establish automated and efficient authentication models.

In terms of benignity authentication, existing methods mainly focus on the static benignity of entities, with limited scalability in dynamic environments. These methods heavily rely on pre-established dependency analysis, equivalence relationships, and protocol state specifications. The trusted formal modeling employed in these methods is predominantly based on cryptographic techniques, which presents several challenges in real-world business scenarios. Cryptographic modeling processes can be time-consuming and hinder the efficiency of authentication procedures. The practicality of these cryptographic methods may be limited due to high hardware requirements for deployment, affecting their efficiency and usability. These methods lack the necessary adaptability and flexibility, which makes it challenging to promptly capture potential risks. For example, coupon clippers may not initially violate regulations, but the large-scale organization of coupon clippers' activities is likely to increase overall transaction and credit costs, ultimately hindering normal business activities. There is a pressing need for more adaptive and flexible approaches in benignity authentication that can overcome these limitations and meet the demands of dynamic environments in real-world business scenarios.

8.3 Preliminaries

Experts in the field of cybersecurity have paid attention to the importance of user behavior research in addressing security issues. Dr. Douglas Maughan, the inaugural Office Head for the National Science Foundation (NSF) Convergence Accelerator, and the former Director of the Cyber Security Division at the Science and Technology Directorate of the US Department of Homeland Security, believes that researchers should prioritize viewing cybersecurity issues from a human factors' perspective [72]. Professor Angela Sasse of the University College London, funded by the British government, is studying cybersecurity issues in the business sector from a novel perspective of social and behavioral science [73]. Professor Stefan Savage of the University of California, San Diego, is also conducting research on behavior analysis for the prevention and control of network fraud [74]. *Nature* published an article titled *How to Hack The Hackers: The Human Side of Cybercrime*, highlighting the progress made in representative research efforts and emphasizing the importance of leveraging behavioral science to understand the behavior patterns of both perpetrators and victims for enhancing cybersecurity [75].

For behavioral identity authentication, devices typically collect behavioral data in the background. The collected data is then utilized for training a machine learning model [76]. Features are extracted through the trained machine learning model based on the collected behavioral data, which form the users' behavior profiles. The identity behavior profile is represented as a matrix U by the model. During the process of behavioral identity authentication, the newly generated credential of kth user is compared against stored identity behavior profile matrix to determine the authenticity of the user's identity. Verification involves comparing a provided behavior record with a stored template to determine a level of similarity. It grants access to legitimate users if their presented behavior record exhibits a similarity measure surpassing a predefined threshold. Let x, $g(\cdot)$, and d_k denote the behavior record, the behavioral identity authentication model, and the predefined threshold of user k, respectively. The result of behavioral identity authentication model $R = True$ if $g(x) > d_k$ and $False$ otherwise.

For behavioral conformity authentication, it refers to the process of ensuring that the internal behavior of users and business activities comply with relevant laws, regulations, and industry standards. Let $\mathscr{X} = \{x_1, x_2, \cdots, x_N\}$ denote the set of behavior records. Behavioral conformity authentication aims to learn a decision function $\phi(\cdot)$: $\mathscr{X} \rightarrow \mathbb{R}$ that assigns behavioral conformity scores. The goal is to effectively distinguish noncompliant behavior records from compliant behavior records in the space defined by the behavioral conformity score decision function. By inputting behavioral data into $\phi(\cdot)$, it can directly infer the conformity scores. Larger outputs of $\phi(\cdot)$ indicate a greater degree of noncompliance, which requires the system to take appropriate actions timely to prevent or rectify such behaviors. Certainly, it is necessary to characterize and map the original features of behavior records before inputting them, which is beneficial for obtaining more accurate scores of behavioral compliance.

Researchers have already begun exploring security and safety issues within the scope of behavioral benignity authentication. Relevant research works primarily focus on the following aspects, including traceability of behavior, predictability of risk, consistency of execution, and integrity of record. The traceability of behavior refers to the tracking of individual or entity behavior activities to obtain detailed historical behavioral data, which is helpful for verifying the legitimacy of user behavior. Predictability of risk involves a comprehensive evaluation of individual or entity behavior activities to timely identify potentially risky behaviors and update the scope of compliance. Consistency of execution evaluates credibility and detects potential risks without affecting consistent operations in cross-domain interactions over heterogeneous networks. Integrity of record entails protecting user privacy and ensuring consistency in the certification process. These aspects of behavioral benignity authentication provide additional protection for systems.

The behavioral authentication method has the following advantages based on its technical principles:

- **Frictionlessness.** This method operates as a backend program, eliminating the need for user intervention. The collection of data and verification are seamlessly performed without requiring any explicit actions from the user. This frictionlessness of behavioral authentication ensures a smooth and convenient experience.
- **Continuousness.** The method transforms cybersecurity authentication from a one-time process to a continuous one. It allows for ongoing analysis of user behavior over time, ensuring that access remains granted as long as the behavior patterns align with the established user profile. This continuous authentication approach enhances security by detecting any anomalous or suspicious behavior in real time.
- **Synthesizability.** The method directly captures the projection of multiple subspaces of behavior, which exhibits synthesizability among the subspaces. By capturing various dimensions of behavior, it creates a comprehensive and reliable behavior profile. Synthesizability of behavior enhances the accuracy and robustness of the authentication process.
- **Inherence.** The data used for behavioral authentication typically originates from users' inherent characteristics and patterns. Therefore, intruders are unlikely to perfectly mimic the user's behavior patterns, as individual behavior is unique and difficult to imitate convincingly. Furthermore, intruders typically exhibit intrusive actions aimed at stealing benefits, which often deviate from the user's normal behavior.

We consider typical scenarios to illustrate the connections between different authentication methods. For example, credit loan services are typical data-intensive services, and the synthetic and intrinsic characteristics of behavioral authentication provide support for enhancing representational capabilities and breaking down isolated data attributes of credit loan services. During the initial stages of loan application, some criminals may steal others' information through illegal means for fraudulent loan applications. The specific process is illustrated in Fig. 8.2a. It shows

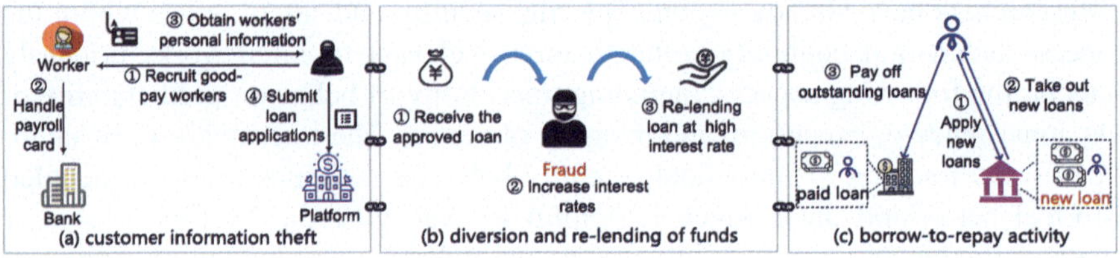

Fig. 8.2 A typical example demonstrating how different levels of behavioral authentication ensure the safety and security of credit loan services

criminals collect information from victims and submit loan applications, and it can be addressed by behavioral identity authentication. The evaluation of whether the information submitted by a borrower matches the true circumstances falls under the level of behavioral identity authentication, which protects the security of the loan application for financial institutions. While behavioral identity authentication can ensure the legitimacy of a borrower's identity, there are instances of noncompliant behavior after the loan is approved. Diversion and re-lending of funds is one such noncompliant behavior where the borrower does not use the loan funds as intended during the loan application. Instead, after receiving the loan from a financial institution, the borrower engages in illegal lending activities. As shown in Fig. 8.2b, the fraudster further increases the loan interest rates approved by financial institutions and lends the funds to others to profit from the interest rate differential. Detecting such frauds falls under the level of behavioral conformity authentication. It uses methods like behavior tracking to ensure the security of loan flows. Furthermore, in real business processes, there are cases where a borrower's identity is legitimate, and the borrower's behavior complies with platform regulations, but failure to restrict such behavior in a timely manner can increase overall system risks. Figure 8.2c illustrates the concept of borrow-to-repay activity, which is detrimental to the stable operation of the financial platform. Initially, the borrower only applies for loans on a single platform and can adhere to the repayment policies of that platform. However, later on, due to financial difficulties, the borrower cannot repay the loans on one platform on time. As a result, the borrower applies for new loans on other platforms to repay the outstanding loans. In reality, the borrower's ability to repay loans has declined. The collaborative audit across multiple platforms to promptly detect such behavior not only ensures the security of the business but also further ensures platform safety by timely refining loan strategies and making the controllability of platform risks. They are crucial aspects of behavioral benignity authentication. Safeguarding user identities from theft, monitoring the compliant use of loans, and continuously improving the approval processes provide essential protection for credit loan services. The progressive enhancement of financial market security and safety through the expansion of behavioral authentication sub-functions with the level of identity, conformity, and benignity promotes the development of inclusive finance.

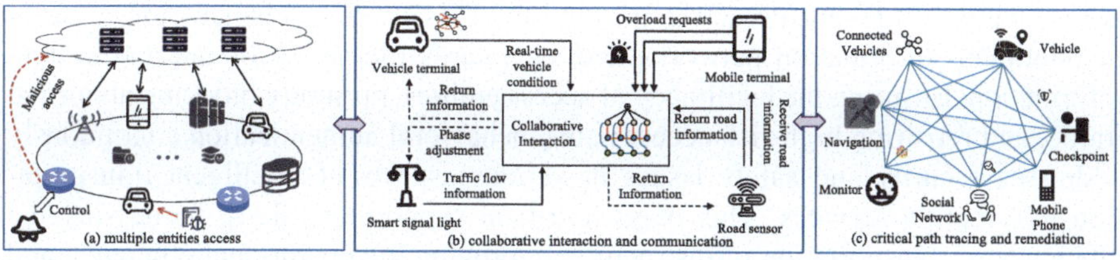

Fig. 8.3 An example illustrating the security and safety issues at different levels of behavioral authentication

Intelligent transportation information services also involve the extension of different levels of behavioral authentication. As communication and computation-intensive services, the frictionless and continuous characteristics of behavioral authentication are helpful in addressing emerging security and safety challenges, where entities' access is dynamic, devices interact frequently, and cross-domain paths are concealed. During the access phase of different entities, behavioral identity authentication prevents different devices from falling victim to phishing attacks and identity theft, ensuring data and access security for intelligent devices, which is shown in Fig. 8.3a. It describes the entity access phase where various entities may face threats such as identity theft and phishing attacks, and it belongs to the level of behavioral identity authentication. While some entities have successfully verified their legal identities and registered successfully within the intelligent transportation system, during the process of heterogeneous entity interaction, some entities may not adhere to the standards and regulations of services. Figure 8.3b illustrates typical noncompliant behavior in intelligent transportation information services. During the collaborative development of intelligent transportation information system, for the convenience of information sharing and resource coordination, different entities open certain interfaces to other interacting entities, based on additional standards. These entities must conform to the specified usage of external cooperative entities' API interfaces according to these additional standards to ensure efficient and stable information sharing and resource allocation in the system. There are some access entities that have verified the legitimacy of their identity but generate traffic requests that exceed service standards. This directly leads to network congestion and increases the system's response time. Behavioral conformity authentication is responsible for monitoring the traffic of API interfaces, promptly detecting noncompliant request entities, and safeguarding the overall security and stability of the multi-entity collaborative process. Additionally, some entities have legitimate identities and generate traffic that complies with standards when using open API interfaces. However, in the process of cross-domain access, there may exist potential high-risk access paths. These paths can give rise to undisclosed security vulnerabilities and potentially expose sensitive internal system information. Such vulnerabilities might inadvertently offer attackers opportunities to establish back-doors within the system, as shown in Fig. 8.3c. Behavioral benignity authentication uses traceability analysis to locate the minimum-cost repair points for high-risk access paths, achieving proactive prevention. This further enhances the security of intelligent information networks and safeguards the safety of internal system information. Ensuring the legitimacy of access entities, promptly blocking abnormal traffic, and repairing high-risk access paths, behavioral authentication continuously addresses security and safety issues in various aspects of intelligent transportation information services. This plays a critical supporting role in harnessing the efficiency of transportation infrastructure, improving the operational efficiency and management level of transportation systems, and facilitating smooth public travel.

8.4 Behavioral Authentication Studies

8.4.1 Behavioral Identity Authentication

According to the different characteristics of behavioral data, this part categorizes behavioral identity authentication into five major types: keystroke, touch gesture, motion, intrinsic signaling behavior, and user interaction behavior. In addition, in certain specific scenarios, the above different behavioral feature types can be combined into multifactor approach to reflect the user behavior, as shown in Fig. 8.4.

Keystroke-based authentication is one of the earlier behavioral identity authentication methods that verifies a user's identity by analyzing the characteristics of the user's input characters on the keyboard, such as keystroke force, keystroke speed, keystroke frequency, keystroke sequence, and keystroke time intervals. Early research into keystroke-based authentication focused on analyzing keystroke force and frequency. Zhu et al. [77] proposed a novel approach to authenticate users

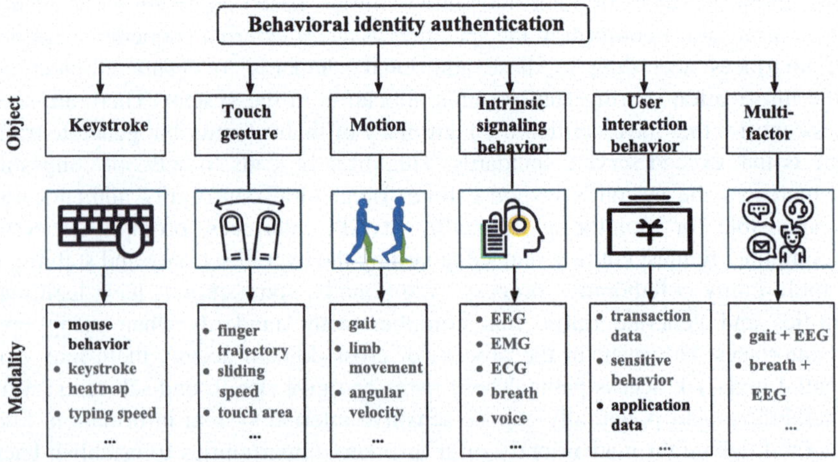

Fig. 8.4 Components of behavioral identity authentication

based on keystroke dynamics while entering passwords. The proposed method leverages a keystroke feature vector that consists of the user's keystroke force and frequency to confirm the user's identity. Subsequently, researchers introduced extra keystroke features to continuously improve the accuracy and security of keystroke-based authentication. Primo et al. [78] investigated the effect of music on keystroke behavioral data using two validation techniques: relative and absolute measurements. The researchers analyzed the data based on three dimensions of keystroke behavior: key hold, key press, and inter-key intervals, to understand how music affects user verification performance. Their findings confirmed that a music environment can improve the accuracy of user authentication based on keystroke dynamics. Ho et al. [79] introduced a user authentication approach based on keystrokes, which integrates a one-class naive Bayes algorithm (ONENB) and a typing speed inspection in typing skills (SITS) algorithm. The ONENB algorithm computes the probability of keystroke behavioral data, while the SITS algorithm incorporates keystroke patterns to define user keystroke characteristics and builds an authentication model to differentiate legitimate users from impostors. Lee et al. [17] designed a parameterized model that utilizes a feature selection method based on the median and interquartile range to extract keystroke dynamics features. The model binds the calculated security measure (FAR) obtained from these keystroke dynamics features for user identity verification. Furthermore, with the development of understanding of keystroke dynamics, many input devices have been used for authentication in similar ideas to ensure the security of user's identity, such as mice and virtual keyboard. Mao et al. [80] developed an innovative real-time identity authentication technique using mouse behavior learning. This approach entails collecting the mouse behavioral feature vectors of each new user during their initial login, both in dynamic and static scenarios, and it compares them with the feature vectors gathered during subsequent login authentication in simulated abnormal scenarios to attain user identity verification. Shen et al. [81] proposed a user identification method that involves analyzing mouse behavior. In a controlled environment, data collection is conducted on the mouse operations performed by the user for specific mouse operations. Subsequently, the features are extracted and categorized to accomplish user identity authentication by analyzing the results of the classification. Kang et al. [82] expanded the applicability of a dynamic user authentication approach that relies on keystroke dynamics. This approach now enables the authentication of users typing various textual strings across multiple input devices while distinguishing between legitimate and potential imposter users. Inguanez et al. [83] developed a user authentication technique for smart touch screen devices. This method utilizes a multilayer perceptron (MLP) neural network to classify the graph of the user's keystroke behavior and compares it with the characteristics of the keystroke heatmap in order to achieve user authentication. However, keystroke-based authentication still has certain environmental dependencies and limitations. Variable factors such as different input devices and keyboard layouts may require re-establishing authentication models. The behavioral feature information provided by the keystroke patterns is relatively limited, so more identity authentication methods using other behavioral features have been developed.

With the popularization of smartphones, touch gesture-based authentication methods have been proposed, which make up for the shortcomings of the insufficient use of behavioral characteristics in the keystroke-based authentication methods. Touch gesture refers to the interaction behavior between users and devices through gestures or touches. Existing authentication schemes usually adopt finger trajectory, touch pressure, sliding speed, touch area, and gesture pattern as behavioral features that need to be extracted and analyzed. Cao et al. [84] devised a continuous user authentication scheme that uses vibration response as an implicit biometric feature. It applies a high-pass filter to remove noise and segments the vibration data for each touch event to achieve accurate data matching and recognition and proposes a novel centroid vector method to infer the touch position accurately. Shen et al. [85] proposed a method to use gesture pattern, touch operation sequence, and other behavioral data for identity verification. It uses a Markov classifier to model the motion sensor data sequence and capture the temporal and dynamic features of the sensor events, which improves the security and applicability of identity verification. Mao et al. [86] proposed an implicit continuous authentication model for user authentication on mobile devices through touch behavior. The model uses data from sensors such as accelerometers and gyroscopes to create feature vectors that include both macroscopic and microscopic features. Yang et al. [19] designed a touch-based behavioral biometric recognition technique using a single-class support vector machine and independent random forest training models. The technique evaluates the accuracy of each type and then applies Bayes' theorem to estimate the confidence level of each type, which provides a safe and continuous authentication method for mobile applications. Xu et al. [3] combined the physical features of finger touch and the biometric features of touch behavior by a feature fusion authentication framework. It utilizes a particular training sample selection strategy to convert signal features into behavior-agnostic features and subsequently applies knowledge distillation in constructing a touch user authentication scheme. Yang et al. [87] presented an integrated identity verification scheme that combines passwords and touch behavioral factors, which include touch pressure and sliding speed. The scheme utilizes a novel algorithm to differentiate fine-grained finger input and supports different forms of passwords in the frequency domain, which improves the security of authentication. The abovementioned methods, without exception, are only for the authentication of a single user. In actual scenarios, there may be multiple users on the same device. The aforementioned studies struggle to address this type of authentication issue.

In contrast to keystroke-based and touch gesture-based authentication, motion-based authentication methods emerged later due to the requirement for a multitude of sensors to collect user motion data for analysis. With the development of wearable and mobile devices, more sensors, such as accelerometers and gyroscopes, are built into the devices to meet the needs of users, which also meets the basic conditions of motion-based authentication methods. Existing motion-based authentication schemes typically utilize motion features such as gait, limb movement, acceleration, and orientation to perform identity verification to ensure the security of users' identities. Chen et al. [88] proposed a framework for continuous user authentication based on motion behavioral characteristics, including direction, acceleration, and angular velocity. The framework utilizes smartphones' built-in sensors to collect typical daily motion data of users such as time, frequency, and wavelet domains, which achieves relatively accurate user authentication. Lee et al. [89] introduced an advanced smartphone authentication system called iAuth, which leverages the capabilities of multiple sensors embedded within smartphones, Bluetooth connectivity, and wearable devices equipped with sensors. By utilizing machining learning methods to capture the user's unique gait behavior pattern in sensor data from different devices, iAuth enables seamless and ongoing authentication of end-users. Zou et al. [18] developed a hybrid deep neural network method for identity verification that extracts robust gait features from time series data. The network combines the convolutional output features with the temporal properties of the data, which enhances the gait features and increases the authentication accuracy. He et al. [1] developed a gait feature extractor using transfer learning to reduce time costs and enhance the model's robustness. The extractor has undergone pre-training for user identification tasks. Song et al. [90] developed an authentication system based on eye movement. The system verifies the user's identity by capturing the features of human eye movement from the front camera of the smartphone. In terms of multiuser authentication, the authentication schemes need to solve the problem of changing motion features to ensure the system's safety. Kong et al. [91] presented an optimized method for user identification based on gait features. To accurately capture the behavioral characteristics of gait data, their method employs a spatial transformation algorithm to optimize coordinate drift and utilizes a support vector machine algorithm to address the issue of gait feature changes when switching between users. Compared to keystroke-based authentication methods, motion-based authentication methods involve multiple sensor data that may be leaked during transmission, thereby compromising the safety of the system. Wang et al. [4] proposed a novel behavioral authentication framework for user motion characteristics to address issues including behavioral dynamics, data privacy, and side-channel leakage. The framework accelerates feature transfer speed on mobile devices and mitigates potential side-channel leakage and improves security during transmission. In summary, motion-based authentication schemes introduce more sensors to capture behavioral features, which makes full use of user behavioral features but also brings greater computational overhead and sensor data leakage problems that necessitate optimization and resolution.

Compared with touch gesture and motion, intrinsic signals, such as EEG, EMG, ECG, breath, facial expression, voice, and others, are highly unique in behavioral recognition, making it more difficult for attackers to accurately replicate. Moreover, existing research also focuses on these sequences of behavioral signals which typically reflect an individual's behavior patterns. Chauhan et al. [92] proposed the ContAuth system, which targets inherent behavioral signals of users, such as breath and EEG, obtained from sensors using a class-incremental learning method. It combines deep learning models with online learning models to enhance the robustness of behavior-based authentication. Perera et al. [15] designed a sparse representation-based multiuser mobile active authentication scheme according to the dynamic facial expressions of users that automatically adjusts the parameters using the extremum distribution mechanism. It also includes an extension algorithm for applying the scheme in a single-user scenario. Lu et al. [93] studied the acoustic Doppler effect of user speech and created a lip-reading-based user authentication system called LipPass that operates in noisy environments. Ji et al. [94] introduced a position-sensitive identity verification mechanism called NAuth with nonlinear enhancement. This verification scheme ensures consistent device identity verification by extracting acoustic nonlinear patterns (ANP). Implementing intrinsic signal authentication often requires specialized hardware, which may not be available on all devices. Also, acquiring these signals often requires stable physical and emotional states, and changes in signals caused by emotions or illness may affect the performance of the system.

In addition to the abovementioned methods relying on devices or sensors for identity authentication, some methods can achieve identity authentication only through user interaction data. With the advancement of big data, machine learning, artificial intelligence, and other technologies, these methods have received more and more attention and research. User interaction behavior, such as social networking, financial transactions, and information browsing, has generated a series of data that includes but is not limited to search history data, transaction data, user profile data, sensitive behaviors, and social behaviors. Ruan et al. [95] introduced a user authentication scheme that relies on social behavioral characteristics. They extracted and classified social behavioral features such as user browsing and clicking on OSN websites, determined metrics for each feature, and constructed user behavior profiles. Finally, they validated the accuracy of distinguishing genuine users from impostors using user behavior profiles. Shi et al. [96] presented an implicit identity verification scheme based on user behavior patterns, leveraging history data on smartphones and movement data collected by sensors to extract behavioral features for user authentication. Skravcic et al. [97] presented an implicit authentication scheme centered around user behavior patterns. This approach employs classification models that are built using vast amounts of user transaction data, call records, and email correspondences from various systems, such as banking and social networks, to identify legitimate and malicious users. By leveraging these behavior patterns, the scheme effectively distinguishes between these two user categories. Yang et al. [12] designed a wind vane module to achieve lightweight implicit authentication. This module determines the amount of data needed to be collected at different times based on the user's identity legitimacy and interactive behavior habit and adjusts the sampling rate accordingly, which provides an energy-efficient solution for real-time implicit authentication on mobile devices. Shi et al. [13] developed an end-point identity authentication technology based on the analysis of user-associated behaviors. The approach employs an interactive behavior common-subsequence similarity algorithm, which extends the traditional behavior common-subsequence (BCS) sequence pattern, and considers the maximum overlap of user behavior sequences and the short sequence overlaps at different time intervals to better identify any anomalies during each user's login session. Wu et al. [14] proposed a hidden Markov model (HMM) to detect malicious user interactive behavior in network systems by extracting relevant features and defining the observation symbols and hidden states based on the user's access behavior patterns. It maps the user's behavior to an HMM chain and identifies any abnormal or harmful actions to ensure the security of network systems. Such methods need to collect, analyze, and store user interaction behavioral data, especially for behaviors containing sensitive data, which may involve privacy issues. Furthermore, such methods are prone to adversarial attacks. When the attacker adds fake historical data to the training samples, the model will not be able to perform correct authentication.

With the further development of authentication technology, some studies realized that multifactor authentication systems, which rely on multiple factors to provide robust and accurate results, have stronger security compared with systems that only consider a single behavioral feature. Dasgupta et al. [98] proposed a multifactor authentication system that considers various combinations of different data features through a subset of available authentication modalities, which ensures efficiency in dynamic environments. They conducted tests on one-time and continuous authentication for smartphone users and confirmed that there is complementarity between different signals, which can enhance the performance of the authentication system. Zhang et al. [2] designed a multimodal biometric authentication system that combines EEG and gait data and leverages their unreplicated characteristic. This system uses a doubly authenticated method to improve the anti-counterfeiting and security of the authentication process. Wazzeh et al. [99] devised an authentication scheme for mobile devices utilizing federated learning (FL). This scheme allows each user to keep their private data locally for safety and trains models to capture their multimodal behavioral data with a server for global aggregation. Liu et al. [100] presented a user authentication method for smartphones by analyzing user interaction behavior. The approach establishes a behavioral characteristic classification model using data such as the user's touch screen interaction method, motion, and phone power consumption to enable continuous user identity verification. However, when designing multifactor authentication methods, suitable algorithms are needed to solve problems such as multifactor data fusion and data imbalance, and it is technically challenging to seamlessly integrate various factors.

Based on the aforementioned analysis, characteristics and issues of different objects of behavioral identity authentication are summarized in Table 8.1.

8.4.2 Behavioral Conformity Authentication

Behavioral conformity authentication aims to identify potential security and safety risks within a system that fall outside the scope of behavioral identity authentication. The risky objects detected by behavioral conformity authentication mainly include five typical categories, i.e., fraud, malicious intrusion, insider threat, unfair discrimination, and privacy leakage, as shown in Fig. 8.5.

Fraud risks commonly occur in industries such as telecommunications, healthcare, and finance, with a particularly significant impact observed in the financial

Table 8.1 Summary of behavioral identity authentication

Object	Description	Characteristics	Issues
Keystroke [17, 77–83]	Keystroke refers to the action of a user inputting information through a keyboard, a keypad, or even a mouse	(1) Without reliance on additional devices. (2) Extra layer of security for password. (3) Noninvasive continuous authentication	Data quality easily affected by environmental and user conditions
Touch gesture [3, 19, 84–87]	Touch gesture refers to the interaction between users and devices through gesture or touch	(1) Covering multiple dimensional features to improve the security of verification. (2) Ensuring frictionless authentication	(1) Difficulty in distinguishing and identifying multiple behavior patterns. (2) Cross-platform versatility limited by inconsistent behavioral data formats
Motion [1, 4, 18, 88–91]	Motion refers to the various postures and movements made by users while using wearable or mobile devices.	(1) Utilization of broader behavioral features. (2) Noninvasive continuous authentication. (3) Less prone to variation caused by external factors	(1) Multi-device dependency. (2) Privacy risks posed by sensor attacks
Intrinsic signaling behavior [15, 92–94]	Intrinsic signaling behavior refers to the signals emitted by human organs during the interaction between users and devices	High biometric uniqueness	(1) High device dependency. (2) Data quality largely affected by emotions, diseases, etc.
User interaction behavior [12–14, 95–97]	User interaction behavior refers to the behavior of users interacting with applications.	(1) No requirement for sensor data collection and conversion. (2) A wider range of behaviors beyond keystroke or motion	(1) Suffering from data privacy problem and breach risk. (2) The increasing probability of misjudgments of the model due to adversarial attacks
Multifactor [2, 98–100]	Multifactor refers to the use of multiple categories of behavioral data or a combination of conventional methods with behavioral authentication to verify a user's identity	(1) Flexible combination of authentication methods. (2) The addition of cross-validation for an extra layer of security. (3) Reducing reliance on a single piece of sensitive data	(1) The challenge of seamlessly integrating various behavioral features and authentication technologies. (2) Imbalanced data from various behavioral features

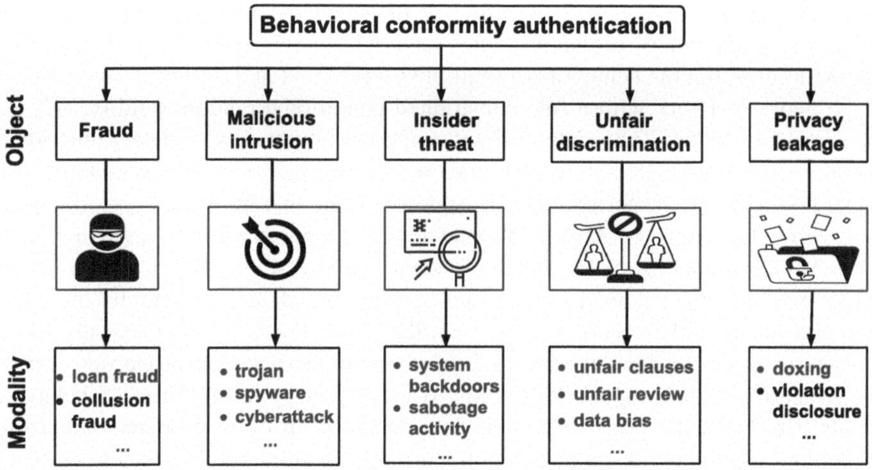

Fig. 8.5 Components of behavioral conformity authentication

field. In the financial domain, user identity authentication is typically subjected to heightened scrutiny. This is attributed to the involvement of substantial funds and sensitive information in financial transactions, necessitating the assurance of transactional security and accuracy. Despite the stringent measures, instances of loan fraud still occur, even when identity authentication processes are in place. For fraudulent behavior in the financial sector, some methods can detect noncompliant activities and transactions by monitoring customer transactions and behavior patterns. Jiang et al. [101] developed a more comprehensive network for embedding location information, called the Fuller Location Information Embedding Network. The network employs self-supervised learning to characterize the address features of users, with a focus on analyzing the relationship between address information, behavioral information, and customer fraudulent behavior in loan applications, effectively improving the performance of loan fraud detection models. Wu et al. [102] proposed a two-stage detection model for identifying fraudulent agents on online large-scale loan platforms. The model extracts 26 features from activity records such as communication logs and application activity histories of agents and borrowers, as well as loan histories. Based on these 26 features, the model characterizes the behavior patterns of the classification objects in detail and achieves high accuracy in identifying fraudulent agents. Awotunde et al. [103] proposed an artificial neural network-based detection method for the fraud in bank loan management. They extracted a series of borrower-related and loan-related information data as features for identification and classification, in order to detect fraudulent behavior in loan transactions. Chang et al. [104] presented a general model for detecting financial fraud using natural language processing technology to accurately detect and classify fraud. As an instance of the model, they implemented an anti-fraud chatbot on a widely used social network service. Nevertheless, certain advanced fraudulent activities transcend the actions of solitary individuals, involving collaboration among multiple users or intermediary agents. In view of these scenarios, a number of research studies have been dedicated to investigating the correlations within the behaviors of multiple users. Xu et al. [105] proposed a novel graph neural network with a role-constrained conditional random field (GRC) for loan fraud detection. The model utilizes graph neural network to detect individual user loan fraud and collusion fraud based on borrower role information and network social relationships. Experimental results demonstrate that the model performs well in detecting loan fraud on Alipay. Wang et al. [23] proposed a graph-based approach for behavior modeling called behavior identification graph (BIG). This method delves into the property-level associations in behavioral data and integrates the inter- and intra-behavioral correlations into a unified space. Furthermore, they introduced a property graph to describe fine-grained correlations between properties, where the structure of the graph corresponds to the topology information of behavior events. Based on the property graph, they designed an event-property composite model and used network representation learning algorithms to extract fine-grained associations at the behavioral property-level. The behavior patterns are represented in a multidimensional spatial distribution of behavioral properties. The effectiveness of this method has been verified in various network threat detection scenarios, particularly in the fraud detection scenario. In addition, some research has focused on purchase fraud and misinformation. He et al. [106] proposed a novel malicious user detection system, called Datingswc, that aims to address fraudulent activities such as misinformation or illegal information on dating applications. The system utilizes user profiles, comments, and other relevant information to establish behavior patterns using machine learning models, such as MLP and LSTM. These behavior patterns are then combined and inputted into an attention module to automatically detect suspicious activities following these behavior patterns. Wang et al. [107] designed a novel graph-based fraud detection framework to detect fraudulent orders placed by employment fraud teams. The framework comprises two parts: the DPP module, which extracts feature sequences of user click locations from the website, and the GSR module, which performs neighborhood sampling and information aggregation. The effectiveness of this framework has been validated through purchase fraud detection on the JD platform. Wang et al. [5] developed an interactive fraud detection dialogue system, which actively engages in conversations with clients by means of intelligent voice interactions. The system employs imitation learning to master the dialogue strategy and accurately assesses the risk of actual payment, thereby reducing the likelihood of misjudging payment behavior without actual risk. However, behavior-based fraud detection methods still exhibit certain shortcomings and face a set of challenges. Constructing a fraud model necessitates the inclusion of user privacy data, which requires the consent and authorization of both the regulatory platform and the user. Once privacy data is acquired, the model confronts the difficulty of addressing the inherent imbalance between instances of fraudulent and normal behavior during the training phase. Furthermore, the trained model must contend with the ever-evolving deceptive techniques, thereby necessitating regular updates to accommodate the identification of emerging fraud patterns.

Different from fraud with deceptive features, malicious intrusion has evident attack-oriented features. This type of behavior poses substantial risks, not only to personal computers but also to smartphones, to Internet of Things (IoT) devices, and even to network security as a whole. The category of malicious intrusion spans a broad spectrum, including trojans, ransomware, spyware, phishing attacks, and cyberattacks. Existing intrusion detection frameworks often utilize machine learning algorithms to extract features from network behavior, aiding in the differentiation between normal and malicious activities. Chen et al. [108] presented an efficient network-based anomaly detection (NBAD) algorithm using a combination of deep belief network (DBN) and long short-term memory (LSTM) network for cyberattack detection. Firstly, on the premise of maintaining the accuracy, the DBN method is utilized to automatically extract the features of the original data nonlinearly, so as to express the features of the original data with a lower dimension. Furthermore, the classification results, as the basis for identifying anomalous network behavior, are obtained through a lightweight LSTM network. Chen et al. [109] presented an automated ransomware pattern extraction and early detection tool. The tool analyzes discovered malware samples and generates a log, from which it extracts sequences of events triggered by ransomware. It also ranks the features of ransomware and detects malicious activity from the learned behavior, which effectively enhances the security of infrastructures. Hamid et al. [110] designed a hybrid feature selection method for detecting phishing attack behavior, which combines content-based and behavior-based analysis. The method analyzes the content and ID tags of phishing emails to identify characteristics of attacker behavior and subsequently detects phishing attacks. Qin et al. [111] devised a novel unsupervised network behavior anomaly detection framework, which combines real-time high-order host state in a dynamic interactive environment with dialogue patterns between hosts. It automatically generates high-order features from a series of basic features extracted from the graph neural network (GNN) and identifies various cyberattack behaviors more effectively. Jiang et al. [112] proposed a behavior-based method for intelligent recognition and security supervision of unmanned aerial vehicles (UAVs). The method uses location tracking and flight data from the onboard black box to collect real-time behavioral data of UAVs. Then, it identifies suspected intrusion and attack behaviors of UAVs through behavior modeling and issues warnings in potential illegal situations. Garg et al. [8] addressed the issue of high false alarm rates in existing real-time anomaly detection by proposing a hybrid detection model that utilizes grey wolf optimization (GWO) and convolutional neural networks (CNNs). The model improves the feature selection and anomaly classification capabilities. In the first stage, improved grey wolf optimization (ImGWO) is used for feature selection to minimize the feature set. In the second stage, optimized convolutional neural network (ImCNN) is used for more effective anomaly classification. Pajouh et al. [113] devised a novel intrusion detection model that uses two layers of dimensionality reduction and two layers of classification modules to identify malicious activities such as user to root and remote to local attacks. The model uses naive Bayes and the deterministic-KNN version to detect suspicious behavior. Wang et al. [114] proposed a security detection system (IoT-Praetor) for malicious attacks on IoT devices. The system uses a novel DUD model to construct norms for the interaction and communication behavior of IoT devices. A behavior rule engine is employed to detect device behavior in real time, enabling the identification of behaviors that damage devices through malicious network communication. Some studies also consider the security of data privacy and the safety of the detection framework while striving for accurate detection of malicious behavior. Pei et al. [115] devised a personalized federated anomaly detection framework in order to take into account privacy protection in the process of detecting anomalous network traffic. It archives a personalized detection model by fine-tuning the model structure of different systems, which improves the data utility on the premise of protecting the privacy and takes into account the safety of the method while improving efficiency. Mothukuri et al. [116] presented a federated learning-based method for detecting malicious attacks in IoT. The method utilizes decentralized device data to proactively identify intrusion behaviors in IoT networks, achieves privacy preservation of terminal devices, and performs better in attack detection than nonfederated learning methods. Kurt et al. [6] devised a data-driven method to detect UDP flooding and spam attacks in IoT networks. To ensure the privacy of node data, the scores are encrypted and perturbed before being sent to the network operator for aggregation of statistical information, followed by anomaly detection through generalized accumulation and algorithms. Intrusion detection methods still have some deficiencies and challenges. Similar to fraud detection, the model of intrusion detection also has the problem of imbalanced training data. When the intrusion detection model needs to be deployed on multiple devices, the problem of heterogeneous data fusion is also considered.

The abovementioned fraud and malicious intrusion are both behaviors that occur outside the system, while insider threats are relatively intuitive risk behaviors inside the system, which may also lead to system collapse or huge economic losses. System backdoors and equipment failures are relatively common insider threats. Ji et al. [117] designed an active anomaly detection network for mobile robots to address system malfunctions caused by outdoor environmental factors. The network effectively integrates multiple sensor signals to ensure robust anomaly detection even in the presence of sensor obstruction in the field environment. Cui et al. [118] introduced a blockchain-supported decentralized and asynchronous FL framework for anomaly detection in IoT systems. This framework ensures data integrity, avoids single-point failures, and improves the security of the system. Luo et al. [119] first introduced autoencoder neural networks to solve anomaly detection problems in wireless sensor networks. By constructing a three-layer autoencoder neural network, they overcame the huge demands for network resources that deep learning requires. This method is mainly used to detect IoT device failures and changes in the environment. Li et al. [120] proposed an active learning and contrastive-based detection model, which monitors the system performance indicators such as CPU utilization, latency to form a multivariate time series. It models the abnormal and normal sequences based on the VAE model to recognize the abnormal sequences and discover the potential security backdoors in the system. In addition, information theft, unauthorized activity, and even sabotage activity by system insiders are also insider threats. Modell et al. [121] devised an incremental approach to analyze anomalous user behavior in event logs. It employs graph embedding to acquire a vector representation of the users, which is updated over time and utilized to model the configuration profiles of user-accessed resources and builds the formation of a dynamic, interactive network comprising users and resources. The method is applicable for identifying suspicious or unauthorized user behavior in enterprise networks. Hou et al. [7] proposed a lightweight framework for detecting abnormal driving behavior. The framework utilizes IoT devices as carriers and captures video and image data using cameras. Based on the analysis of this data, the framework identifies dangerous behaviors such as fatigue driving and erratic driving. Mazzawi et al. [122] proposed a machine learning algorithm for detecting malicious user activity in databases, which is used to detect suspicious behaviors, such as information theft and unauthorized activity. The algorithm consists of two primary components: one is responsible for generating models using user behavior, and the other focuses on clustering similar behaviors to detect abnormal patterns that might be shared among a group of users. The current detection methods rely on high-quality behavioral data to ensure models' performance. When encountering sporadic zero-day backdoors or highly covert unauthorized activities, these existing detection methods have difficulty adapting rapidly to newly emerging threats due to a lack of behavioral data.

Unlike insider threats, the risk of unfair discrimination might not be immediately apparent, but rather, it accumulates gradually within the system over time. For instance, within certain machine learning models, data bias could propagate unfairness to model decisions. In cloud environments, resources might be unfairly distributed due to scheduling or allocation mechanisms. If the resources allocated to maintain the security of the environment fall short, the environment will gradually fall into a perilous situation. Additionally, on some network platforms, there may be unfair user clauses. Existing research mainly focuses on detecting or optimizing problems such as data bias, imbalanced distribution, and unfair clauses. Li et al. [123] designed a q-fair federated learning (q-FFL) optimization method that enables fairer performance allocation among devices in large-scale federated networks. This method can also be applied to other related problems such as meta-learning, which helps in fair initialization across multiple tasks. Mohri et al. [124] developed a novel unbiased federated learning framework to address the issue of model bias among different clients in federated learning scenarios. This framework is also applicable to learning scenarios such as cloud computing, domain adaptation, and data drift. Wei et al. [125] designed a method for optimizing resource balance distribution in cloud services. First, they employed a binary integer programming approach to address the resource allocation optimization problem among independent applicants. Second, they used evolutionary programming to modify the reuse strategy of initial optimal solutions of different applicants. This method provides a solution for resolving the complex issue of resource balance distribution in cloud computing. Lin et al. [126] proposed a combined approach of single-layer dominant and max-min fair (SDMMF) allocation and multilayer dominant and max-min fair (MDMMF) allocation to address the issue of fair resource allocation in intrusion detection systems (IDS) in edge computing. The IDS architecture is divided into six layers, and SDMMF allocation is executed recursively starting from the first layer until resources are assigned to the bottom layer, resulting in the equitable allocation of resources that achieves both single-layer and multilayer fairness in terms of multiple resources. Lippi et al. [127] proposed a machine learning and natural language method for detecting unfair clauses in applications or websites. They defined unfair terms and expanded the corpus of clauses in order to better train the model. In addition, the model can not only perform classification tasks but can also identify more information in the clauses and detect and classify the implicit unfair semantics of clauses in the terms. Dolly et al. [128] designed a scheme for detecting unfair reviews. This scheme uses sentiment analysis algorithms and supervised techniques to determine the overall semantics of customer comments based on the positive or negative emotions reflected in them. All of the above detection or optimization methods do reduce the unfairness of the target, but the selection of fairness indicators, such as α-fairness [123], has certain subjectivity. The fairness indicators vary depending on different models and scenarios. Therefore, universally applicable fairness indicators should be further studied to promote the formulation of universal fairness clauses.

With the emergence of privacy protection standards such as CCPA [129] and GPDR [130], people pay more attention to the security of private data. Privacy leakage has become a prevalent security concern. Specifically, privacy leakage may arise due to not only malicious intrusion or insider threat but also user disclosure. All these activities will lead to the result of system or user privacy leakage. Current research predominantly centers on two key aspects: the detection of privacy leakage and the enhancement of privacy protection measures. In terms of detection, several studies have explored issues related to privacy leakage. Li et al. [131] proposed a static taint analyzer for detecting sensitive data leaks during data propagation between application components. This method focuses on contextual information regarding data propagation between multiple components, which allows for higher performance. Liu et al. [132] conducted a study on the problem of privacy leakage caused by background access to location in location-based service applications. Their research shows that accessing locations in the background of an application can generate user movement trajectory data, thereby identifying personal information and resulting in privacy leakage. Mehdy et al. [133] proposed a hybrid neural network model with multiple inputs and outputs for detecting privacy leaks. The model incorporates pre-trained language models, semantic analysis, linguistics, and other knowledge to accurately identify personal information related to health, finance, and social relationships that Twitter users may disclose when posting tweets. This enables the detection of privacy disclosures by users. Shokri et al. [134] conducted a study on the privacy implications of machine learning models, specifically focusing on membership inference attacks that can lead to data leakage. They conducted experiments using a dataset that contained hospital-related information and analyzed various factors that influence the privacy leakage. Additionally, they evaluated different strategies to mitigate these privacy risks. Karimi et al. [135] proposed an automated method for detecting privacy violations on Twitter. The approach utilizes contextualized string embedding to detect sensitive information in tweets, specifically targeting second-party and third-party doxing and malicious information disclosure, while excluding instances of self-disclosure by users and privacy disclosures not targeting any specific identity. In addition, some researches are dedicated to the improvement of privacy protection. Wang et al. [136] developed a new solution for the challenges faced in the collection of multidimensional data under local differential privacy (LDP), including high communication costs and noise issues. The solution includes a multivariate k-ary randomized response (kRR) mechanism called multi-KRR to reduce communication cost, a Markov-based dynamic privacy budget allocation mechanism called Markov-kRR to mitigate the impact of noise, and an improvement on Markov-kRR flipping times threshold to optimize data utility. Rahat et al. [137] developed a convolutional neural network-based privacy policy classification model to assess compliance with the privacy policies of various websites. They used General Data Protection Regulation (GDPR) as a standard and identified 18 labels from it to annotate and classify the privacy policy dataset. Their experiments demonstrated that very few source websites in the dataset strictly followed the GDPR standards. Li et al. [138] proposed a mobile cloud framework to prevent applications from over-collecting user data. This framework stores all user data in the cloud and restricts application access to user data in the cloud, proactively eliminating instances of data over-collection. However, different application services have different protection levels for private information, and the existing privacy leakage detection models are only established for specific scenarios and lack of sufficient scenario transferability. Confronted with scenarios where there are no clearly defined levels of privacy data, the detection model may struggle to yield satisfactory outcomes. Establishing an efficient detection model under such data constraints is a subsequent challenge.

From the above analysis, characteristics and issues of different objects of behavioral conformity authentication are summarized in Table 8.2.

8.4.3 Behavioral Benignity Authentication

The study of behavioral benignity authentication can be mainly divided into the following four aspects: predictability of risk, consistency of execution, traceability of behavior, and integrity of record, as shown in Fig. 8.6.

Predictability of risks is one of the important attributes of a secure network or system. By analyzing and assessing potential risks, the system can identify possible threats and vulnerabilities in advance and take preventive measures. Even in cases where risk threats cannot be entirely mitigated, early warnings can significantly reduce the impact of risk events. Many studies, aimed at risk prediction, have introduced methodologies encompassing risk quantification and credibility assessment. Hu et al. [139] proposed a new threat identification method and risk quantification model for predicting threats in multimedia communication networks. The threat recognition method uses a dynamic Bayesian attack graph-based threat prediction

Table 8.2 Summary of behavioral conformity authentication

Object	Description	Characteristics	Issues
Fraud [5, 23, 101–107]	Fraud refers to the behavior of fraudsters for obtaining illegal benefits through fraudulent means, such as concealing facts and stealing information	(1) Ensuring process security by detecting suspicious activity during transactions. (2) Uncovering hidden fraud networks through behavioral correlation analysis	(1) Possible leakage and illegal use of user privacy data. (2) Imbalanced training data. (3) Hard to deal with new fraud patterns timely
Malicious intrusion [6, 8, 108–116]	Malicious intrusion refers to attacks from outside the system, posing threats to personal computers, mobile phones and even the entire network.	(1) Ensuring system security by detecting malicious activity. (2) Timely alerts and responses	(1) Difficulty in detecting advanced persistent threats. (2) Imbalanced training data
Insider threat [7, 117–122]	Insider threat refers to the risks existing in the system, affecting the safety of the system from the inside	(1) Enhancing network safety and system availability. (2) Strengthening the compliance of systems, such as enterprise networks	(1) Dependencies on high-quality behavioral data. (2) Difficult to address zero-day backdoors
Unfair discrimination [123–128]	Unfair discrimination refers to unfairness or imbalance within a system, which gradually becomes apparent as the system runs	Detecting patterns of unfair discrimination at an early stage allows system personnel to intervene and correct such practices timely, which avoids the cumulative risk of unfair discrimination	The scarcity of data complicates the comprehensive selection of fairness indicators and hinders the establishment of universally applicable fairness metrics
Privacy leakage [131–138]	Privacy leakage usually refers to the exposure of users' private data due to security issues, but it also includes disclosure by a second party, a third party, or even other users	(1) The discovery of external threats and enhancement of system security through privacy leak detection. (2) The improvement of privacy safety through privacy protection measures	Under data constraints, such as scenarios where there are no clearly defined levels of privacy data, establishing an efficient detection model is a subsequent challenge

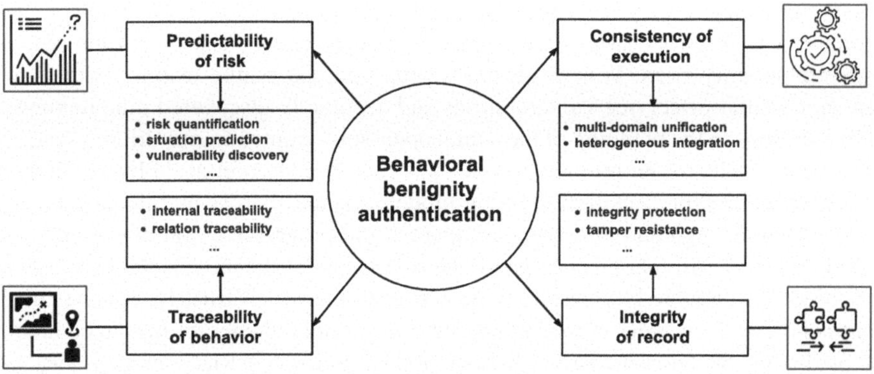

Fig. 8.6 Components of behavioral benignity authentication

algorithm, which aims to predict threat scenarios using complete information. The risk quantification model quantifies the risk status of the entire network and individual hosts by analyzing the security risks at the host and network levels. Wang et al. [140] proposed a network behavior risk measurement method by analyzing network traffic. They considered traffic data as network behavior and characterized the network traffic and network topology information. Additionally, they introduced the theory of differential manifolds to measure the behavior risk of the network system. Li et al. [141] developed a trust-based model for detecting suspicious behavior in network groups. Firstly, the model constructs a trust matrix between network nodes utilizing network topology information. Secondly, it calculates the similarity matrix of nodes based on their trust levels. Finally, the model clusters the nodes utilizing the similarity matrix to identify potential malicious groups. This approach has a high distinction rate for hidden risks in the network. There are also some studies that indirectly conduct risk prediction through network situation prediction methods. Yang et al. [142] proposed a network security situational assessment method based on attack intent distinction. They analyzed the correlation between the attack phase, network configuration information, and attack intent. The method distinguishes the attacking intent and predicts the next attack, which does not rely on historical sequences and is more effective in predicting network security situational assessments. Ghazel et al. [143] proposed a global model involving LC (logistics center) region's railway and road transportation, where each model describes the behavior of components throughout the LC environment. By employing the Monte Carlo principle, the global system behavior can be simulated. They also give some solutions for risk mitigation according to the simulation. Vulnerability discovery plays a crucial role in mitigating potential risks within a system, and it has witnessed significant advancements with the progress of detection technologies. Liu et al. [144] introduced a software vulnerability detection system, which leverages deep learning and domain adaptation techniques to address the challenge of software vulnerability detection. The system harnesses the automatic feature representation capabilities of deep learning and the domain adaptation framework to discover various vulnerabilities in heterogeneous projects and reduce software security risks. Wan et al. [145] proposed a dynamic testing method for semantic denial-of-service vulnerabilities and accurately discovered nine unknown vulnerabilities in the planning of the actual open-source autonomous driving system. Code reuse leads to the propagation of vulnerabilities. Luo et al. [146] developed an intermediate representation function model to achieve cross-architecture binary code search through an entropy-based adapter and progressive search strategy and tested on seven different tasks to prove the robustness of the model. Cui et al. [147] proposed an automated vulnerability detection framework VRust for Solana smart contracts. The framework automatically detects potential vulnerabilities in contracts by analyzing the intermediate representation of translated Rust source code and program data flow. However, in real scenarios, network or system risk prediction faces more difficulties and challenges, and it is necessary to enhance the robustness of risk prediction in terms of consistency of execution, traceability of behavior, and integrity of record. These aspects can significantly improve the reliability and effectiveness of network or system risk prediction.

Consistency of execution helps ensure smooth collaboration among multiple systems or entities involved in cross-domain risk detection. In real scenarios, achieving unified execution operations in cross-domain interactions over heterogeneous networks contributes to informed risk management and resource optimization. Ding et al. [148] designed a resource management algorithm for heterogeneous integrated networks. During the process of collecting and managing heterogeneous resources in the heterogeneous network, the algorithm uses information security transmission technology to ensure the safe collection of resources and uses the improved management algorithm of heterogeneous resources to realize the security management of heterogeneous integrated network resources. Guo et al. [149] introduced a reliable cross-domain authentication mechanism applied to the IoT. To achieve cross-domain authentication between heterogeneous IoT domains, the mechanism uses a master-slave blockchain architecture to ensure cross-domain privacy security. To achieve trusted authentication, an improved Byzantine fault-tolerant device based on the reputation value model (RIBFT) is used to conduct a credible assessment. Xuan et al. [150] devised a certificate-less cross-domain authentication scheme that possesses the capability to distinguish parameters. The scheme relies on the principles of certificate-less public key cryptography and smart contract technology, which allows for the use of differentiated cryptographic system parameters for authentication between heterogeneous IoT networks, thus enhancing the security of cross-domain authentication for heterogeneous IoT networks. Hao et al. [151] proposed a lightweight architecture for consortium blockchain. The architecture utilizes a token accumulation mechanism for authentication of data access control and trust evaluation of requesting nodes. It supports cross-domain data sharing among Internet of Things users from different geographical locations. Li et al. [152] introduced a trust mechanism hierarchy that relies on cooperative detection among blockchain nodes. Firstly, they employed federated learning to train a cross-domain unified behavior detection model which broke down data barriers and achieved cross-domain unified evaluation of device behavior trust. Based on this, they designed a layered trust mechanism based on federated detection combined with the transaction performance of blockchain. By dynamically evaluating devices based on behavior detection and blockchain transaction detection, graded trust management of devices is implemented. Sheff et al. [153] aimed at the problem of safe scheduling in a federated environment and implemented a static detection compiler and a system that introduced a phased commit protocol to ensure the consistency and security of scheduling. Chen et al. [154] proposed an efficient and privacy-preserving cross-domain authentication scheme, named XAuth, to address the cross-domain authentication issue in public key infrastructure (PKI). The scheme utilizes multiple Merkle hash trees to ensure the responsiveness of cross-domain data management and employs zero-knowledge proof algorithms to ensure privacy in cross-domain authentication. Lin et al. [155] designed a time-aware cross-scenario keystroke dynamic authentication mechanism to address the issue of collecting a large amount of data every time the authentication scenario switches. The method improves the quality of data by selectively learning and encoding time information to achieve efficient behavior pattern transfer across scenarios. The method improves data diversity and cross-scenario applicability through a local Gaussian data augmentation method to enable consistent authentication across different scenarios. However, objectives such as cross-domain communication and heterogeneous integration still face several challenges. Cross-domain and heterogeneous networks typically entail distinct security policies and mechanisms. Achieving unified execution operations necessitates overcoming conflicts between diverse security requirements while ensuring that cross-domain operations do not introduce novel security vulnerabilities.

Traceability of behavior plays a key role in risk detection. There is a great deal of uncertainty in the task of risk prediction. In instances of prediction failure, it is necessary to promptly trace the risk behavior to address vulnerabilities and minimize losses. Traceability of behavior is mainly categorized into two types: internal traceability and relation traceability. Internal traceability involves tracing user behavior within the network environment to distinguish potential risk behaviors or to trace the source of suspicious behavior that has already occurred. Relation traceability refers to analyzing a series of behaviors within the network environment, tracing suspicious associated behaviors, or tracing attacks from outside the network environment. Zhang et al. [156] presented a secure s-health system designed for cloud service environments. The system introduces a decryption component that is integrated with the user's information during key retrieval. Once integrated, the component remains fixed and prevents key owners from re-randomizing, thereby establishing a binding between the user's information and enabling the tracking of a series of user behaviors. Lin et al. [157] proposed a method for detecting internal threats in cloud environments based on behavior traceability. First, the method analyzes the call rules of the cloud service interface to construct the complete behavior process of the call. Then, it uses the behavior tree construction algorithm to generate a legitimate behavior tree describing the behavior of cloud users. Subsequently, behavior trace points are set up to capture call behavior information on the service invocation interface of cloud services. Finally, user interface call information is matched with legitimate behavior trees through keyword matching to trace the source of malicious user behavior. Yu et al. [158] proposed a blockchain-enhanced security access control scheme to support traceability and revocability in IoT. Specifically, the scheme involves blockchain-based authentication to store all user information and public keys. Subsequently, system parameters are issued by administrators to users, along with a unique parameter embedded in private keys. The scheme enables tracing malicious behavior by utilizing the parameter in private keys and revoking malicious users accordingly. Wang et al. [159] proposed a method for tracing associated events using a composite blockchain structure. Firstly, a storage structure model for the composite blockchain was constructed to achieve data association storage. Secondly, by obtaining the source entity block, an event association graph was constructed by using a source tracing method based on the Apriori algorithm. Finally, the entities were subjected to risk assessment using reinforcement learning. Zhu et al. [160] proposed an Ethereum attack traceability method based on graph analysis. They applied graph analysis techniques to analyze the behavioral characteristics of attackers and the relationships among them. Additionally, they used RPC mechanisms to trace the related attackers and attack sources. Behavioral tracing also encounters several challenges. More sophisticated attack behaviors excel at concealing their attack features, rendering themselves indistinguishable from normal behavior patterns. As a result, these behaviors can lurk within the system for an extended duration, which is currently difficult to trace.

Integrity of record can provide a reliability and accuracy guarantee for risk prediction and behavior traceability. Integrity of record includes integrity, accuracy, and tamper resistance of data. Current research is mainly focused on data integrity protection and tamper resistance. Li et al. [161] integrated the trusted execution environment SGX with blockchain technology to construct a privacy-preserving multimedia authentication system. The system utilizes SGX to create a trusted execution environment and employs PhotoChain's hybrid storage mode to only store the hash values of the photos on the blockchain. By using blockchain to ensure data integrity, the system does not add a storage burden to the blockchain. Javaid et al. [162] proposed a solution based on physically unclonable functions (PUFs) and blockchain, known as BlockPro, for enhancing the security of data sources and ensuring data integrity in IoT environments. Specifically, the characteristics of PUFs can be used to establish a data source to ensure a unique source, and the data storage method of Ethereum ensures data integrity. Patil et al. [163] proposed an efficient privacy-protecting authentication protocol that combines blockchain technology and PUFs. The protocol employs a decentralized digital ledger using blockchain smart contracts to resist attacks from data tampering, thereby ensuring security in IoT environment. Additionally, by integrating the uniqueness and tamper-proof properties of PUFs with blockchain, the protocol ensures unique device IDs and data integrity in the IoT. Barbareschi et al. [164] proposed a mutual authentication scheme relying on the use of PUFs. The scheme employs PUFs' characteristics of being unclonable, unique, and tamper-evident to protect edge nodes in IoT from physical attacks and data tampering. Wei et al. [165] proposed a solution to address data security concerns in cloud computing by integrating blockchain technology. The approach involves deploying a distributed virtual machine proxy model in the cloud through the use of mobile agents, which ensures reliable data storage. Furthermore, the solution leverages a blockchain-based data integrity protection framework and generates hash values for corresponding files using the Merkle hash tree. Data alterations are then monitored via smart contracts and alerts are triggered in case of any tampering. In the process of data integrity protection for cross-domain transmission, researchers need to find a balance between security and efficiency, so as to provide more reliable support for risk prediction and behavior traceability.

Based on the aforementioned analysis, characteristics and issues of different objects of behavioral benignity authentication are summarized in Table 8.3.

8.5 Challenges of Behavioral Authentication

8.5.1 The Limited Quality of Behavioral Data

How to overcome the limitations of behavioral data quality in the modeling process remains a primary concern. These limitations can be generally attributed to three factors: behavioral data collection and processing, privacy protection, and business attributes. In terms of data collection and processing, inherent difficulties in behavioral data collection or processing can result in low data quality [166]. For example, during the collection of behavioral data, issues such as duplicate or missing data may arise due to unreliability in collection equipment or inconsistencies in data sources, leading to inaccurate or even erroneous results in behavioral data analysis. These undesirable results have a direct impact on the quality and effectiveness of behavior modeling. Therefore, to guarantee the high quality of behavior modeling, it is imperative to have a rigorous process of data cleaning and validation. In terms of business attributes, the inherent characteristics of commercial activities pose challenges in effectively modeling user behavior [167, 168]. In scenarios such as online payment services, the highly imbalanced ratio of fraudulent accounts to legitimate accounts presents significant difficulties in behavior modeling. For interaction data across diverse terminal devices, the distinction between normal and abnormal data is frequently ambiguous. Relying solely on attribute data has limited effectiveness. Designing feasible data augmentation methods is a key prerequisite for achieving effective behavioral authentication and aims to establish high-quality behavior models using low-quality behavioral data.

Table 8.3 Summary of behavioral benignity authentication

Object	Description	Characteristics	Issues
Predictability of risk [139–147]	Predictability of risk refers to the distinction of potential behavioral risks in the network environment	(1) Prevention of potential attacks. (2) Remediation of exploitable vulnerabilities. (3) Reduction of losses through early warning	Due to the inherent uncertainty of evolving threats, ensuring the robustness of risk prediction needs to be complemented with other security technologies
Consistency of execution [148–155]	Consistency of execution primarily refers to the method of credibility evaluation and potential risk detection that can perform consistent operations across complex and diverse networks.	Ensuring the seamless execution of operations across various domains and heterogeneous networks in multiple scenarios facilitates unified risk management and resource optimization	Achieving consistency of execution requires compatibility with different resource forms and operating mechanisms in cross-domain interactions over heterogeneous networks
Traceability of behavior [156–160]	Traceability of behavior refers to the traceability of potential risks or suspicious behaviors in the network environment	(1) Tracing suspicious behavior before it becomes a threat. (2) Tracing malicious behavior back to its source and close loopholes. (3) Tracing the source of related behaviors to reveal hidden attack chains	More sophisticated attackers excel at hiding their attack signatures and making attack behaviors indistinguishable from normal behaviors so that they lurk in systems for extended periods of time to carry out continuous attacks
Integrity of record [161–165]	Integrity of record means that user data in a trusted network environment should remain complete, accurate, and non-tamperable	(1) Improvement of risk prediction accuracy through complete and accurate data. (2) Enhancement of traceability analysis reliability	Protecting record integrity faces multiple data tampering attacks and introduces performance overhead and system complexity like data protection in distributed systems

8.5.2 The Insufficient Behavioral Data Quantity

Behavioral authentication is also data-driven, and the insufficient behavioral data quantity may influence the effectiveness of behavioral authentication systems. With limited behavioral data available for training, it becomes difficult to create robust behavioral authentication models capable of accurately distinguishing between legitimate and malicious behaviors. This scarcity also increases the risk of biased or unfair models, as they may learn from a narrow subset of behaviors, leading to decreased reliability in authentication performance. Furthermore, the attackers may not repeat previously detected or blocked methods to attack the system, which also intensifies the risks associated with the system [169]. The lack of behavioral data inhibits the generalizability of authentication systems, as models may struggle to adapt to new or evolving threats without comprehensive training data. Therefore, addressing the challenge of insufficient behavioral data quantity is crucial for the development of highly reliable behavioral authentication systems.

8.5.3 Privacy Concerns and Communication Costs

Under the stricter privacy protection regulatory constraints, such as CCPA, GPDR, and DSL [129, 130, 170], behavioral data are more tightly controlled. People are also increasingly valuing privacy as they become more aware of the risks associated with data breaches and unauthorized surveillance. Heightened concerns about personal information security have motivated individuals to demand greater transparency and control over how their data is collected and used. These changes directly lead to the difficulty of implementing existing centralized authentication methods for collecting behavioral data from different agents. Despite the advancements in hardware technology and the development of 5G technology, the current computational resources and latency in behavioral authentication also face significant challenges. Behavioral data exists in various forms such as text, voice, and video, and the required models for training are of high complexity [171]. This leads to excessive consumption of computational resources and significant communication costs. For example, when there is a change in the existing behavioral authentication patterns, it becomes necessary to update the entire model, which not only affects the usability of authentication but also results in unnecessary waste of computational resources. It is of great significance to determine models that are compatible with resource constraints and establish efficient knowledge transfer mechanism of multimodal behavioral data. Regarding authentication latency, current mainstream authentication schemes still rely on the transmission of frequent behavioral data packets to achieve centralized behavior modeling. However, there is distance between different terminal devices, and centralized collection takes time to transmit behavioral data. Even a delay of a few milliseconds, while browsing the Internet or attempting to connect to a smart refrigerator, may only affect user experience.

Additionally, the centralized authentication relies on a single centralized entity, leading to excessive concentration of authority, and single-point failures can result in severe system issues. However, in scenarios such as automated remote surgery or autonomous driving [172, 173], a few milliseconds of latency could lead to fatal accidents. Therefore, establishing more flexible deployment schemes for behavioral authentication not only ensures lower latency but also avoids unnecessary network congestion and reduces unnecessary bandwidth costs.

8.5.4 Lack of Theoretical Analysis Basic

The existing foundational theoretical efforts primarily focus on conducting specialized theoretical analysis from specific perspectives, considering the constrained conditions of user behavioral data. On the one hand, these studies excessively rely on the employed models, making it hard to perform prior evaluation and analysis of data utility. This circumstance prevents accurate anticipation of model performance before data utilization and hinders the determination of whether the chosen models possess sufficient applicability for particular tasks or applications. On the other hand, the data-driven paradigm is difficult to cover or traverse all possible scenarios, consequently impeding the attainability of a general applicable theoretical framework. The diverse requirements arising from different data sources and tasks may necessitate distinct methods and models for conducting behavioral authentication. In summary, the existing relevant research lacks theoretical analysis basic of behavioral authentication, which makes it difficult to evaluate the performance of behavioral authentication in practical applications and hampers the provision of theoretical guidance for optimizing approaches. Hence, the exploration of the formalization for behavioral authentication holds significant significance, akin to the guiding role of information theory's fundamental limits in digital communication technology. It will not only offer architectural guidance for high-performance models and algorithm design in behavioral authentication but will also provide systematic metrics for evaluating specific performances. Simultaneously, it will provide a theoretical foundation and analytical methodologies for understanding the mechanisms of behavior modeling in typical services, holding particular importance for estimating data utility in the data handover phase. Efforts toward addressing this issue contribute to a better comprehension of essential issues and drive the advancement and progress of the entire field.

8.6 Future Behavioral Authentication Research

We thoroughly examine the main limitations associated with existing behavioral authentication methods and discuss innovative research directions that have the potential to significantly augment the applicability and effectiveness of behavioral

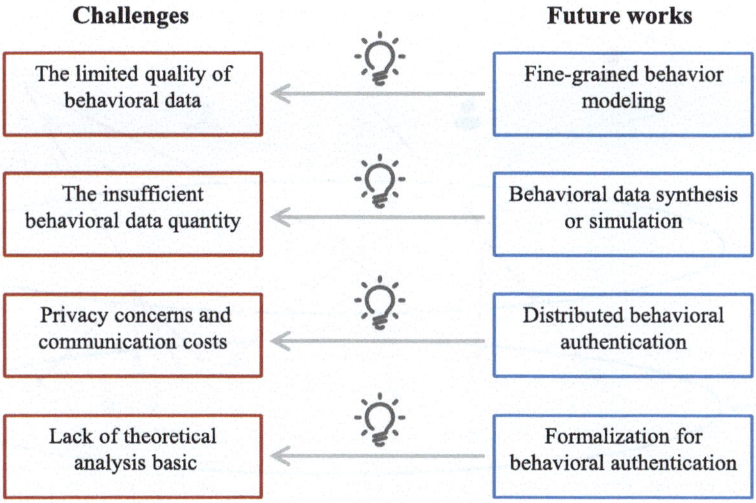

Fig. 8.7 Overview of main challenges and future research directions for behavioral authentication

authentication. A comprehensive overview of limitations and future research directions is shown in Fig. 8.7.

8.6.1 Fine-Grained Behavior Modeling

Future research should focus on developing the behavior modeling method that fully utilizes limited data, taking into account the interaction and collaboration between behaviors from a fine-grained perspective. Fine-grained behavior modeling leverages deep and rich information from the behavior data, which can provide accurate and reliable behavior distribution for behavioral identity, conformity, and benignity authentication.

We are committed to proposing a fine-grained behavior modeling framework to enhance the limited behavior data. It considers the associations of behavior across multiple dimensions such as behavior sequences, subject correlations, and attribute distributions from complex and diverse behavioral data in both virtual and real spaces. Through holistic learning of abundant behavioral information, we can better understand the underlying semantic meanings in behavioral data and subsequently apply the learned semantics for enhancing the authentication performance.

Correspondingly, as shown in Fig. 8.8, a potential solution is to establish a three-layer knowledge graph structure, including the subject correlation layer, behavior sequence layer, and attribute distribution layer. The subject correlation layer reflects the association between subjects and is coupled with a mapping relationship to the behavior sequence layer. There is a time-annotated edge with time information

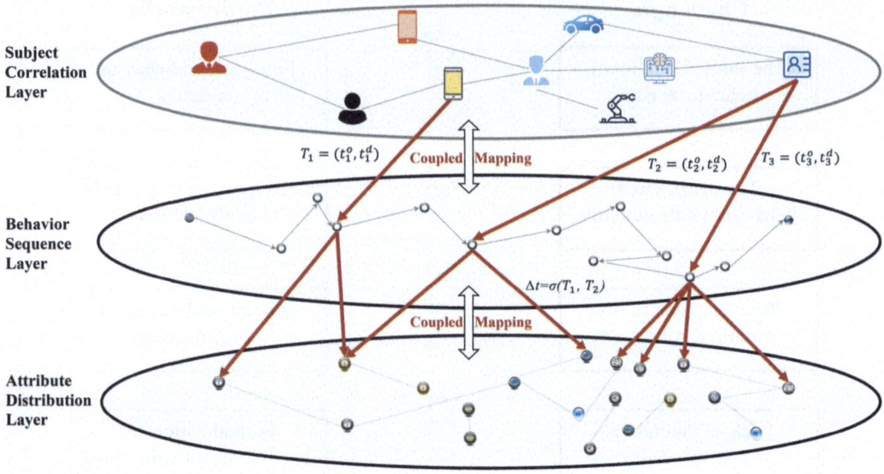

Fig. 8.8 The fine-grained behavior modeling framework

$T = (t^o, t^d)$ between each subject and behavior, where time T indicates the starting time t^o and ending time t^d when the subject initiates the behavior. In the behavior sequence layer, different behaviors occur in a clear sequence, and we denote $\Delta t = \sigma(T_1, T_2)$ as the time difference between two behaviors T_1 and T_2. A behavior is described in detail by several attributes, and their associations are reflected as the coupling mapping between the behavior sequence layer and the attribute distribution layer.

In the subject correlation layer, subject correlation associations are modeled by introducing the similarity between subjects and their social relevancies. Specifically, the behavior sequence layer and attribute distribution layer provide characterization of the latent behavior space encompassing all behaviors, allowing generalization of subjects from instantiated modeling objects in network services to any behavioral attributes (such as population, individual, and class-based subjects). Multidimensional intelligent synthesis strategies are designed to cooperatively generalize the distribution of subject models, thereby mitigating the technical bottleneck of insufficient data in security and safety authentication. In the behavior sequence layer, the temporal relationships of behaviors, such as the order in which behaviors occur, are modeled. We aim to propose a variable-length sequence modeling solution that specifically can be divided into prefix, infix, and suffix partitions based on behavioral order and adaptive partitions based on behavioral windows. Compared to traditional sequence modeling techniques, it combines behavior intent recognition and time dependency learning. By extracting information from behavioral context and intent, it alleviates the inefficiency issues of authentication techniques on limited data. In the attribute distribution layer, co-occurrence associations between attributes from the same behavior are extracted. For example, multiple attributes that appear together in a behavior can serve as co-occurrence relationships between attributes. We conceptualize behavior as a stable system composed of internal associations at the topological level. By designing customized behavioral scanners, we quantify the associations between fine-grained behavioral attributes as measurable graph objects. It excludes confounding information arising from knowledge-driven interference within behaviors, focusing instead on potentially internal associations between fine-grained behavioral attributes as driven by the data. So it can eliminate behavioral noise caused by erroneous or outdated human annotation. Based on the three customized layers, two types of coupled mapping, i.e., subject behavior mapping and behavior attribute mapping, are devised. The former realizes the effect of subject collaborative filtering through the interaction between subjects and behaviors. Subjects with similar preferences are reflected in the subject correlation layer. The correlations between subjects are further described by the interaction with time between subjects and behaviors. The latter realizes the effect of content collaborative filtering through the interaction between behaviors and attributes, which reflects behaviors with similar attributes in the behavior sequence layer. Through hierarchical modeling of behavior, in addition to introducing more associations to enrich the description of behavior, we can perform customized behavioral authentication based on specific associations within each layer. The coupled mapping between layers further allows aggregating information from other layers into a single layer, thereby obtaining high-density behavioral semantics (information stacking and reduced information carriers jointly improve density) under limited behavioral data to support different authentication tasks.

8.6.2 Behavioral Data Simulation

The challenge of insufficient behavioral data consists of two parts. On the one hand, in scenarios with a defined process, behavioral data synthesis methods are required to supplement the datasets, especially samples of illegal behavior. Data synthesis methods can be utilized to expand existing datasets with data of the same distribution, and the gap between normal behavior samples and illegal behavior samples could be narrowed. On the other hand, in scenarios with complex interactive processes, the interactions of individuals lead to chaos so that there does not exist a fixed distribution of illegal behaviors. Therefore, synthesis methods creating behavioral data of the same distribution no longer work, and we suggest behavioral simulation as an effective resolution.

For the synthesis of behavioral data, the existing methods exhibit sensitivity to outliers, which may compromise the quality of generated samples in the latent space. A potential approach to address these limitations is to combine a generator and two discriminators. The generator incorporates an autoencoder to map the data into a high-dimensional space, where the outlier characteristics of abnormal behavior are accentuated through a specially designed mapping function. The interaction between the generator and the first discriminator enhances the similarity between the generated data distribution and the real data distribution, while the interac-

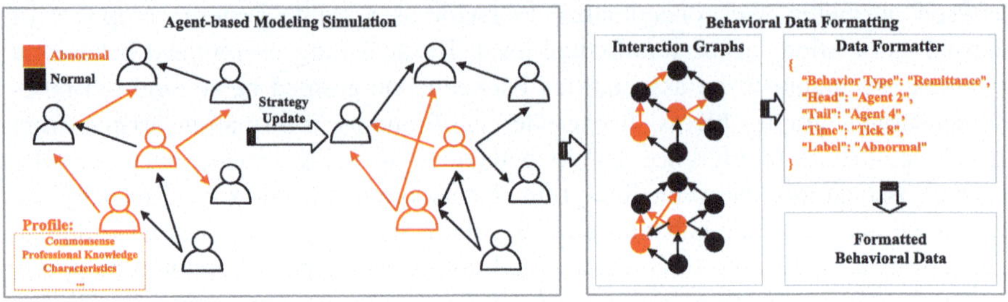

Fig. 8.9 The illustration of behavioral data simulation

tion between the generator and the second discriminator enhances the separation between the distributions of normal and illegal behaviors. The adversarial training between the first discriminator and the second discriminator further enhances the distinction of illegal behaviors. Simulation technologies have developed for many decades, consistently regarded as a powerful tool for analyzing complex systems. Agent-based modeling simulation (ABMS) represents an advanced simulation approach, characterized by the generation of individual-level behaviors for both items and agents [174]. In the era of data science, deep learning models have been harnessed to enhance the authenticity of individual behavior generation. This paradigm falls short in addressing the challenge of dynamic distributions, as agents empowered by deep learning models remain constrained by historical data and limited behavior sets. To enhance ABMS for simulating diverse behavioral data, we suggest a novel simulation framework centered around agents endowed with reasoning and learning capabilities. Notably, large language models (LLMs) have demonstrated open-domain reasoning abilities as agents. Leveraging these agents for simulating security scenarios (as depicted in Fig. 8.9), we provide them with security-related knowledge and detailed profiles encompassing motivation, behavior preferences, and other relevant factors, using a variety of models. First, agents with open-domain reasoning ability and detailed profiles are launched in an ABMS system. Second, the interactions between agents are converted into interaction graphs. Third, these graphs are formatted into the required form through a data formatter. Finally, these data are regarded as simulated behavioral data to supplement the origin dataset. Through the combination of foundational models with smaller-scale models, each agent is effectively characterized to respond thoughtfully to situational cues. Furthermore, these agents possess learning abilities, enabling them to devise novel strategies beyond historical data, similar to real-world attackers and defenders. In summary, an ABMS built upon this new type of agents equipped with hierarchical knowledge structures and adaptive learning mechanisms offers a promising resolution for simulating behavioral data in complex scenarios.

8.6.3 Distributed Behavioral Authentication

In general, the effectiveness of behavioral authentication methods often relies on centralized data collection to construct classification models. However, with the increasing emphasis on stricter privacy protection regulations and the need for local data processing in distributed environment, the control and transmission of behavioral data become more challenging. It is necessary to develop distributed behavioral authentication models across different devices while minimizing the costs associated with transmitting privacy-sensitive behavioral data between devices. The concept of computing power network has been implemented as the infrastructure continues to improve, enabling behavioral data to transition from single central deployment to diverse distributed deployment. This shift from centralized scheduling to distributed collaborative scheduling has provided a favorable opportunity for the advancement of behavioral authentication in near-real-time processing. Sensitive behavioral data can be processed and stored locally, while other behavioral data can be decoupled, allowing for selective interaction with servers in different physical locations. Even if the parties involved in data sharing do not have identical definitions of sensitive behavioral information, negotiations can be conducted to determine the optimal approach for decoupling the sensitivity levels of behavioral information. Distributed behavioral authentication systems can bring enhanced processing capabilities and reduce response time, which improves effectiveness and reliability of the authentication process significantly.

Therefore, as an essential computational paradigm, distributed learning holds great potential in designing viable architectures for behavioral authentication to achieve a balance among authentication performance, user privacy, and communication costs. An objective is to improve cloud-edge-terminal collaboration to fully leverage the centralized and intensive deployment capabilities of the central cloud, the low latency and high flexibility of edge servers, and the local processing power of the terminal devices. The collaboration approach enhances the safety of behavioral data interactions. Through the collaboration of cloud computing, edge verification, and local data processing, it reduces the processing burden on the cloud center, lowers bandwidth load and authentication latency, and mitigates behavioral data transmission costs.

Correspondingly, as shown in Fig. 8.10, a distributed behavioral authentication framework that leverages cloud-edge-device collaboration has been established, further expanding the application scenarios of behavioral authentication. The overall architecture of distributed behavioral authentication. The central server is responsible for collaborating with the terminals to construct feature extractors and coordinating the unified scheduling of edge servers. Edge servers are responsible for deploying specific authentication credentials and partitioning user clusters, as well as responding to specific authentication requests. Terminals are responsible for storing private sensitive data and initiating relevant requests during the verification process. To achieve this, an auxiliary task is defined to obtain behavioral embedding vectors based on a deep neural network. During this process, terminal data remains

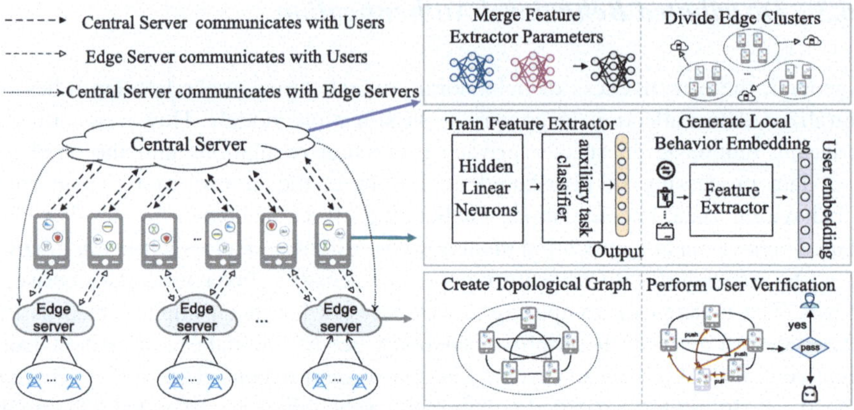

Fig. 8.10 The overall architecture of distributed behavioral authentication

to be stored locally without the risk of privacy leakage. The local devices share the partial model parameters with the central server to ensure the measurability of different behavioral embedding vectors. If the learning rate of the device is l when the global parameters are updated in the t round, each device will perform parameter updates locally through ${w_{t+1}}^k = {w_t}^g - l\nabla g_k$. The central server uses the operation of ${w_{t+1}}^g = \sum_{k=1}^{K} \frac{n_k}{\sum_{k=1}^{K} n_k} \cdot {w_{t+1}}^k$ to aggregate parameters and returns ${w_{t+1}}^g$ to different devices. The device updates the local model with the new parameters for the next training round. Once the feature extractor has been built and deployed to various devices, corresponding behavioral vectors are generated locally by different devices, and then different devices send these vectors to the central server. When the central server receives behavioral representation vectors from various terminals, it calculates the similarity between these vectors. Subsequently, the authentication service is deployed to the optimal edge locations based on demand. For the deployment of edge servers, the optimization objective is centered around minimizing latency by: $\min \mathscr{L}(r) = \sum_{s_k \in S} \sum_{a_i \in Z_k} q(a_i, s_k; r)$, where r represents one certain feasible edge deployment scheme, Z_k represents all base stations covered by current authentication server s_k, S represents all edge authentication servers, and $q(a_i, s_k)$ signifies the latency from base station a_i to edge authentication server s_k. Furthermore, each edge server establishes a topological graph. For terminals with similar patterns, corresponding behavior profiles are created, and based on the results, a topological graph $G = (V, E)$ is constructed. The node set is denoted as $V = \{x|x \in X\}$, where X represents the terminals within that cluster, and $E = \{(x, y)|x, y \in V\}$, where (x, y) represents an edge between two terminals. The connection strength of credentials between different terminals is calculated and stored on the edge server. During the verification phase, the k-th terminal generates new behavioral data. For the privacy-sensitive data of this terminal, the input data is processed through the trained feature extractor to generate a vector representation, denoted as Φ_{k_i}, which is then sent to the edge server. The final authentication result is returned by the edge server. This approach achieves distributed behavioral authentication by the collaboration of cloud, edge, and devices, which significantly reduces authentication latency while ensures authentication performance.

The proposed framework represents initial efforts toward enabling distributed behavioral authentication, which possesses good flexibility and compatibility. In scenarios where datasets from different clients are highly heterogeneous and non-independently distributed, local model structures and model training strategies can be replaced in our framework. Personalized feature extractors can be trained through techniques such as model decoupling and global model personalization [175, 176]. Additionally, the framework can be used in conjunction with many other technologies. When addressing privacy concerns, our framework can employ technologies such as differential privacy [177] to alleviate privacy issues. Simultaneously, it can utilize existing robust adversarial defense techniques [178, 179] to defend against security threats during the behavior modeling process. Furthermore, when there is distrust among multiple clients, our framework can address this issue by introducing blockchain technology [180, 181]. It is orthogonal to our framework and thus can be incorporated into our framework for storing behavior profiles.

8.6.4 Formalization for Behavioral Authentication

Inspired by the formal hypotheses and principles in information theory, we analogize behavioral authentication to concurrent network communication since both the problems of behavioral authentication and concurrent network communication share the common goal of eliminating uncertainty. The fundamental problem in communication is for one end of communication to accurately or approximately reproduce the message selected by the other end.

We give the formalization method for behavioral authentication, which satisfies the formal hypotheses. As shown in Fig. 8.11, we encode the complex features of data to obtain behavior representations, which extract meaningful information and behavior event structures. Inspired by the formal hypotheses and principles in information theory, the ultimate essence of behavioral authentication is the elimination of behavioral uncertainty. This uncertainty can be quantified using

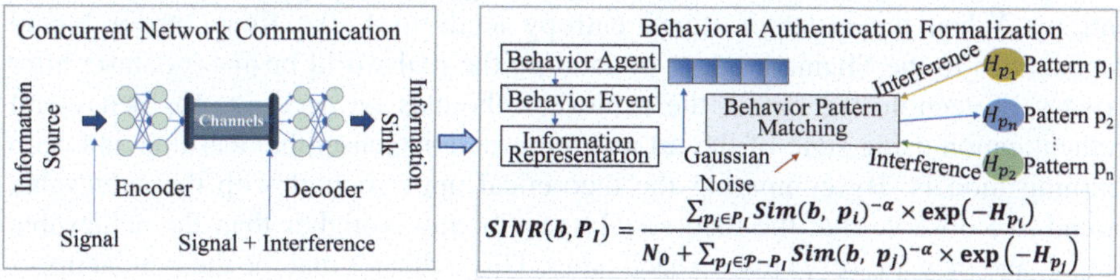

Fig. 8.11 The formalization for behavioral authentication

a metric similar to the signal-to-interference-plus-noise ratio (*SINR*) commonly found in communication. The impact on behavioral authentication performance arises from interference by users with similar behavior patterns and the noise inherent in the data itself. Ultimately, behavioral authentication aims to eliminate the uncertainty of behaviors. Similar to the signal-to-interference-plus-noise ratio (*SINR*) in communication, this uncertainty can be measured as follows:

$$SINR(b, P_I) = \frac{\sum_{p_i \in P_I} Sim(b, p_i)^{-\alpha} \cdot \exp(-H_{p_i})}{N_0 + \sum_{p_j \in \mathscr{P} - P_I} Sim(b, p_j)^{-\alpha} \cdot \exp(-H_{p_j})},$$

where P_I denotes the set of behavior patterns of the user to be verified, $\mathscr{P}$ represents the set of all behavior patterns, and H_{p_i} represents the behavior entropy of the current matched user. The function $Sim()$ measures the similarity between behaviors. The impact on behavioral authentication performance arises from interference by users with similar behavior patterns and the noise inherent in the data itself. For different levels of behavioral authentication, the elements contained in P_I have distinct meanings. Taking behavioral identity authentication as an example, the set P_I degenerates into a singleton set, containing only the user with a deterministic behavior pattern. The signal strength of behavioral identity authentication can be measured by calculating the stability and similarity between the given behavior and the behavior pattern of the user to be verified.

Building upon the formalization for behavioral authentication described above, we conduct a preliminary exploration of the fundamental limits of data utility for behavioral authentication [182]. Specifically, from the perspective of data distribution, we introduce a data utility function based on conditional entropy. We analogize behavioral authentication to the problem of signal transmission in communication channels. By establishing the relationship between conditional entropy and authentication accuracy, we derive upper bounds on accuracy for behavioral authentication: the Shannon upper bound and Rényi upper bound. The former is based on Shannon entropy and is obtained by transforming the Fano inequality; the latter utilizes conditional entropy defined on Rényi entropy and is derived by applying Jensen's inequality and the principle of maximum discrete entropy. When the order of Rényi entropy is greater than 1, the Rényi upper bound can be expressed as follows: where $\boldsymbol{X}$ denotes the attribute combination, N refers to the number of categories of the label Y, e denotes the average authentication error probability, H_α denotes the Rényi entropy, and H_S denotes the Shannon entropy. When the order of Rényi entropy tends to 1, the Rényi upper bound degenerates to the Shannon upper bound. On the real-world business dataset after privacy protection, we obtain the achievable bounds on accuracy for behavioral authentication using state-of-the-art and representative ensemble learning and deep learning models. By comparing the theoretical upper bound with the achievable bound, we observe that the theoretical upper bound is higher than the achievable bound, and they are very close to each other. This indicates that the theoretical upper bound provides valuable insights for optimizing the achievable bound.

Furthermore, in the field of communication, there have been initial advancements in semantic communication paradigms [183, 184], which improve the transmission efficiency and reduce the latency of communication systems. In the future, incorporating semantic information into our proposed formalization method for behavioral authentication holds the potential to push the fundamental limits of behavioral authentication.

8.7 Discussion

Behavioral authentication has already been applied in practical applications such as financial risk control, healthcare, and intelligent transportation.

In financial transaction services, behavioral authentication can deduce invariant behavioral patterns from changing behavioral data. By integrating multiple factors such as the financial chain, consumption patterns, timing, and location of ordinary consumers, fragmented information is used to accomplish behavioral modeling of consumers. This forms the basis for constructing a comprehensive risk prevention and control system. Behavioral authentication technologies have been implemented in the risk prevention systems of banks and Ant Group [23, 185]. It enables early detection and prevention of fraudulent activities.

In healthcare applications, smart wearable devices from companies like Apple and Huawei use your behavioral data to assess your physical health status [186, 187]. Behavioral authentication provides continuous monitoring and important guidance for the user's health. The industry has witnessed the emergence of behavioral capture and analysis products, such as fall detection vests for the elderly. When the walking patterns, accelerations, and other behavioral features of elderly individuals deviate from their normal patterns, the vest automatically activates a protective mode to mitigate the impact of a potential fall and prevent injuries [188, 189].

In intelligent transportation services, leading automotive companies such as Byd and Tesla have implemented real-time behavior detection of drivers in their onboard systems [190, 191]. When non-benign behaviors are detected, such as abrupt lane changes without signaling or tailgating at an unsafe distance, the onboard system intervenes to mitigate potential risks. This intervention can take the form of audible warnings, visual alerts, or even automated corrective actions, such as gentle steering corrections or adaptive cruise control adjustments. The implementation of behavioral authentication in intelligent transportation services helps to cultivate safer driving habits and reduce the likelihood of accidents.

8.8 Conclusions

Technological advancements have ushered humanity into an unprecedented era of artificial intelligence. Behavioral authentication has emerged as a promising authentication solution in many scenarios and been successfully and widely applied in practice. It has gradually become a fundamental problem in the field of authentication. This work summarizes the background and applications of behavioral authentication. In particular, it introduces concept of behavioral authentication including behavioral identity authentication, behavioral conformity authentication, and behavioral benignity authentication. The paper presents a comprehensive review of work conducted in these three levels of behavioral authentication, establishes a clear framework for categorization, and summarizes their corresponding characteristics and issues. The main challenges in current behavioral authentication are analyzed, and key research directions for the future are pointed out including fine-grained behavior modeling, distributed behavioral authentication, and formalization for behavioral authentication. Obviously, with the increasing sophistication of artificial intelligence in various fields and a greater emphasis on user experience, behavioral authentication will display even stronger vitality and play an increasingly important role.

References

1. L. He, C. Ma, C. Tu, Y. Zhang, in *25th IEEE International Conference on Computer Supported Cooperative Work in Design, Hangzhou, 2022*. pp. 280–285
2. X. Zhang, L. Yao, C. Huang, T. Gu, Z. Yang, Y. Liu, ACM Trans. Intell. Syst. Technol. **11**(4), 1 (2020)
3. X. Xu, J. Yu, Y. Chen, Q. Hua, Y. Zhu, Y.C. Chen, M. Li, in *Proceedings of the 26th Annual International Conference on Mobile Computing and Networking, London, 2020*. pp. 1–13
4. C. Wang, Y. Xiao, X. Gao, L. Li, J. Wang, IEEE Trans. Mobile Comput. **22**(1), 19 (2021)
5. Z. Wang, M. Yang, C. Jin, J. Liu, Z. Wen, S. Liu, Z. Zhang, in *Proceedings of the AAAI Conference on Artificial Intelligence, 2021*. pp. 16117–16119
6. M.N. Kurt, Y. Yılmaz, X. Wang, P.J. Mosterman, IEEE J. Sel. Areas Commun. **40**(3), 982 (2022)
7. M. Hou, M. Wang, W. Zhao, Q. Ni, Z. Cai, X. Kong, Comput. Commun. **184**, 128 (2022)
8. S. Garg, K. Kaur, N. Kumar, G. Kaddoum, A.Y. Zomaya, R. Ranjan, IEEE Trans. Netw. Service Manag. **16**(3), 924 (2019)
9. S.G. Lyastani, M. Schilling, S. Fahl, M. Backes, S. Bugiel, in *27th USENIX Security Symposium, Baltimore, 2018*. pp. 203–220
10. C. Wang, H. Zhu, B. Yang, IEEE Trans. Comput. Soc. Syst. **9**(2), 428 (2022)
11. Z. Li, X. Yang, C. Wang, K. Ma, C. Jiang, IEEE Internet Things J. **9**(2), 1622 (2022)
12. Y. Yang, J. Sun, in *IEEE INFOCOM 2017-IEEE Conference on Computer Communications, Atlanta, 2017*. pp. 1–9
13. C. Shi, M. Du, W. Lu, X. He, S. Lu, Mobile Netw. Appl. 1–13 (2022)
14. Z. Wu, L. Tian, Z. Wang, Y. Wang, in *2021 IEEE International Conference on Information Communication and Software Engineering (ICICSE), Chengdu, 2021* (2021), pp. 76–82
15. P. Perera, V.M. Patel, IEEE Trans. Inform. Forensics Secur. **14**(5), 1240 (2018)
16. C. Shen, Y. Chen, X. Guan, Inform. Sci. **430**, 538 (2018)
17. H. Lee, J.Y. Hwang, S. Lee, D.I. Kim, S.H. Lee, J. Lee, J.S. Shin, Pervasive Mobile Comput. **54**, 45 (2019)
18. Q. Zou, Y. Wang, Q. Wang, Y. Zhao, Q. Li, IEEE Trans. Inform. Forensics Secur. **15**, 3197 (2020)
19. Y. Yang, B. Guo, Z. Wang, M. Li, Z. Yu, X. Zhou, Ad Hoc Netw. **84**, 9 (2019)
20. C. Wang, H. Zhu, R. Hu, R. Li, C. Jiang, IEEE Trans. Big Data **9**(2), 758 (2023)

21. C. Wang, S. Chai, H. Zhu, C. Jiang, IEEE Trans. Depend. Secure Comput. **20**(3), 2565 (2022)
22. C. Wang, C. Wang, H. Zhu, J. Cui, IEEE Trans. Depend. Secure Comput. **18**(5), 2122 (2021)
23. C. Wang, H. Zhu, IEEE Trans. Inform. Forensics Secur. **17**, 2703 (2022)
24. M. Ring, S. Wunderlich, D. Scheuring, D. Landes, A. Hotho, Comput. Secur. **86**, 147 (2019)
25. K. Shu, D. Mahudeswaran, S. Wang, D. Lee, H. Liu, Big Data **8**(3), 171 (2020)
26. G. Charles, N. Mori, Int. J. Dev. Issues (2017)
27. C. Bailey, R. Brody, M. Sokolowski, J. Financ. Crime **29**(2), 519 (2022)
28. C. Wang, H. Zhu, IEEE Trans. Depend. Secure Comput. **19**(1), 301 (2022)
29. 52th Statistical Report on the Development of the Internet in China (2023). https://www.gov.cn/yaowen/liebiao/202308/content_6900651.htm
30. X. Chen, C. Wang, Q. Yang, T. Hu, C. Jiang, Sci. China Inform. Sci. **66**(1) (2023)
31. X. Chen, C. Wang, J. Cui, Q. Yang, T. Hu, C. Jiang, IEEE Trans. Big Data **9**(2), 499 (2023)
32. W. Youyou, M. Kosinski, D. Stillwell, Proc. Natl. Acad. Sci. **112**(4), 1036 (2015)
33. M. Kosinski, D. Stillwell, T. Graepel, Proc. Natl. Acad. Sci. **110**(15), 5802 (2013)
34. C. Song, Z. Qu, N. Blumm, A.L. Barabasi, Science (2010)
35. H. Yin, Z. Hu, X. Zhou, H. Wang, K. Zheng, Q.V.H. Nguyen, S. Sadiq, in *2016 IEEE 32nd International Conference on Data Engineering (ICDE), Helsinki, 2016*. pp. 942–953
36. S.H. Hashemi, J. Kamps, in *Proceedings of the 25th Conference on User Modeling, Adaptation and Personalization, Bratislava, 2017*. pp. 50–58
37. C. Zhou, J. Bai, J. Song, X. Liu, Z. Zhao, X. Chen, J. Gao, in *Proceedings of the AAAI Conference on Artificial Intelligence, New Orleans, 2018*. pp. 4564–4571
38. Q. Wang, H. Yin, H. Wang, Z. Huang, in *Databases Theory and Applications: 29th Australasian Database Conference (ADC), Gold Coast, 2018*. pp. 211–223
39. W. Nai, L. Liu, S. Wang, D. Dong, Algorithms **11**(2), 15 (2018)
40. J. Yang, Y. Qiao, X. Zhang, H. He, F. Liu, G. Cheng, IEEE Trans. Emer. Topics Comput. **3**(1), 95 (2014)
41. C. Huo, Y. Zhao, W. Ren, in *Proceedings of the 3rd International Conference on Communication and Information Processing, Tokyo, 2017*. pp. 164–169
42. Y. Zhu, H. Li, Y. Liao, B. Wang, Z. Guan, H. Liu, D. Cai, in *Proceedings of the 26th International Joint Conference on Artificial Intelligence (IJCAI), Melbourne, 2017*. pp. 3602–3608
43. Q. Cao, X. Yang, J. Yu, C. Palow, in *Proceedings of the 2014 ACM SIGSAC Conference on Computer and Communications Security, Scottsdale, 2014*. pp. 477–488
44. Y. Zhang, S. Lv, D. Fan, J. Comput. Res. Dev. **38**(10), 2011 (2015)
45. Z. Wang, C. Jiang, Y. Ding, X. Lyu, Y. Liu, Electron. Commerce Res. Appl. **27**, 74 (2018)
46. S. Liu, S. Wang, F. Zhu, IEEE Trans. Knowl. Data Eng. **27**(7), 2005 (2015)
47. Q. Wang, D. Shen, S. Feng, Y. Kou, T. Nie, G. Yu, J. Softw. **29**(3) (2018)
48. R. Zafarani, H. Liu, in *Proceedings of the 19th ACM SIGKDD International Conference on Knowledge Discovery and Data Mining, Chicago, 2013*. pp. 41–49
49. S. Pearman, J. Thomas, P.E. Naeini, H. Habib, L. Bauer, N. Christin, L.F. Cranor, S. Egelman, A. Forget, in *Proceedings of the 2017 ACM SIGSAC Conference on Computer and Communications Security, Dallas, 2017*. pp. 295–310
50. G. Ye, Z. Tang, D. Fang, X. Chen, K.I. Kim, B. Taylor, Z. Wang, in *Proceedings of the 24th Network and Distributed System Security Symposium (NDSS), California, 2017*
51. A. Constantinides, M. Belk, C. Fidas, A. Pitsillides, in *Proceedings of the 25th International Conference on Intelligent User Interfaces, Cagliari, 2020*. pp. 33–37
52. A. Constantinides, M. Belk, C. Fidas, A. Pitsillides, in *Proceedings of the 27th ACM Conference on User Modeling, Adaptation and Personalization, Larnaca, 2019* (2019), pp. 201–205
53. U. Pal, R.K. Roy, K. Roy, F. Kimura, in *2009 10th International Conference on Document Analysis and Recognition, Barcelona, 2009*. pp. 456–460
54. J. Bonneau, C. Herley, P.C. Van Oorschot, F. Stajano, Commun. ACM **58**(7), 78 (2015)
55. S. Turner, R. Housley, *Implementing Email and Security Tokens: Current Standards, Tools, and Practices* (John Wiley & Sons, 2008)

56. S. Hallsteinsen, I. Jorstad, et al., in *Proceedings of the Second International Conference on Systems and Networks Communications (ICSNC), Cap Esterel, 2007*. pp. 68–68
57. J. Bonneau, C. Herley, P.C. Van Oorschot, F. Stajano, in *IEEE Symposium on Security and Privacy, San Francisco, 2012*. pp. 553–567
58. M.V. Ruiz-Blondet, Z. Jin, S. Laszlo, IEEE Trans. Inform. Forensics Secur. **11**(7), 1618 (2016)
59. Z. Sitová, J. Šeděnka, Q. Yang, G. Peng, G. Zhou, P. Gasti, K.S. Balagani, IEEE Trans. Inform. Forensics Secur. **11**(5), 877 (2015)
60. M. Bicego, A. Lagorio, E. Grosso, M. Tistarelli, in *IEEE Conference on Computer Vision and Pattern Recognition Workshop (CVPRW), New York, 2006*. pp. 35–35
61. T.C. Clancy, N. Kiyavash, D.J. Lin, in *Proceedings of ACM SIGMM Workshop on Biometrics Methods and Applications, Berkley, 2003*. pp. 45–52
62. A. Kumar, A. Passi, Pattern Recogn. **43**(3), 1016 (2010)
63. A. Das, J. Bonneau, M. Caesar, N. Borisov, X. Wang, in *21st Annual Network and Distributed System Security Symposium (NDSS), San Diego, 2014*. pp. 23–26
64. H.M. Sun, Y.H. Chen, Y.H. Lin, IEEE Trans. Inform. Forensics Secur. **7**(2), 651 (2011)
65. J. Xu, J. You, F. Liu, in *Proceedings of IEEE Networking, Sensing and Control, 2005*. pp. 44–48
66. L. Cao, IEEE Trans. Knowl. Data Eng. **22**(6), 755 (2010)
67. G. Tandon, P.K. Chan, in *Proceedings of ICDM Workshop on Data Mining for Computer Security, 2003*
68. D. Pan, D. Liu, J. Zhou, G. Zhang, Microelectron. Reliab. **55**(9–10), 2082 (2015)
69. X. Li, J. Han, S. Kim, H. Gonzalez, in *Proceedings of the 17th SIAM International Conference on Data Mining, Minneapolis, 2007*. pp. 273–284
70. X.X. Lin, P. Lin, E.H. Yeh, IEEE Netw. **35**(1), 212 (2020)
71. G. Tandon, P.K. Chan, in *Proceedings of the 13th ACM SIGKDD International Conference on Knowledge Discovery and Data Mining, San Jose, 2007*. pp. 697–706
72. M.M. WALDROP, Nature **533** (2016)
73. K. Krol, J.M. Spring, S. Parkin, M.A. Sasse, in *The LASER Workshop: Learning from Authoritative Security Experiment Results (LASER), San Jose, 2016*. pp. 21–31
74. P. Pearce, V. Dave, C. Grier, K. Levchenko, S. Guha, D. McCoy, V. Paxson, S. Savage, G.M. Voelker, in *Proceedings of the 2014 ACM SIGSAC Conference on Computer and Communications Security, Scottsdale, 2014*. pp. 141–152
75. M.M. Waldrop, Nature **533**(7602) (2016)
76. M. Abuhamad, A. Abusnaina, D. Nyang, D. Mohaisen, IEEE Internet Things J. **8**(1), 65 (2020)
77. M. Zhu, J. Zhou, J. Wang, Comput. Eng. **28**(10), 138 (2002)
78. A. Primo, in *8th IEEE Annual Ubiquitous Computing, Electronics and Mobile Communication Conference (UEMCON), New York, 2017*. pp. 217–225
79. J. Ho, D.K. Kang, Appl. Intell. **48**, 1547 (2018)
80. C. Mao, Y. Cheng, W. Yu, Chin. J. Netw. Inform. Secur. **2**(3), 76 (2016)
81. C. Shen, Z. Cai, X. Guan, Y. Du, R.A. Maxion, IEEE Trans. Inform. Forensics Secur. **8**(1), 16 (2012)
82. P. Kang, S. Cho, Inform. Sci. **308**, 72 (2015)
83. F. Inguanez, S. Ahmadi, in *IEEE 6th International Conference on Consumer Electronics-Berlin (ICCE-Berlin), Berlin, 2016*. pp. 193–197
84. H. Cao, H. Jiang, D. Liu, J. Xiong, in *41st IEEE International Conference on Distributed Computing Systems (ICDCS), Washington DC, 2021*. pp. 1020–1030
85. C. Shen, Y. Li, Y. Chen, X. Guan, R.A. Maxion, IEEE Trans. Inform. Forensics Secur. **13**(1), 48 (2017)
86. R. Mao, H. Ji, D. Cheng, X. Wang, Y. Wang, D. Sun, in *IEEE Symposium on Computers and Communications (ISCC), Rhodes, 2022*. pp. 1–7
87. X. Yang, S. Yang, J. Liu, C. Wang, Y. Chen, N. Saxena, IEEE Trans. Mobile Comput. **21**(10), 3565 (2021)

88. Y. Chen, C. Shen, Z. Wang, T. Yu, in *IEEE International Conference on Identity, Security and Behavior Analysis (ISBA), New Delhi, 2017*. pp. 1–6
89. W.H. Lee, R. Lee, in *Proceedings of the Hardware and Architectural Support for Security and Privacy 2016* (2016), pp. 1–8
90. C. Song, A. Wang, K. Ren, W. Xu, in *35th Annual IEEE International Conference on Computer Communications (INFOCOM), San Francisco, 2016*. pp. 1–9
91. J. Kong, Y. Guo, C. Liu, Y. Wang, J. Comput. Appl. **39**(6), 1747 (2019)
92. J. Chauhan, Y.D. Kwon, P. Hui, C. Mascolo, Proc. ACM Interact. Mobile Wearable Ubiquitous Technol. **4**(4), 1 (2020)
93. L. Lu, J. Yu, Y. Chen, H. Liu, Y. Zhu, Y. Liu, M. Li, in *IEEE INFOCOM 2018-IEEE Conference on Computer Communications, Honolulu, 2018*. pp. 1466–1474
94. X. Ji, X. Zhou, C. Yan, J. Deng, W. Xu, IEEE Trans. Mobile Comput. **21**(4), 1155 (2020)
95. X. Ruan, Z. Wu, H. Wang, S. Jajodia, IEEE Trans. Inform. Forensics Secur. **11**(1), 176 (2015)
96. E. Shi, Y. Niu, M. Jakobsson, R. Chow, in *Information Security 13th International Conference, Boca Raton, 2010*. pp. 99–113
97. K. Skračić, P. Pale, Z. Kostanjčar, Comput. Secur. **67**, 107 (2017)
98. D. Dasgupta, A. Roy, A. Nag, Comput. Secur. **63**, 85 (2016)
99. M. Wazzeh, H. Ould-Slimane, C. Talhi, A. Mourad, M. Guizani, IEEE Netw. (2022)
100. X. Liu, C. Shen, Y. Chen, in *Biometric Recognition: 13th Chinese Conference, CCBR 2018, Urumqi, 2018*. pp. 669–677
101. J. Jiang, B. Ni, C. Wang, in *Companion Proceedings of the Web Conference 2021, 2021*. pp. 238–246
102. Y. Wu, Z. Xie, S. Ji, Z. Liu, X. Zhang, C. Lin, S. Deng, J. Zhou, T. Wang, R. Beyah, IEEE Trans. Depend. Secure Comput. **20**(2), 1169 (2022)
103. J.B. Awotunde, S. Misra, F. Ayeni, R. Maskeliunas, R. Damasevicius, in *Hybrid Intelligent Systems: 21st International Conference on Hybrid Intelligent Systems (HIS 2021), 2021* (2022), pp. 463–472
104. J.W. Chang, N. Yen, J.C. Hung, J. Ambient Intell. Hum. Comput. **13**(10), 4663 (2022)
105. B. Xu, H. Shen, B. Sun, R. An, Q. Cao, X. Cheng, in *Proceedings of the AAAI Conference on Artificial Intelligence, 2021*. pp. 4537–4545
106. X. He, Q. Gong, Y. Chen, Y. Zhang, X. Wang, X. Fu, IEEE Trans. Depend. Secure Comput. **18**(5), 2193 (2021)
107. S. Wang, Y. Liu, C. Zheng, R. Lin, in *Artificial Intelligence: Second CAAI International Conference, CICAI 2022, Beijing, China, August 27–28, 2022* (2023), pp. 112–123
108. A. Chen, Y. Fu, X. Zheng, G. Lu, Comput. Secur. **114**, 102600 (2022)
109. Q. Chen, S.R. Islam, H. Haswell, R.A. Bridges, in *Science of Cyber Security: Second International Conference, SciSec 2019, Nanjing, 2019* (2019), pp. 199–214
110. I.R. A. Hamid, J. Abawajy, in *Algorithms and Architectures for Parallel Processing: 11th International Conference, ICA300 2011, Melbourne, Australia, October 24–26, 2011* (2011), pp. 266–275
111. Z.Q. Qin, H.Z. Xu, X.K. Ma, Y.J. Wang, Secur. Commun. Netw. (2022)
112. C. Jiang, Y. Fang, P. Zhao, J. Panneerselvam, IEEE Trans. Ind. Inform. **16**(10), 6652 (2020)
113. H.H. Pajouh, R. Javidan, R. Khayami, A. Dehghantanha, K.K.R. Choo, IEEE Trans. Emer. Topics Comput. **7**(2), 314 (2016)
114. J. Wang, S. Hao, R. Wen, B. Zhang, L. Zhang, H. Hu, R. Lu, IEEE Internet Things J. **8**(2), 927 (2020)
115. J. Pei, K. Zhong, M.A. Jan, J. Li, Comput. Netw. **209**, 108906 (2022)
116. V. Mothukuri, P. Khare, R.M. Parizi, S. Pouriyeh, A. Dehghantanha, G. Srivastava, IEEE Internet Things J. **9**(4), 2545 (2021)
117. T. Ji, A.N. Sivakumar, G. Chowdhary, K. Driggs-Campbell, IEEE Robot. Autom. Lett. **7**(2), 4975 (2022)
118. L. Cui, Y. Qu, G. Xie, D. Zeng, R. Li, S. Shen, S. Yu, IEEE Trans. Ind. Inform. **18**(5), 3492 (2021)

119. T. Luo, S.G. Nagarajan, in *2018 IEEE International Conference on Communications (ICC), Kansas City, 2018* (2018), pp. 1–6
120. Z. Li, Y. Zhao, Y. Geng, Z. Zhao, H. Wang, W. Chen, H. Jiang, A. Vaidya, L. Su, D. Pei, IEEE J. Sel. Areas Commun. **40**(9), 2746 (2022)
121. A. Modell, J. Larson, M. Turcotte, A. Bertiger, in *2021 IEEE International Conference on Big Data (Big Data), Orlando, 2021*. pp. 2650–2655
122. H. Mazzawi, G. Dalal, D. Rozenblatz, L. Ein-Dorx, M. Niniox, O. Lavi, in *IEEE 33rd International Conference on Data Engineering (ICDE), San Diego, 2017* (2017), pp. 1140–1149
123. T. Li, M. Sanjabi, A. Beirami, V. Smith, arXiv preprint arXiv:1905.10497 (2019)
124. M. Mohri, G. Sivek, A.T. Suresh, in *Proceedings of the 36th International Conference on Machine Learning, Long Beach, 2019* (2019), pp. 4615–4625
125. G. Wei, A.V. Vasilakos, Y. Zheng, N. Xiong, J. Supercomput. **54**, 252 (2010)
126. F. Lin, Y. Zhou, X. An, I. You, K.K.R. Choo, IEEE Consum. Electron. Mag. **7**(6), 45 (2018)
127. M. Lippi, P. Pałka, G. Contissa, F. Lagioia, H.W. Micklitz, G. Sartor, P. Torroni, Artif. Intell. Law **27**, 117 (2019)
128. M. Dolly Nithisha, B. Divya Sri, P. Lekhya Sahithi, M. Suneetha, in *High Performance Computing and Networking: Select Proceedings of CHSN 2021* (Springer, 2022), pp. 295–306
129. W. Stallings, IEEE Secur. Privacy **18**(1), 61 (2020)
130. M. Godinho de Matos, I. Adjerid, Manag. Sci. (2021)
131. L. Li, A. Bartel, T.F. Bissyandé, J. Klein, Y. Le Traon, S. Arzt, S. Rasthofer, E. Bodden, D. Octeau, P. McDaniel, in *2015 IEEE/ACM 37th IEEE International Conference on Software Engineering, Florence, 2015*. pp. 280–291
132. D. Liu, X. Gao, H. Wang, in *2017 IEEE 37th International Conference on Distributed Computing Systems (ICDCS), Atlanta, 2017* (2017), pp. 2423–2429
133. A. Mehdy, H. Mehrpouyan, arXiv preprint arXiv:2108.08483 (2021)
134. R. Shokri, M. Stronati, C. Song, V. Shmatikov, in *2017 IEEE Symposium on Security and Privacy (SP), San Jose, 2017*. pp. 3–18
135. Y. Karimi, A. Squicciarini, S. Wilson, Proc. ACM Hum.-Comput. Interact. **6**(CSCW2), 1 (2022)
136. X. Chen, C. Wang, Q. Yang, T. Hu, C. Jiang, ACM Trans. Manag. Inform. Syst. **14**(1), 1 (2023)
137. T.A. Rahat, T. Le, Y. Tian, arXiv preprint arXiv:2111.04224 (2021)
138. Y. Li, W. Dai, Z. Ming, M. Qiu, IEEE Trans. Comput. **65**(5), 1339 (2015)
139. H. Hu, H. Zhang, Y. Yang, Multimedia Tools Appl. **77**, 21693 (2018)
140. Q. Wang, X. Zhao, J. Guo, J. Xue, B. Zhao, et al., Secur. Commun. Netw. (2023)
141. Q. Li, H. He, B. Fang, H. Zhang, Y. Wang, J. Comput. Sci. Technol. **37**(1), 1 (2014)
142. H. Yang, Z. Zhang, L. Xie, L. Zhang, Int. J. Intell. Syst. **37**(10), 6909 (2022)
143. M. Ghazel, IEEE Trans. Intell. Transp. Syst. **10**(4), 668 (2009)
144. S. Liu, G. Lin, L. Qu, J. Zhang, O. De Vel, P. Montague, Y. Xiang, IEEE Trans. Depend. Secure Comput. **19**(1), 438 (2020)
145. Z. Wan, J. Shen, J. Chuang, X. Xia, J. Garcia, J. Ma, Q.A. Chen, in *29th Annual Network and Distributed System Security Symposium (NDSS), San Diego, 2022* (2022)
146. Z. Luo, P. Wang, B. Wang, Y. Tang, W. Xie, X. Zhou, D. Liu, K. Lu, in *30th Annual Network and Distributed System Security Symposium (NDSS), San Diego, 2023* (2023)
147. S. Cui, G. Zhao, Y. Gao, T. Tavu, J. Huang, in *Proceedings of the 2022 ACM SIGSAC Conference on Computer and Communications Security (CCS), Los Angeless, 2022* (2022), pp. 639–652
148. L. Ding, Z. Wang, X. Wang, D. Wu, Comput. Commun. **155**, 32 (2020)
149. S. Guo, F. Wang, N. Zhang, F. Qi, X. Qiu, J. Netw. Comput. Appl. **172**, 102812 (2020)
150. S. Xuan, H. Xiao, D. Man, W. Wang, W. Yang, Wirel. Commun. Mobile Comput. **2021**, 1 (2021)

151. X. Hao, W. Ren, Y. Fei, T. Zhu, K.K.R. Choo, IEEE Trans. Services Comput. **16**(2), 773 (2022)
152. C. Li, H. Yang, Z. Sun, Q. Yao, B. Bao, J. Zhang, A.V. Vasilakos, IEEE Internet Things J. **10**(1), 447 (2022)
153. I. Sheff, T. Magrino, J. Liu, A.C. Myers, R. Van Renesse, in *Proceedings of the 2016 ACM SIGSAC Conference on Computer and Communications Security, Vienna, 2016* (2016), pp. 229–241
154. J. Chen, Z. Zhan, K. He, R. Du, D. Wang, F. Liu, IEEE Trans. Depend. Secure Comput. **19**(5), 3301 (2021)
155. C. Lin, J. He, C. Shen, Q. Li, Q. Wang, IEEE Trans. Depend. Secure Comput. **20**(3), 2314 (2022)
156. Y. Zhang, J. Li, D. Zheng, X. Chen, H. Li, Pers. Ubiquitous Comput. **21**, 815 (2017)
157. L. Lin, S. Li, X. Lv, B. Li, in *2021 IEEE Intl Conf on Parallel & Distributed Processing with Applications, Big Data & Cloud Computing, Sustainable Computing & Communications, Social Computing & Networking (ISPA/BDCloud/SocialCom/SustainCom), New York* (2021), pp. 344–351
158. K. Yu, L. Tan, M. Aloqaily, H. Yang, Y. Jararweh, IEEE Trans. Ind. Inform. **17**(11), 7669 (2021)
159. J. Wang, S. Li, J. Wanting, D. Li, B. Song, Peer-to-Peer Networking and Applications pp. 1–20 (2023)
160. H. Zhu, W. Niu, X. Liao, X. Zhang, X. Wang, B. Li, Z. He, Secur. Commun. Netw. **2022**, 3448950:1 (2022)
161. X. Li, L. Wei, L. Wang, Y. Ma, C. Zhang, M. Sohail, Int. J. Intell. Syst. **37**(5), 3050 (2022)
162. U. Javaid, M.N. Aman, B. Sikdar, in *Proceedings of the 1st Workshop on Blockchain-enabled Networked Sensor Systems, Shenzhen, 2018*. pp. 13–18
163. A.S. Patil, R. Hamza, A. Hassan, N. Jiang, H. Yan, J. Li, Comput. Secur. **97**, 101958 (2020)
164. M. Barbareschi, A. De Benedictis, E. La Montagna, A. Mazzeo, N. Mazzocca, Future Gener. Comput. Syst. **101**, 246 (2019)
165. P. Wei, D. Wang, Y. Zhao, S.K.S. Tyagi, N. Kumar, Future Gener. Comput. Syst. **102**, 902 (2020)
166. C. Jing, C. Wang, C. Yan, in *Financial Cryptography and Data Security: 23rd International Conference, FC 2019, Frigate Bay, St. Kitts and Nevis* (2019), pp. 588–604
167. R.A. Bauder, T.M. Khoshgoftaar, Health Inform. Sci. Syst. **6**, 1 (2018)
168. P. Kaur, A. Gosain, in *ICT Based Innovations: Proceedings of CSI 2015* (2018), pp. 23–30
169. L. Eisenberg, T.H. Noe, Manag. Sci. **47**(2), 236 (2001)
170. J. Chen, J. Sun, Int. Cybersecur. Law Rev. **2**(2), 209 (2021)
171. T. Phan-Minh, E.C. Grigore, F.A. Boulton, O. Beijbom, E.M. Wolff, in *Proceedings of the IEEE/CVF Conference on Computer Vision and Pattern Recognition (CVPR), Seattle, 2020*. pp. 14074–14083
172. E. Derman, A.A. Salah, in *13th IEEE International Conference on Automatic Face & Gesture Recognition (FG 2018), Xi'an* (2018), pp. 577–584
173. R. Gupta, S. Tanwar, S. Tyagi, N. Kumar, IEEE Netw. **33**(6), 22 (2019)
174. E.R. Smith, F.R. Conrey, Pers. Soc. Psychol. Rev. **11**(1), 87 (2007)
175. A.Z. Tan, H. Yu, L. Cui, Q. Yang, IEEE Trans. Neural Netw. Learn. Syst. (2022)
176. L. Collins, H. Hassani, A. Mokhtari, S. Shakkottai, in *International Conference on Machine Learning* (PMLR, 2021), pp. 2089–2099
177. K. Wei, J. Li, M. Ding, C. Ma, H.H. Yang, F. Farokhi, S. Jin, T.Q. Quek, H.V. Poor, IEEE Trans. Inform. Forensics Secur. **15**, 3454 (2020)
178. P. Chen, J. Yang, J. Lin, Z. Lu, Q. Duan, H. Chai, in *2023 IEEE International Conference on Data Mining (ICDM)* (IEEE, 2023), pp. 41–50
179. A. Huang, Y. Liu, T. Chen, Y. Zhou, Q. Sun, H. Chai, Q. Yang, ACM Trans. Intell. Syst. Technol. **12**(4), 1 (2021)
180. M. Shayan, C. Fung, C.J. Yoon, I. Beschastnikh, IEEE Trans. Parallel Distrib. Syst. **32**(7), 1513 (2020)

181. Y. Lu, X. Huang, Y. Dai, S. Maharjan, Y. Zhang, IEEE Trans. Ind. Inform. **16**(6), 4177 (2019)
182. Q. Yang, C. Wang, C. Wang, H. Teng, C. Jiang, IEEE Trans. Comput. Soc. Syst. **8**(2), 398 (2020)
183. C. Dong, H. Liang, X. Xu, S. Han, B. Wang, P. Zhang, IEEE J. Sel. Areas Commun. **41**(1), 202 (2022)
184. J. Dai, S. Wang, K. Tan, Z. Si, X. Qin, K. Niu, P. Zhang, IEEE J. Sel. Areas Commun. **40**(8), 2300 (2022)
185. L. Roa, A. Correa-Bahnsen, G. Suarez, F. Cortés-Tejada, M.A. Luque, C. Bravo, Expert Syst. Appl. **169**, 114486 (2021)
186. S. Kunchay, S. Abdullah, in *Adjunct Proceedings of the 2020 ACM International Joint Conference on Pervasive and Ubiquitous Computing and Proceedings of the 2020 ACM International Symposium on Wearable Computers* (2020), pp. 488–493
187. X. Xiao, L. Li, L. Zeng, X. Liu, P. Wei, K. Xue, in *Seventh International Conference on Traffic Engineering and Transportation System (ICTETS 2023)*, vol. 13064 (SPIE, 2024), pp. 370–376
188. A. Ramachandran, A. Karuppiah, et al., BioMed Res. Int. **2020** (2020)
189. K. Pei, S. Wang, Y. He, L. Qin, in *2021 IEEE 5th Information Technology, Networking, Electronic and Automation Control Conference (ITNEC)*, vol. 5 (IEEE, 2021), pp. 413–418
190. Y. Xun, J. Qin, J. Liu, IEEE Trans. Veh. Technol. **70**(6), 6172 (2021)
191. X.c. Ma, J. Lu, Y.D. Wong, et al., J. Adv. Transp. **2023** (2023)

Chapter 9
Sophisticated Behavioral Simulation

Abstract As artificial intelligence becomes increasingly prevalent in scientific research, data-driven methodologies appear to overshadow traditional methods in resolving scientific problems. We revisit a classic classification of scientific problems and rethink the evolution of scientific paradigms from the standpoint of data, algorithms, and computational power. We observe that the strengths of new paradigms have expanded the range of resolvable scientific problems, but the continued advancement of data, algorithms, and computational power is unlikely to bring a new paradigm. To tackle unresolved problems of organized complexity in more intricate systems, we argue that the integration of paradigms is a promising approach. Consequently, we propose behavioral rehearsing, checking what will happen in such systems through multiple times of simulation. One of the methodologies to realize it, sophisticated behavioral simulation (SBS), represents a higher level of paradigm integration based on foundational models to simulate complex social systems involving sophisticated human strategies and behaviors. SBS extends beyond the capabilities of traditional agent-based modeling simulation (ABMS) and, therefore, makes behavioral rehearsing a potential solution to problems of organized complexity in complex human systems.

9.1 Introduction to Behavioral Simulation

As Warren Weaver elucidated in "Science and Complexity" [1], scientific problems can be classified into three categories: problems of organized simplicity, problems of disorganized complexity, and problems of organized complexity, each distinguished by their unique levels of intrinsic complexity and randomness. The quest for resolving more problems has led to the continual proposal of new methodologies, which is a destined result of the comprehensive progress in data, algorithms, and computational power.

Data, algorithms, and computational power are the cornerstones of scientific development, and these three elements are closely related to *the four paradigms* [2]. Based on the observation of natural phenomena and experimental summary, the integration of data and algorithms formed the first paradigm—empirical science [3].

C. Wang, H. Zhu, *Universal Behavior Computing for Security and Safety*,
https://doi.org/10.1007/978-981-97-9014-2_9

With the improvement of scientists' ability in algorithmic design, the integration of algorithms and computational power to explain the principles behind certain natural laws formed the second paradigm—model-based theoretical science. Subsequently, the algorithm development and increase in computational power gave rise to the third paradigm—computational science [4]. Finally, with the further development of computers, the ability to collect, store, and process large amounts of data has dramatically changed science. A high degree of integration of data, algorithms, and computational power achieved the fourth paradigm—data-driven science [5].

Subsequent to the publication of Weaver's work in 1948, the advent of the third and fourth paradigms has yielded significant advancements in tackling problems of organized complexity. Methods of either paradigm partially possess the ability to analyze the uncertainty beyond completely randomness which follows some specific distribution instead of typical distributions like normal distribution. However, the comprehensive development of the three elements indicates the stagnation of paradigms [6]. To better research problems concerning higher levels of uncertainty, the introduction of new mechanism is imperative. We suggest integration as such a mechanism, enabling the exploration of these problems through collaborative efforts of multiple paradigms to compensate for each other's limitations. Consequently, when targeting problems of organized complexity within more complex systems, we posit that the combination of simulation and machine learning (ML) stands out as a promising approach [7], providing an efficacious platform to actualize the concept of paradigm integration.

Inspired by the explosion of foundation models and LLM-based agent simulation [8, 9], we propose a novel technology utilizing paradigm integration, termed *behavioral rehearsing*. This term is a good metaphor for researching a complex human system through more sophisticated simulation technologies to verify expected phenomena and discover unexpected risks. *Sophisticated behavioral simulation* (SBS) is the methodology to realize behavioral rehearsing, where "sophisticated" refers to complex experience-related features of human behaviors. It is an integration of the second paradigm, the third paradigm, and the fourth paradigm. Within SBS, methods of the second paradigm formulate the fundamental rules of a system; methods of the fourth paradigm realize authentic individual behaviors in systems through our proposed model tower. Methods of the third paradigm combine these two elements making heterogeneous simulations. We believe that SBS will be unprecedentedly powerful in analyzing problems of organized complexity in complex human systems that are always of high-level uncertainty.

We rethink the existent scientific paradigms through the lens of data, algorithms, and computational power. We discuss the challenges of addressing problems of organized complexity in complex human systems from a standpoint of uncertainty. Through pertinent examples, we argue that the paradigm integration is a possible solution to problems of organized complexity. Accordingly, we propose behavioral rehearsing and demonstrate the implementation and utilization methods of SBS.

9.2 Organized Complexity

Throughout the evolution of science, complexity and randomness have emerged as two pivotal criteria for characterizing the nature of scientific problems [1]. Complexity refers to the number of variables implicated in a problem, while randomness refers to the unpredictability or disorderliness of each variable within a system. The classification based on these criteria offers a clear and robust framework for comprehending the essence of scientific problems. The correlations between complexity, randomness, and systems are depicted in Fig. 9.1.

The problems of organized simplicity and the problems of disorganized complexity represent two extremes, with a category known as the problems of organized complexity existing between them [10]. Problems of organized simplicity can be analyzed with formulas and equations, thanks to the limited variables and simple relationship [11]. Problems of disorganized complexity can be analyzed with statistical methodologies because large quantities of variables exhibit stable features in such systems [12]. The number of variables involved, although fewer than that of problems of disorganized complexity, far exceeds that of problems of organized simplicity. In contrast to disorganized problems, each variable and its interactions in problems of organized complexity exhibit a certain degree of organization and regularity. This implies that, despite the numerous factors within the system, the relationships between these factors are not completely random but possess a certain order and predictability.

In reality, problems of organized simplicity and problems of disorganized complexity represent only a small fraction of scientific problems. Many problems in human society fall into the category of organized complexity, with economics being a particularly prominent example. Considering the seemingly straightforward issue

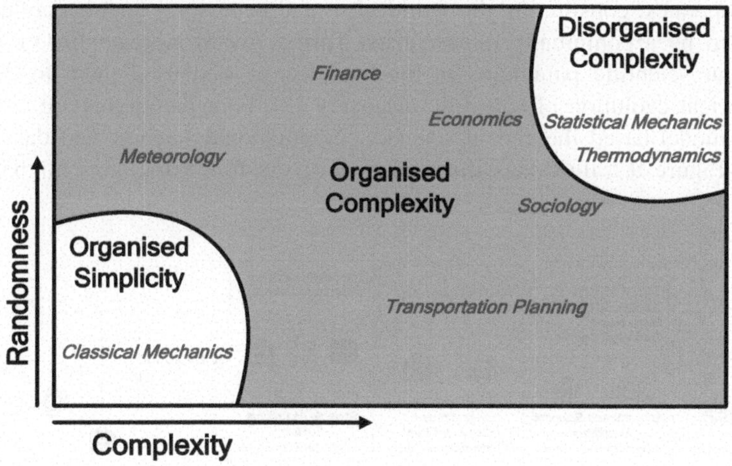

Fig. 9.1 Three classes of systems and their examples

of agricultural product pricing, this involves the influences of various factors such as policies, production levels, and the market, with these factors interconnecting in a complex and organic manner. At the theoretical level, Hausmann et al. [13] introduced two fundamental concepts of economic complexity: relatedness and complexity. Measurement of economic complexity not simply involves adopting aggregation (corresponding to simple systems) or distribution (corresponding to complex systems). It involves the use of dimensionality reduction techniques to preserve the characteristics of the involved elements and consider their interactions.

Besides economics, there are also problems of organized complexity in sociology, transportation, etc. In sociology, intricate elements such as interactions among social actors and cultural exchanges form complex social networks [14–16]. In transportation, factors such as traffic flow and road planning interact, leading to the emergence of complex characteristics within the transportation system [17, 18].

These instances underscore the pervasive presence of organized complexity within complex human systems. Addressing these complexities necessitates a comprehensive consideration of numerous factors and a profound understanding of the interdependencies among these factors, thereby posing new challenges to scientists. Warren Weaver asserted that resolving such problems demands scientific advancements surpassing those necessitated by the other two categories of problems. The subsequent evolution of scientific paradigms confirmed his assertion.

9.3 Evolution of Scientific Paradigms

The concept of scientific paradigm was originally introduced by the renowned American philosopher of science, Thomas Samuel Kuhn [19], in 1962. Paradigm refers to the theoretical foundation and practical norms upon which conventional science operates, constituting the world view and behavioral patterns collectively adhered to by a community of scientists. Turing Award laureate Jim Gray proposed four scientific paradigms in his final lecture and used them to describe the historical evolution of scientific discovery [2]. Four paradigms are empirical science, model-based theoretical science, computational science, and data-driven science. Figure 9.2 illustrates the evolution of the four paradigms aligned with

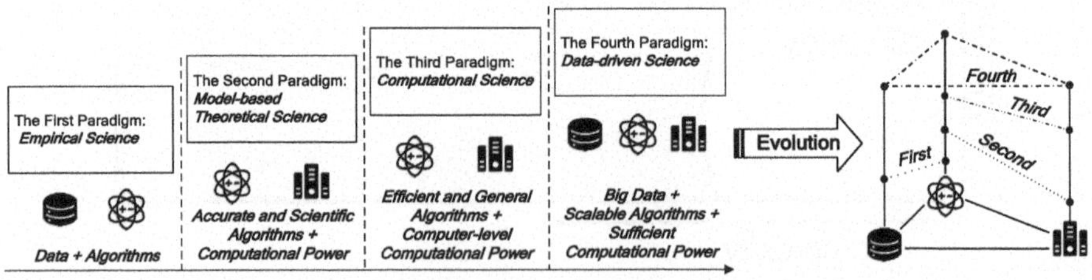

Fig. 9.2 The evolution of the four paradigms aligned with the three elements

the development of three elements—data, algorithms, and computational power. The four paradigms, each adopting distinct perspectives and methodologies, have facilitated addressing a wide range of scientific problems.

The first paradigm, termed empirical science [3], is fundamentally empirical, characterized by the recording and description of natural phenomena. It hinges upon direct observation and hands-on experimentation within the natural world. Although numerous apparent rules are observed, there lacks systematic methods to capture or express them. However, the disadvantage in all three elements makes it inefficient and challenging to achieve a more precise understanding of phenomena.

With the improvement of scientists' capabilities of algorithm, induction, and deduction, the second paradigm, known as model-based theoretical science, emerged. In comparison with the first paradigm, which relies on observation and experimentation to discern rules, the scientific theories of the second paradigm require principled explanations behind certain natural rules. Examples include Newton's laws of motion from the seventeenth century [20] or Maxwell's electrodynamics equations from the nineteenth century [21]. These theoretical models are inductively derived from empirical observations and can be generalized to broader situations to deduce more theory. However, when it comes to problems of organized complexity where complex interactions cannot be ignored, the difficulty of validating theories increases, diminishing the effectiveness of the second paradigm.

In the twentieth century, the rapid development of computers brought about a significant increase in computational power and algorithm capabilities, giving rise to the third paradigm—computational science [4]. Around the mid-twentieth century, John von Neumann proposed modern electronic computer architecture, which resulted in the widespread adoption of simulation and emulation for scientific experiments using electronic computers [22]. As computer simulation increasingly replaced experiments, the third paradigm gradually became a conventional method in scientific research. Methodologies of the third paradigm are a better solution to some problems of organized complexity in natural science. Numerical analysis realized through mathematical models and enough computational power eliminates the bottleneck of dealing with large amounts of complex interactions in a system. However, mathematical models are insufficient to simulate all the patterns of the real world.

In the early twenty-first century, the development of computers once again transformed science. The ability of computers to collect, store, and process large amounts of data gave rise to the data-intensive fourth paradigm—data-driven science [5]. Methodologies of the fourth paradigm further resolve some problems of organized complexity through pattern recognition, such as image classification and temporal prediction. The complex interactions in a specific problem are represented as a complicated distribution.

The four scientific paradigms reveal an ongoing evolutionary trajectory in scientific research. Each paradigm holds a significant position within its respective era, expanding the scope of resolvable problems. Although they progressively reduce the area of unresolved problems, the comprehensive advancement of the three elements indicates a limited progression in addressing further problems of organized complexity. The scaling of data, algorithms, and computational power can only yield more powerful methodologies within the fourth paradigm [6], thereby failing to fundamentally resolve problems that cannot be transformed into specific distribution-finding tasks on big data. The four paradigms are now facing this bottleneck.

9.4 Bottleneck Arising from Uncertainty

Uncertainty refers to the unpredictability of individual and systemic states. It is an ubiquitous characteristic in systems of organized complexity or disorganized complexity. Not all problems of uncertainty are intractable. For example, in thermodynamics, temperature is a physical quantity that quantitatively expresses the characteristics of random motions of microscopic particles. Its effectiveness results from the fact that random motions consistently follow a normal distribution, and these motions can be regarded as a whole when scaled up. Similarly, it is the relatively fixed distribution inherent in the uncertainty that makes the methods of the fourth paradigm, data-driven science, effective for tasks like image classification and generation.

However, in most systems of organized complexity, there is no fixed distribution. The absence of a fixed distribution leads to higher levels of uncertainty. The theory of chaos serves as a pertinent demonstration. In 1962, Lorenz [23] observed a phenomenon where a finite deterministic system, representing forced dissipative hydrodynamic flow through ordinary nonlinear differential equations, exhibited instability. This instability manifested as nonperiodic solutions that diverged significantly from bounded solutions with minor differences in initial states. Subsequently, May [24] explored systems described by different equations and identified similar instabilities while studying biological populations. Two years later, May [25] further demonstrated the potential for creating nondeterministic systems even with simpler mathematical models. In essence, chaos is the union of the transience of stability, the sensitivity of initial states, the general existence of nonperiodic state sequences and the very-long-range unpredictability of non-idealized systems. Consequently, complex systems of this property are called chaotic systems. While some problems of this level of uncertainty have been partially resolved, such as improved weather forecasting in recent decades through the advancements in simulation and deep learning (DL), bottleneck remains.

After demonstrating the concept of chaos, we can explore its extension, high-order chaos. Different from individual simulacra of specific expression or fixed distribution in systems such as meteorology, human behaviors are notably intricate. Research conducted in 1990 [26] revealed that human decision-making within classical game theory systems is heavily influenced by others' behaviors, often deviating from the anticipated outcomes in Nash equilibrium scenarios [27]. This indicates human behaviors are dynamic in complex decision-making scenarios.

Existent methods of the third paradigm or the fourth paradigm primarily rely on theories derived from induction, idealized hypotheses, and fixed distribution of behaviors [28–32], limiting their capacity to authentically simulate human behaviors in these scenarios. When applied to complex systems such as stock markets and social networks, they may suddenly encounter failures due to the oversimplification of human behavioral process.

The higher level of uncertainty within human strategy-related systems can be regarded as a form of high-order chaos, where every individual within the chaotic system is itself chaotic. There exists a bottleneck that the four paradigms are not capable of resolving problems concerning high-order chaos.

9.5 Integration of Scientific Paradigms

Confronted with the bottleneck arising from high-level uncertainty, existing methods of the second paradigm, the third paradigm, and the fourth paradigm prove inadequate. In these scenarios, which include fields like sociology and economics, researchers have to resort to the first paradigm that emphasizes observing the real world and inducing laws. The limitations of the first paradigm are evident. Real-world experiments are often limited, inefficient, and costly. Since the three elements have well developed, we suggest the integration of the four paradigms as a new paradigm to eliminate the bottleneck.

With the scientific problems under research having become more complex, researchers employing research methods of different paradigms often find that a single method cannot fully address these challenges. This necessitates collaborative efforts, integrating research methods of different paradigms to capitalize on their respective strengths and compensate for limitations, which demonstrates the advantages of complementation among scientific paradigms. The deciphering of the deoxyribonucleic acid (DNA) structure marks a significant milestone in the history of biology. Its discovery process involves exploration through research methods of different paradigms. Scientists represented by Rosalind Franklin and Maurice Wilkins captured and observed the DNA structure adopting an X-ray diffractometer, i.e., the first paradigm research method through observation and analysis of experiments [33]. Scientists like James Watson and Francis Crick explored the DNA structure by deriving models based on chemical rules, utilizing the second paradigm through model deduction. After extensive research on their methods, neither side fully revealed the truth. Eventually, collaboration ensued: Watson and Crick [34] utilized reports from Wilkins and Franklin along with high-resolution DNA photos taken by Franklin to remodel and deduce the DNA double helix. Simultaneously, Franklin and Raymond Gosling [35] provided crystallographic evidence supporting the double helix structure, while Wilkins further validated the model's reliability through DNA crystal experiment data. Besides, AlphaFold2 predicted the distance distribution between each pair of amino acids in a protein, along with the angles between the chemical bonds connecting them, and then compiled all the measured results of amino acid pairs into 2D distance histograms. Convolutional neural networks learned from these images to build 3D structures of the protein [36]. The combination of this first paradigm research method with the fourth paradigm research method effectively predicts protein structures. The combination of two paradigms, as evidenced by the three examples mentioned, offers a complementary approach that leverages strengths to compensate for weaknesses. This synergy proves essential.

In conclusion, the four paradigm research methods each hold their unique value and, therefore, can be integrated and complemented. We strive to achieve a higher degree of integration among the four paradigms, thereby establishing a novel research methodology that leverages their collective strengths to resolve problems of organized complexity. Drawing upon the success of the third and fourth paradigms, we posit that a new simulation technology that harnesses the power of paradigm integration holds promise. This technology is *behavioral rehearsing*.

9.6 Behavioral Rehearsing

Currently, there have been methodologies integrating the third paradigm and the fourth paradigm to generate data for research purposes. In the realm of the Internet of things (IoT), a knowledge-enhanced generative adversarial model effectively produced network traffic data, addressing the significant challenge of acquiring large-scale IoT traffic data [31]. Similar methods were utilized in personal activity trajectory data [32] and other daily life research [37]. Moreover, they are also applied to simulate groups and infer system changes. A simulation framework based on data-driven entities could simulate crowds for disaster prevention [31, 38], urban management [39], and social systems research [40].

Although these methodologies can recognize patterns from data, its prediction or classification performance often suffers when encountering individual-level samples not consistent with the historical data distribution, i.e., the uncertainty of dynamic distribution in problems of organized complexity.

Therefore, we propose a technology that integrates paradigms to a deeper extent, and it is a possible solution to problems of organized complexity: *behavioral rehearsing*, scrutinizing a system through more authentic individual simulation to gather feedback. Behavioral rehearsing draws inspiration from the term "behavioral economics" [41]. According to the terming of "behavioral economics," the modifier behavioral has historical significance, originating from behavioral decision research (BDR) [42]. BDR, in essence, is dedicated to studying the intricacies of how individuals make decisions. Historically and theoretically, it focuses on comparing actual decision-making processes with the principles of rationality [43]. It is from this foundation that behavioral economics emerged as a distinct field. Behavioral economics focuses on the study of irrational human behavior, recognizing that individuals often deviate from purely rational decision-making processes. The term "behavioral" highlights the departure from traditional economic models that assume strict rationality. By incorporating "behavioral" into the term, we aim to create simulations that authentically represent the complexities, biases, and nuances inherent in human decision-making to scrutinize complex systems of high-order chaos.

According to the conclusions of Lorenz [23] and May [24], the intrinsic nature of chaotic dynamical systems suggests an inherent inability to fully comprehend the intricate correlations among individuals and the latent patterns within chaos. Consequently, the recourse to understanding chaotic system states primarily lies in simulation, aside from the limited scope of observing real-world systems. Moreover, the more authentic one simulation is, the more informative its results will be. If we can better simulate sophisticated human beings, problems concerning high-order chaos will be better analyzed. Advanced to conventional agent-based modeling simulation (ABMS), the simulation methodologies in behavioral rehearsing can be in the form of a higher degree of paradigm integration to simulate individual-level behaviors of dynamic distributions. Therefore, we propose SBS in complex human systems where sophistication like mental factors and learning mechanisms have to be considered.

The agents in existent ABMS generate their behaviors based on historical data following a fixed pattern. Human behaves according to his observation, experience, learning, and culture, all of which account for the sophistication. Therefore, its generation have to be more interpretable and follows dynamic patterns to be consistent with human, a typical characteristic of organized complexity. Contrary to agents in ABMS, the agent in SBS should generate its behaviors based on its *knowledge architecture*, which can be updated through a *learning mechanism*.

9.7 Realizing Sophisticated Behavioral Simulation

To realize the so-called knowledge architecture and learning mechanism, we propose the sophisticated agent. Within the sophisticated agent, different levels of knowledge are included by models of different paradigms. Through the fusion of models, the knowledge can form a hierarchical knowledge architecture. With the sophisticated agent as the core, importing learning mechanism, we design the architecture of SBS as illustrated in Fig. 9.3. The framework of SBS is separated into three layers:

- **Functions**
  - The process of simulation is the basic function in SBS.
  - Simulation results are further utilized for past event production and future prediction.
- **Models**
  - The core is our proposed *model tower* which consists of three floors. After the model tower empowers instances of the sophisticated agents, these agents

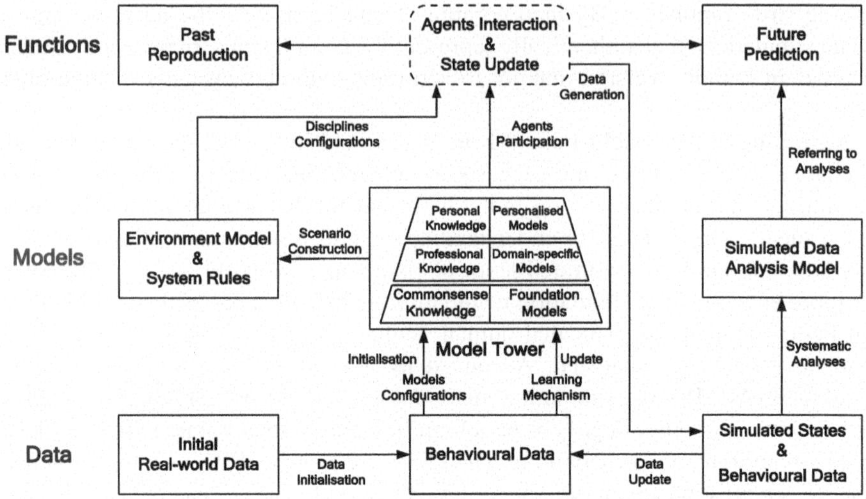

Fig. 9.3 The framework of SBS

become a part of the simulation scenario and participate in interactions within the system.
- Environment analysis model and system update rules confirm the essence of the system which is the disciplines of agents' interactions and states update.
- The simulated data analysis model is designed to analyze systemic data so that its output can be the reference for future prediction.

- **Data**
  - Initial real-world data are collected to initialize the model tower as the seed of SBS.
  - Simulated states/behavioral data are generated in the system states' update and interactions among agents after every tick. Simultaneously, a learning mechanism affects models in the model tower to update agents' hierarchical knowledge systems.
  - Initial data and simulated data constitute all behavioral data in SBS.

9.7.1 *Hierarchical Knowledge Architecture: Model Tower*

The sophisticated agent adopts an architecture we name *model tower*, comprising self-organized multi-floor models. We aim to generate authentic behaviors by initially feeding current system states and other input streams into the bottom floor in a specific format. As the data ascends through each floor, it undergoes bottom-up refinements, culminating in the generation of finely nuanced individual-level behaviors. The model tower is considered a research tool of paradigm integration where DL model is the basis, and upper floors could be the second, third, and fourth paradigm models. This is what we call a higher degree of integration in SBS than in existent ABMS methods. The model tower builds a sophisticated behavior generation framework and hierarchical knowledge architecture containing commonsense knowledge, professional knowledge, and personal knowledge.

9.7.1.1 Commonsense Knowledge

The bottom floor conducts general reasoning based on commonsense knowledge and establishes a preliminary direction for subsequent refinements. As a foundation model, it typically serves as the base for this purpose. Among existing research [9, 44–46], LLM stands out as a predominant choice due to its remarkable reasoning capabilities and extensive inherent commonsense knowledge. Through prompting engineering, the LLM agents are able to apply their knowledge and think as human. To make the most use of commonsense knowledge, there are plenty of tricks behind the design of prompts like templates and cognitive architectures [47], and it is closely related to the application scenarios. As the rapid upgrade of LLMs and other forms of foundation models, they will be doomed to possess more and more powerful commonsense reasoning ability. What the basis of the model tower is required to do is align prior knowledge with human in specific scenarios through well-designed prompts.

9.7.1.2 Professional Knowledge

(1) Internal Specialized Generative Module As depicted in Fig. 9.4, the sequence of dice rolls generated by LLM is intuitively peculiar. To ensure the independence of each roll, we clear the context for every roll request. Despite these rolls being mutually independent, the number "4" emerges as the most frequent outcome. Notably, with either LLaMA-2-7b-chat or LLaMA-2-13b-chat [48], the numbers "1," "2," and "6" never appear. Even when considering that this sequence holds a probability the same as the sequence 123456123456... (which stands at $\frac{1}{6^n}$ when using a standard dice), the observed likelihood remains peculiar.

In the dice rolling problem, if the probabilities assigned to the six sides appearing in the candidate tokens differ, the model struggles to accurately simulate real dice rolling, which typically follows an equal probabilities model.

In comparison with the hallucination of LLM, the generated behavior is not logically incorrect but incorrect in probability. While obtaining a sequence where "4" consistently appears in a finite number of dice rolls is probable, the likelihood of it being a standard dice is notably low.

The specialized generative models within the model tower will effectively address this issue. Introducing an additional floor above the foundational model floor facilitates a two-floor architecture, separating reasoning from generation

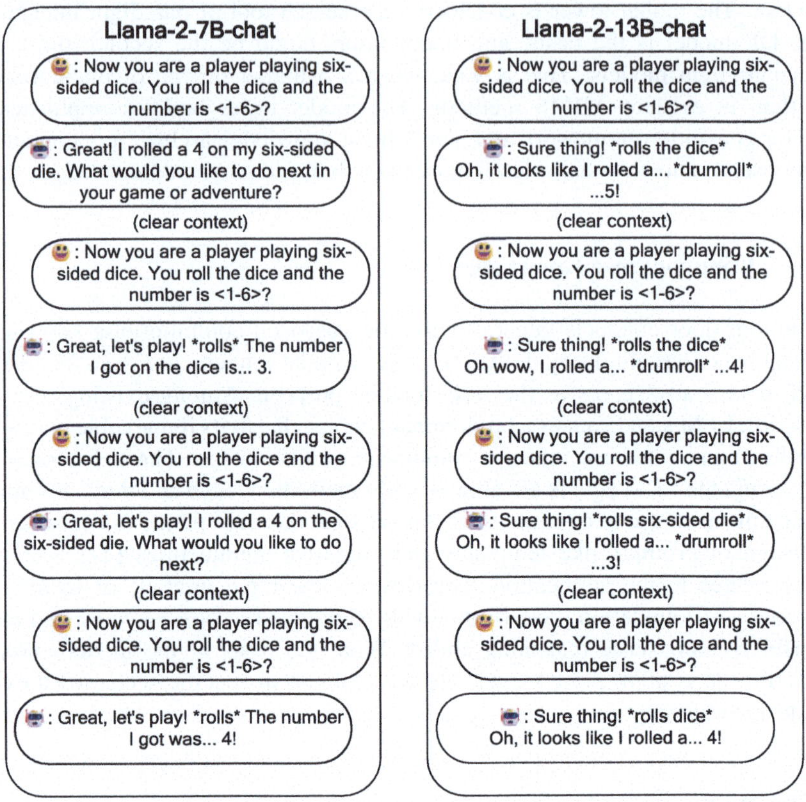

Fig. 9.4 The dice rolling experiments was conducted on LLaMA-2-7B-chat and LLaMA-2-13B-chat

into distinct processes. This significantly mitigates the dice rolling problem. In the dice rolling scenario, we first prompt the LLM to generate "{'action': 'dice rolling'}." Then, the specific generative module processes the request, calling a generative model whose distribution is the same as a standard six-sided dice to generate the final action "{'dice rolling': 4}." Theoretically, the result sequence generated in this pattern can be far more authentic than the basic LLM agent. This pattern allows LLM to perform commonsense reasoning, while invoking next-floor domain-specific generative models, for reliable behavior generation supplemented by professional knowledge in the form of distributions.

(2) External Application Programming Interface The division of reasoning and generation across floors within the aforementioned model tower architecture is similar to the application programming interface (API) calls observed in online language assistant agents such as Copilot and retrieval-augmented generation (RAG) of LLMs [49]. Numerous efforts, as highlighted by Patil et al. [50], have focused on enhancing the comprehensive abilities of LLM agents. Therefore,

beyond the specialized generative module, leveraging various external APIs, such as calculators, databases, and others, holds promise in mitigating hallucinations and improving reasoning, which is also a good choice when the time and hardware costs are acceptable.

In summary, the internal specialized generative module and external APIs collectively form the second floor within the model tower, relieving the first floor (foundation model) from professional inference. As a result, most task-specific professional knowledge is obtained in the second floor through agents calling APIs. Plenty of work about professional LLM agent has verified the feasibility and necessity of the import of professional knowledge [51–53].

9.7.1.3 Personal Knowledge

Dramaturgy proposed by Erving Goffman [54] states that everyone's social behaviors heavily rely on societal norms. Besides external factors, various internal elements distinctly influence one's behaviors, including character, experience, gender, occupation, mood, etc. Therefore, to achieve the sophisticated agent, a comprehensive profile of an agent containing the aforementioned external and internal information is required to differentiate agents driven by a two-floor model tower. Above aforementioned impersonal knowledge floors, a floor provides group and individual bias is required. We call this floor personalized models floor. Rather than relying solely on textual profiles [44], the more sustainable resolution is adding a personality model upper the two-floor model. This addresses the issue between the capacity of the LLM context and the intricate nature of personality. In other words, the personalized models can be realized in the form of rules, which belongs to the second paradigm. A better resolution is using personality models, which belongs to the fourth paradigm.

9.7.2 Behavioral Learning Mechanism: Reinforcement Learning

Simultaneously, a learning mechanism is utilized to update agents' hierarchical knowledge systems. The learning mechanism empowers agents with flexibility, enabling them to adjust their policies and behaviors in response to challenges and changes. This mechanism enables agents to learn from mistakes, self-improve, and continuously optimize their decision-making and behavioral patterns. Through continuously accumulating experiences and refining their performance, agents enhance their efficiency and accuracy in task execution. The learning mechanism is crucial for agents, and the reinforcement learning (RL) method is widely selected as the learning mechanism for agents owing to its high flexibility, self-learning, experience accumulation, etc. LLMs with CoT prompts have demonstrated remarkable advancements in tackling intricate language processing tasks [55, 56]. Post the agent's generation of pertinent prompts derived from its current observations, CoT engages in intricate reasoning to determine the optimal behavioral scenarios.

9.8 Utilizing Behavioral Rehearsing

To deal with the problems of organized complexity, behavioral rehearsing needs to scrutinize issues about system states. Given that accurate prediction of states within chaotic systems is impossible, it implies that simulations may not perfectly correspond with the actual evolution of these systems. As a result, it becomes crucial to ensure the generation of diverse initial states that yield varied "world lines" for meaningful analysis. The more simulations are conducted, the higher the likelihood of capturing the possible future trajectory of a system will be. By leveraging the interpretability inherent in logs of agents' behaviors and thoughts generated by foundational models such as LLMs, we primarily present some analysis patterns for two functions in SBS.

9.8.1 Past Reproduction

SBS makes past reproduction possible. Through the details of simulated events, causality analyses on abnormal behaviors and events can be conducted. This has the potential to be utilized in scenarios such as security event investigation and review in complex human systems. When some unexpected states or events arise in a real-world system (represented by e_0), simulating the system from a particular past moment becomes a viable approach.

As is shown in Fig. 9.5a, conducting numerous simulations yields a spectrum of diverse states or events, denoted as $E = \{e_1, \cdots, e_n\}$, occurring when the simulated system progresses to the same point as e_0. Among these states or events, we filter those resembling e_0, forming the set $E_\tau = \{e_i, e_j, \cdots, e_k\}$. Subsequently, we can investigate the individual behaviors $B_i = \{b_{i,0}^{a_1}, b_{i,0}^{a_2}, \cdots, b_{i,T}^{a_x}\}$ ($b_{i,T}^{a_m}$ represents the behavior of agent a_m at moment T before e_i) leading to the $e_i \in E_\tau$. Employing manually checking the common features in $B_\tau = \{B_i, B_j, \cdots, B_k\}$, meaningful conclusions regarding the causality hardly noticed in complex systems underlying the appearance of e_0 may be drawn.

9.8.2 Future Prediction

SBS can be instrumental in predicting future risks, utilized in scenarios such as safety events warning and avoiding. Similar to the methodology employed

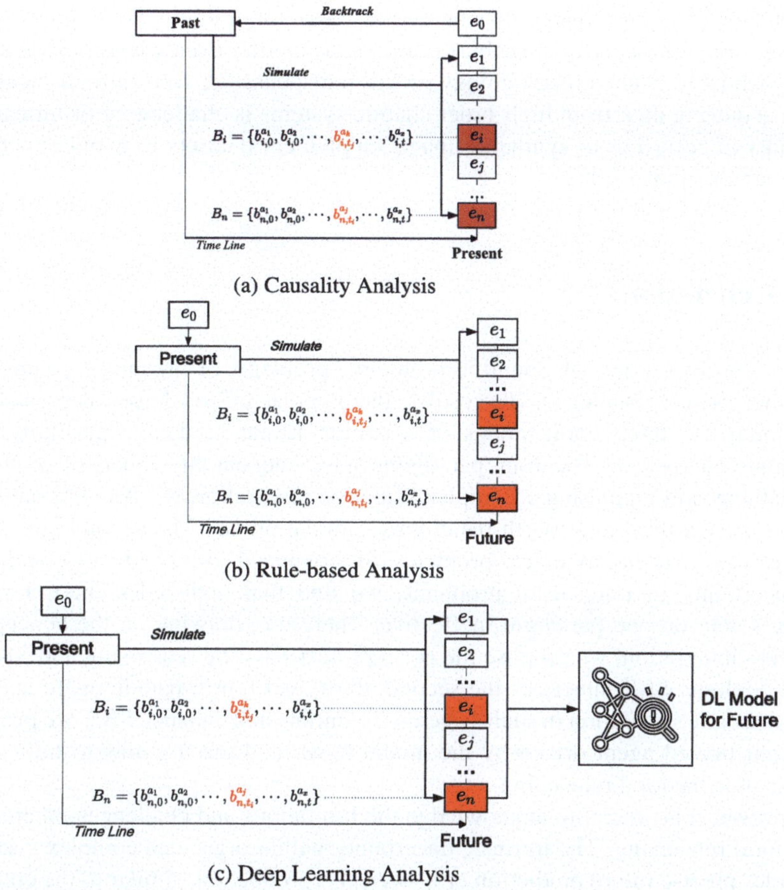

Fig. 9.5 The illustration of causality analysis, rule-based analysis, and DL analysis

in causality analysis, the simulation process initiates from the present moment. As is shown in Fig. 9.5b, as the simulation progresses, diverse states or events $E = \{e_1, \cdots, e_n\}$ are observed at a specified future moment t. In this scenario, predefined rules serve as benchmarks for expectations at moment t. These rules facilitate the identification of unexpected future occurrences, denoted as E_β. If $E_\beta = \emptyset$, indicating no unexpected outcomes, it signifies that the future aligns with expectations. However, if $E_\beta \neq \emptyset$, it indicates that the simulation predicts deviations from expectations due to certain specific and notable behaviors. In such cases, manual intervention becomes imperative to prevent these behaviors.

Besides human-assist analyses, a completely autonomous approach is feasible. As is shown in Fig. 9.5c, regarding a system as a graph where nodes are agents, and edges are interaction behaviors between agents, each simulation generates a series of graphs. By aggregating these graphs from multiple simulations, a comprehensive dataset emerges. This dataset becomes the basis for training DL models. This process is revolutionary as it enables training DL models even when no real-world data is available. Such a methodology proves indispensable, especially in scenarios where acquiring data from high-order chaotic systems is challenging or unfeasible [51], just as resorting to synthetic image supplement datasets to avoid expensive dataset annotations [57].

9.9 Conclusions

Among the three types of scientific problems, problems of organized complexity pose the greatest challenge. To resolve them, great efforts have been made in developing the three cornerstones of scientific research: data, algorithms, and computational power. The four paradigms have sequentially emerged to tackle the challenges in increasingly complex scientific research fields. As representative methods of the third and fourth paradigms, simulation and ML technologies have become the solution to many problems of organized complexity. Considering the uncertainty of unresolved problems, we find that high-order chaos renders relying solely on one paradigm ineffective. Therefore, drawing on the concept of paradigm integration, we propose the concept of behavioral rehearsing and SBS as a methodology. SBS integrates the second, third, and fourth paradigms to achieve more authentic simulation in such systems. To initiate and enhance SBS, we propose the sophisticated agent driven by the model tower and discuss how to utilize the outputs of behavioral rehearsing.

However, it is crucial to acknowledge the limitations and challenges inherent in behavioral rehearsing. The intrinsic uncertainty within organized complex systems precludes precise future prediction or historical reproduction, similar to the concept of Laplace's demon [58], a concept deemed unrealistic. Therefore, behavioral rehearsing serves as a valuable reference rather than an absolute predictor, while DL models excel at specific prediction and classification tasks. Challenges persist, including determining the required width of the context window, achieving personalized modeling without text profiles, implementing experiential learning throughout the model tower, and estimating the resource demands of large-scale SBS.

We are looking forward to the application and other methodologies of behavioral rehearsing. Their advancement, coupled with the benefits derived from the development of other paradigms, positions behavioral rehearsing as a promising and risk-avoiding methodology. This ensures it is indispensable in cost-sensitive social science scenarios such as policy piloting, financial strategy, and systematic risk prediction, where real-world experiments may not be feasible.

References

1. W. Weaver, Am. Sci. **36**(4), 536 (1948)
2. T. Hey, *The Fourth Paradigm* (United States of America, 2009)
3. R.T. Thibault, M. Lifshitz, A. Raz, Cortex **74**, 247 (2016)
4. M. Gahegan, Int. J. Geograph. Inform. Sci. **34**(1), 1 (2020)
5. A. Agrawal, A. Choudhary, Apl Mater. **4**(5) (2016)
6. B. Zhang, Z. Liu, C. Cherry, O. Firat, in *The Twelfth International Conference on Learning Representations* (2023)
7. R. Haasl, in *Nature in Silico: Population Genetic Simulation and Its Evolutionary Interpretation Using C++ and R* (Springer, 2022), pp. 1–20
8. R. Bommasani, D.A. Hudson, E. Adeli, R. Altman, S. Arora, S. von Arx, M.S. Bernstein, J. Bohg, A. Bosselut, E. Brunskill, et al., arXiv preprint arXiv:2108.07258 (2021)
9. J.S. Park, J. O'Brien, C.J. Cai, M.R. Morris, P. Liang, M.S. Bernstein, in *Proceedings of the 36th Annual ACM Symposium on User Interface Software and Technology* (2023), pp. 1–22
10. G.J. Klir, *Foundations of Generalized Information Theory* (2006)
11. I. Newton, *Philosophiae Naturalis Principia Mathematica*, vol. 1 (G. Brookman, 1833)
12. R.K. Pathria, *Statistical Mechanics* (Elsevier, 2016)
13. C.A. Hidalgo, R. Hausmann, Proc. Natl. Acad. Sci. **106**(26), 10570 (2009)
14. L. Freeman, Study Sociol. Sci. **1**(687), 159 (2004)
15. M. Emirbayer, J. Goodwin, Am. J. Sociol. **99**(6), 1411 (1994)
16. F. Vega-Redondo, *Complex Social Networks*, vol. 44 (Cambridge University Press, 2007)
17. B. Chen, Y. Xie, W. Tong, C. Dong, D. Shi, B. Wang, Phys. A: Stat. Mech. Appl. **391**(8), 2730 (2012)
18. J. Tang, Y. Wang, F. Liu, Phys. A: Stat. Mech. Appl. **392**(18), 4192 (2013)
19. T.S. Kuhn, *The Structure of Scientific Revolutions* (University of Chicago Press, 2012)
20. I. Newton, *The Principia: Mathematical Principles of Natural Philosophy* (University of California Press, 1999)
21. J.C. Maxwell, Philos. Trans. Roy. Soc. Lond. **155**, 459 (1865)
22. J. von Neumann, IEEE Ann. Hist. Comput. **15**(4), 27 (1993). https://doi.org/10.1109/85.238389
23. E.N. Lorenz, J. Atmos. Sci. **20**(2), 130 (1963)
24. R.M. May, Science **186**(4164), 645 (1974)
25. R.M. May, Nature **261**(5560), 459 (1976)
26. D. Richards, Behav. Sci. **35**(3), 219 (1990)
27. J. Nash, Ann. Math. 286–295 (1951)
28. Y. Yuan, J. Ding, H. Wang, D. Jin, ACM Trans. Intell. Syst. Technol. (2023)
29. P. Chen, in *Proceedings Visualization '93* (1993), pp. 397–401. https://doi.org/10.1109/VISUAL.1993.398900
30. S. Kang, S. Kahan, J. McDermott, N. Flann, I. Shmulevich, Bioinformatics **30**(21), 3101 (2014)
31. S. Hui, H. Wang, Z. Wang, X. Yang, Z. Liu, D. Jin, Y. Li, in *Proceedings of the ACM Web Conference 2022* (2022), pp. 3336–3346
32. Y. Yuan, J. Ding, H. Wang, D. Jin, Y. Li, in *Proceedings of the 28th ACM SIGKDD Conference on Knowledge Discovery and Data Mining* (2022), pp. 4752–4762
33. R.E. Franklin, R.G. Gosling, Nature **171**(4356), 740 (1953)
34. J.D. Watson, F.H. Crick, Nature **171**(4356), 737 (1953)
35. M. Cobb, N. Comfort, Nature **616**(7958), 657 (2023)
36. J. Jumper, R. Evans, A. Pritzel, T. Green, M. Figurnov, O. Ronneberger, K. Tunyasuvunakool, R. Bates, A. Žídek, A. Potapenko, et al., Nature **596**(7873), 583 (2021)
37. Y. Yuan, H. Wang, J. Ding, D. Jin, Y. Li, in *Proceedings of the ACM Web Conference 2023* (2023), pp. 906–916
38. H. Shi, Q. Yao, Y. Li, in *Proceedings of the ACM Web Conference 2023* (2023), pp. 4200–4209

39. G. Zhang, Z. Yu, D. Jin, Y. Li, in *Proceedings of the 28th ACM SIGKDD Conference on Knowledge Discovery and Data Mining* (2022), pp. 2439–2449
40. F. Lorig, L. Vanhée, F. Dignum, in *ECCAI Advanced Course on Artificial Intelligence* (Springer, 2021), pp. 391–414
41. E. Angner, G. Loewenstein, *Handbook of the Philosophy of Science: Philosophy of Economic* (2007), pp. 641–690
42. J.W. Payne, J.R. Bettman, E.J. Johnson, Ann. Rev. Psychol. **43**(1), 87 (1992)
43. R.M. Dawes (1998)
44. C. Gao, X. Lan, Z. Lu, J. Mao, J. Piao, H. Wang, D. Jin, Y. Li, arXiv preprint arXiv:2307.14984 (2023). https://arxiv.org/abs/2307.14984
45. G. Wang, Y. Xie, Y. Jiang, A. Mandlekar, C. Xiao, Y. Zhu, L. Fan, A. Anandkumar, arXiv preprint arXiv:2305.16291 (2023). https://arxiv.org/abs/2305.16291
46. G. Li, H.A.A.K. Hammoud, H. Itani, D. Khizbullin, B. Ghanem, arXiv preprint arXiv:2303.17760 (2023). https://arxiv.org/abs/2303.17760
47. T.R. Sumers, S. Yao, K. Narasimhan, T.L. Griffiths, arXiv preprint arXiv:2309.02427 (2023). https://arxiv.org/abs/2309.02427
48. H. Touvron, L. Martin, K. Stone, P. Albert, A. Almahairi, Y. Babaei, N. Bashlykov, S. Batra, P. Bhargava, S. Bhosale, et al., arXiv preprint arXiv:2307.09288 (2023). https://arxiv.org/abs/2307.09228
49. P. Lewis, E. Perez, A. Piktus, F. Petroni, V. Karpukhin, N. Goyal, H. Küttler, M. Lewis, W.t. Yih, T. Rocktäschel, et al., Adv. Neural Inform. Process. Syst. **33**, 9459 (2020)
50. S.G. Patil, T. Zhang, X. Wang, J.E. Gonzalez, arXiv preprint arXiv:2305.15334 (2023). https://arxiv.org/abs/2305.15334
51. C. Wang, S. Zeng, C. Wang, arXiv preprint arXiv:2403.18230 (2024)
52. Z. Chu, Y. Wang, F. Zhu, L. Yu, L. Li, J. Gu, arXiv preprint arXiv:2402.03628 (2024)
53. Y. Zhu, S. Qiao, Y. Ou, S. Deng, N. Zhang, S. Lyu, Y. Shen, L. Liang, J. Gu, H. Chen, arXiv preprint arXiv:2403.03101 (2024)
54. E. Goffman (1959)
55. X. Wang, J. Wei, D. Schuurmans, Q. Le, E. Chi, S. Narang, A. Chowdhery, D. Zhou, arXiv preprint arXiv:2203.11171 (2022). https://arxiv.org/abs/2203.11171
56. J. Wei, X. Wang, D. Schuurmans, M. Bosma, F. Xia, E. Chi, Q.V. Le, D. Zhou, et al., Adv. Neural Inform. Process. Syst. **35**, 24824 (2022)
57. A. Shrivastava, T. Pfister, O. Tuzel, J. Susskind, W. Wang, R. Webb, in *2017 IEEE Conference on Computer Vision and Pattern Recognition (CVPR)* (2017), pp. 2242–2251. https://doi.org/10.1109/CVPR.2017.241
58. P.S. Laplace, *A Philosophical Essay on Probabilities* (Courier Corporation, 2012)

Correction to: Universal Behavior Computing for Security and Safety

Correction to:
C. Wang, H. Zhu, *Universal Behavior Computing for Security and Safety*,
https://doi.org/10.1007/978-981-97-9014-2

In the original version of the book, the copyright year has been changed to 2024. The book and the chapter have been updated with the change.

The updated version of this book can be found at
https://doi.org/10.1007/978-981-97-9014-2

C. Wang, H. Zhu, *Universal Behavior Computing for Security and Safety*,
https://doi.org/10.1007/978-981-97-9014-2_10

The manufacturer's authorised representative in the EU is Springer Nature Customer Service Centre GmbH, Europaplatz 3, 69115 Heidelberg, Germany. If you have any concerns regarding our products, please contact ProductSafety@springernature.com

Printed and bound by CPI Group (UK) Ltd, Croydon, CR0 4YY
01/10/2026
02225858-0008